AARON KLUG – A LONG WAY FROM DURBAN
A BIOGRAPHY

The atomic structures of macromolecules provide the key to understanding how life works. Aaron Klug led the way to the development
of methods for solving such structures and is one of the pioneers of
structural molecular biology. He was awarded a Nobel Prize in 1982 for
his work.

Illuminating both his personal life and scientific achievements, this
unique biography begins with Klug's youth in Durban and his studies
at Johannesburg, Cape Town and then Trinity College, Cambridge.
Holmes proceeds to explore Klug's career from his work on the
structure of viruses with Rosalind Franklin at Birkbeck College,
London, to his time as Director of the MRC Laboratory of Molecular
Biology (LMB) in Cambridge and as President of the Royal Society.

Drawing on their long-term collaboration, interviews and unique
access to Klug's archives, Holmes provides a fascinating account of
an innovative man and his place in the history of structural molecular
biology.

Kenneth C. Holmes is Emeritus Director of the Department of
Biophysics at the Max Planck Institute (MPI) for Medical Research,
Heidelberg, Germany. He is a Fellow of the Royal Society and recipient
of the Aminoff Prize, the Gabor Prize and the European Latsis Prize.
Holmes's long-term collaboration with Aaron Klug began in the
late 1950s at Birkbeck College and continued at the Laboratory of
Molecular Biology in Cambridge before his move to MPI in 1968.

AARON KLUG
A LONG WAY FROM
DURBAN
A BIOGRAPHY

∼

Kenneth C. Holmes
Max Planck Institute (MPI) for Medical Research,
Heidelberg, Germany

CAMBRIDGE
UNIVERSITY PRESS

CAMBRIDGE
UNIVERSITY PRESS

University Printing House, Cambridge CB2 8BS, United Kingdom

Cambridge University Press is part of the University of Cambridge.

It furthers the University's mission by disseminating knowledge in the pursuit of education, learning and research at the highest international levels of excellence.

www.cambridge.org
Information on this title: www.cambridge.org/9781107147379

© Kenneth C. Holmes 2017

First published 2017

A catalogue record for this publication is available from the British Library

Library of Congress Cataloging-in-Publication Data
NAMES: Holmes, K. C. (Kenneth Charles), author.
TITLE: Aaron Klug : a long way from Durban : a biography / Kenneth C. Holmes, Max-Planck-Institut für Medizinische Forschung, Heidelberg, Germany.
DESCRIPTION: Cambridge, United Kingdom ; New York, NY : Cambridge University Press, 2016. | Includes bibliographical references and index.
IDENTIFIERS: LCCN 2016007900 | ISBN 9781107147379 (hardback) | ISBN 1107147379 (hardback)
SUBJECTS: LCSH: Klug, A. (Aaron), Sir, 1926– | Chemists–Great Britain–Biography. | Biophysicists–Great Britain–Biography.
CLASSIFICATION: LCC QD22.K572 H65 2016 | DDC 540.92 [B] –dc23 LC record available at https://lccn.loc.gov/2016007900

ISBN 978-1-107-14737-9 Hardback

Additional resources for this publication at www.cambridge.org/9781107147379

Contents

CONTENTS

Colour plates can be found between pages 180 and 181.

Foreword

It is unusual for a brilliant scientist to write a biography about another great scientist and even more unusual for the foreword to this wonderful biography to be written by yet another scientist who has known them both for almost fifty years. A fan of scientific biographies, I had read those of all my towering heroes of science including Max Perutz, Francis Crick, Fred Sanger and Sydney Brenner. These great scientists were all from the same Laboratory of Molecular Biology in Cambridge as Aaron Klug and indeed Ken Holmes. While these other biographies are beautifully written, they were not researched with the remarkable level of rigour demonstrated here by Ken Holmes. This biography of Sir Aaron Klug is so jam-packed with detailed observations that it serves to document not only the life and work of one great scientist, but indeed a period and place of unparalleled scientific discovery.

On a personal level, it revealed my many connections with Aaron Klug. Aaron and my father were both born in Lithuania. Aaron and my mother both grew up in Durban, South Africa. More generally, this book is not just a biography, it is encyclopaedic in its scope, serving to shed light on the history of some of the greatest discoveries including the structure of DNA, macromolecular crystallography, electron microscopy image reconstruction and more. And if this was not enough, the Appendix presents a crystal-clear easily approached introduction to diffraction, which continues to play a central and crucial role in the determination of the detailed structure of the tiny macromolecules that are at the heart of the secret of life.

Even better than the description of the science behind Klug's work provided by Ken Holmes is what he reveals about such icons of British science as Rosalind Franklin, who was very close to both Aaron and Ken; Francis Crick, who closely followed Klug's work on tRNA and chromatin; as well as Max Perutz, John Kendrew and Sydney Brenner, another South African, who also studied at Witwatersrand University in the 1940s and then subsequently came to England. Klug, Crick and Brenner really are for me the giants of twentieth-century science and they all played a huge role in my career.

Although Sydney Brenner holds the unique distinction of having been scientific father to more Nobel Laureates than any other Nobelist, with five other winners who worked with him, I worked most closely with Aaron Klug and Francis Crick. Their styles were very different. I remember joking that if you had a new idea and went to tell Francis about it, you left his office quite sure that the idea had indeed been his. On the other hand, if you had no new ideas and went to talk to Aaron, he helped a new idea germinate and you left his office quite sure that his new idea was yours. In either case, science as a whole benefited, but Aaron's approach made people want to work for him. In this way, Aaron Klug brought out the best in others. His self-effacing manner, coupled to his reluctance to step in, made one feel empowered by what Holmes, in one humorous story, terms 'one of Aaron's endearing qualities: the ability to get someone else to take over onerous practical work by indicating a hesitancy that may be construed as a lack of competence'. In the context of scientific collaboration, it served to empower all those fortunate enough to work for him. The endorsers of this book also describe the influence of Aaron on collaborators and younger scientists. Klug was unique with his constructive ego-free approach to science.

Professor Michael Levitt
November 2016

Acknowledgements

My primary debt is to Aaron Klug's wife, Liebe Klug, who has remained an inspiration and fountain of information throughout the genesis of this work. I am grateful that Aaron Klug was party to a number of most informative interviews: in particular with the National Life Stories (interviewed by Katherine Thompson, British Library) and the Laboratory of Molecular Biology's own recorded interviews (interviewed by John Finch and Tony Crowther). I am particularly indebted to Steve Harrison for his critical reading of the whole manuscript and for his many helpful suggestions for improvement. I am also indebted to Tony Crowther, Daniela Rhodes, Don Caspar, Michel Goedert, Jacqueline Morgan, Debbie Davis, Richard Henderson, Greg Winter, Van Moundrianakis, Hugh Pelham, Andrew Travers, Peter Collins, Adrian Dixon, Jenifer Glynn, Kevin Leonard and Chris Calladine for critically reading large chunks of the manuscript and for making many helpful suggestions and corrections. I am most grateful to Allen Packwood and the staff of the Churchill Archive Centre for allowing me to dredge through the Klug papers before they were catalogued. Annette Faux, Archivist of the MRC Laboratory of Molecular Biology, has given me much support and help. John Finch's *A Nobel Fellow on Every Floor* (Icon Books, 2008) has been a valuable crib. I have drawn on a number of biographies as indicated by footnotes. Wikipedia has been of considerable help. I am deeply indebted to Mary Holmes for critically reading the whole manuscript. I thank the editorial staff of Cambridge University Press for their expert and patient guidance.

Introduction

The first half of the twentieth century saw the discovery of the chemical processes of life, including a sense of wonderment that yeast and humans used the same metabolic processes to burn sugar; the second half saw the unravelling of the processes whereby chromosomal deoxyribonucleic acid (DNA) is duplicated, the DNA sequence determines protein sequence and structure, and protein structure explains the magic of enzyme action. Together, these discoveries constitute the secret of life: the processes of life were revealed in their underlying physico-chemical simplicity but cloaked in a bewildering complexity.

In the 1930s, John Desmond Bernal thought that the secret of life would be revealed by using X-ray crystallography to solve the structures of crystalline proteins. Max Perutz was Bernal's student in Cambridge, England. Twenty-five years later, Perutz and his co-worker John Kendrew realised Bernal's dream by determining the atomic structures of the oxygen-carrying proteins haemoglobin and myoglobin. In the end, the secret of life entailed more than protein structures, but structures are indeed an essential part of learning that secret. Proteins fulfil many functions: hair, skin, enzymes, pumps, or the multitude of nano-machines and motors that make muscles move and cells divide. Structure, together with the genetic approach founded by Salvador Luria and Max Delbrück, became the twin pillars of a new science, Molecular Biology.

Aaron Klug developed methods for structure determination of macromolecular assemblies at the atomic level – a prerequisite for understanding how life's clockwork actually operates. Aaron entered the field of structural molecular biology just as it was beginning. Thus our account perforce embraces many historical aspects of the development of Molecular Biology.

Aaron Klug was born in Zelva (other spellings include Zhelva or Želva), Lithuania, in 1926. His family emigrated to Durban, South Africa, when he was two. He attended the Durban High School and won a scholarship to Witwatersrand University in Johannesburg at 15. He attained his degree from Wits when he was 19 and moved to Cape Town University, where he took a master's degree. There, he worked with the crystallographer R. W. James, who had an important formative effect. During this time, Aaron's first scientific publication appeared in the scientific periodical *Nature*. Two more papers published in *Acta Crystallographica* earned him an 1851 British Empire Fellowship and a Trinity College (Cambridge) Rouse Ball scholarship. He married Liebe Bobrow, a dancer and musician, in 1948, and in 1949 the young couple repaired to Cambridge, England, where Aaron did a PhD with Douglas R. Hartree on the kinetics of the formation of steel. For an ensuing year in Cambridge he did theoretical studies for Jack Roughton on the kinetics of oxygen uptake by the blood pigment haemoglobin. This experience reawakened his interest in biological phenomena.

Armed with a Nuffield Fellowship, Aaron joined John Desmond Bernal's Biomolecular Research Laboratory at Birkbeck College, London. Here he met Rosalind Franklin: after working on the structure of DNA at King's College London, Franklin had moved to Bernal's laboratory to lead a small group working on virus structure by X-ray diffraction. Meeting Franklin at Birkbeck College was Aaron's epiphany. This encounter determined his scientific future: solving the structures of macromolecular assemblies such as viruses. After Franklin's untimely death, he took over direction of the virus group. Together with John Finch, his collaborator for 40 years, he showed that poliovirus and small spherical plant viruses have closely related structures. A fruitful collaboration with Donald Caspar established the geometric rules for assembling 'spherical' viruses. In 1960 the virus group was invited to join the newly founded Medical Research Council's Laboratory of Molecular Biology (LMB) in Cambridge.

At the LMB, while working out the structures of spherical viruses by electron microscopy, Aaron developed the first applicable method for computing three-dimensional images from a set of two-dimensional projections of a structure (tomography). He and his collaborators solved the structure of tobacco mosaic virus. His group worked out the first atomic structure of an RNA-containing macromolecule, tRNA – a molecule containing thousands of atoms. Aaron also carried out structural analysis of the large macromolecular complexes involved in packaging DNA in chromatin (nucleohistones) and mapped out the organisation of the nucleohistone core. He discovered zinc fingers, protein domains that bind to specific DNA sequences, and he pioneered their application in gene therapy.

In 1982, Aaron Klug was awarded the Nobel Prize for Chemistry. Between 1985 and 1995, he was Director of the Laboratory of Molecular Biology. As Director he was also instrumental in getting the British part of the Human Genome project started. From 1995 to 2000, he was President of the Royal Society of London. He was knighted in 1988 and awarded the Order of Merit in 1995.

Science is much held back because scientists concern themselves with that which is not worth knowing, and that which cannot be known[1]. Aaron was careful never to fall into either of these traps – he chose subjccts that were at once topically important and likely to yield answers. Moreover, he never let himself be beholden to a technique. Aaron moved fluently between structure determination by X-ray diffraction or electron microscopy, and biochemistry. His endeavours were guided by an ability to choose topics that would yield to a sustained investigation. His consummate skill as a teacher and leader enabled him to draw forth exceptional performance from his co-workers and collaborators.

---

[1] *Die Wissenschaft wird dadurch sehr zurückgehalten, daß man sich abgibt mit dem, was nicht wissenwert, und mit dem, was nicht wißbar ist.* Goethe, J. W. in *Maximen und Reflexionen* (ed. H. Koopmann) p. 83 Deutsche Taschenbuch Verlag München (2006)

Part I

From Durban to Birkbeck College

1

Durban and Johannesburg

Towards the end of her reign, Catherine the Great, observing that Jews were filling the ranks of the nascent Russian middle class, banished the Russian Jews to the Pale of Settlement. The Pale, established in 1791, corresponded to the borders of the Polish–Lithuanian Commonwealth, which had been forcibly annexed by Imperial Russia. It included much of present-day Lithuania, Belarus, Poland, the Ukraine and parts of western Russia. More than 90% of the Russian Jews were deported to the Pale. Even within the Pale, Jews were discriminated against; they paid double taxes and were forbidden to own land, run taverns or receive higher education. Moreover, all young men were forcibly conscripted into the Imperial Russian Army for 25 years. Despite these heavy burdens, the Jewish population in Russia grew from 1.6 million in 1820 to 5.6 million in 1910.

Thousands of Jews fell victim to pogroms in the 1870s and 1880s. This led to mass emigration, particularly to the United States (two million between 1880 and 1914), as well as to other developments, such as the spread of the Haskalah (Jewish enlightenment). Zionism also took hold in the Pale. Only after the overthrow of the Czarist regime in 1917 was the Pale of Settlement abolished. Even after the Revolution, the relationship of the Jewish communities with the local governments remained strained and emigration continued.

The Pale was large and by no means culturally homogeneous. There were local laws and customs. Moreover, the Jewish population used its

3

ingenuity to circumvent the harsh regulations. Thus in the Lithuanian North, renowned for its sober view of life, in a *shtetl* called Zelva, Aaron's grandfather Benjamin Klug owned quite a large farm. The Klugs may have immigrated into Lithuania from Alsace. Benjamin was a cattle dealer, buying cattle from local farmers, fattening them up and herding them to market in the local town Ukmerge, some 20 miles away. He was helped by his three sons Yudel, Lazar (who was of unusually short stature, on account of having contracted typhus during his adolescence) and Isaac. Essentially drovers, they were accomplished horsemen. Lazar married Bella (née Silin; the Silins ran the local shop), and they had two sons, Benjamin (1924) and Aaron (11th August 1926). Lazar had a traditional Jewish education and secular schooling. The family language was Yiddish, but Lazar was also fluent in Russian and Lithuanian. Although he never had the opportunity to acquire a higher education, Lazar had a natural gift for writing, and he published a number of articles in the newspapers of Kaunas in Lithuania, for which he acted as a freelance correspondent.

As the second son, Lazar would not inherit his father's farm. Furthermore, he had had some very unpleasant experiences with the Bolshevik army: in defence of his brother, it is said that he killed a Bolshevik soldier and was very lucky to have survived the ensuing fracas (alternative accounts also circulate in the family[1]). Members of the Gevisser family, who were cousins of Bella, had emigrated to Durban in South Africa around 1900. Durban welcomed white immigrants, which included a trickle of Jewish families. About 90% of South Africa's Jews came from Lithuania. Given all these factors, in 1927 Lazar decided it was time to seek a new home. He emigrated to Durban where he was accommodated by the Gevisser family. He had been apprenticed as a saddler and therefore knew about leather, and thus he entered the family business of Moshal-Gevisser dealing in hides and general stores. Bella, with her two sons Benjamin and Aaron and her sister Rose, followed to Durban in 1929. Isaac also emigrated to South Africa and settled in Johannesburg.

On the death of his father, Yudel became a landowner. However, as a consequence of the Molotov–Ribbentrop pact, in June 1940 Soviet

---

[1] See Appendix A wherein Mark Gevisser recounts his visit to Zelva.

military forces occupied Lithuania and set up a puppet government consisting mainly of left-wing poets, which made itself popular among the general public by nationalising the land. The landowners were summarily deported. Thus poor Yudel was shipped off to Siberia, where he died in exile. The death was only reported to the family after the Second World War through a charitable organisation that investigated the fate of deportees. His son was conscripted into the Russian army and died fighting the Germans. Yudel's youngest daughter Janina spent the war with the partisans in the forest and then married and emigrated to Israel. Years later, she complained to Lazar that he had taken the easy way out. To Lazar this complaint seemed to be rather misplaced. After all, he had received no inheritance and had to fend for himself.

In January 1834, Sir Benjamin d'Urban had taken office as governor and commander in chief of the South African Cape Colony. He occupied Natal and named it as a new colony for the British Empire. To commemorate this, the name of the principal port was changed from Port Natal to Durban. Durban was a very English microcosm of the Empire – *Civis Britannicus sum*. Facing the Indian Ocean at latitude 30° S, Durban has a subtropical climate varying between warm in winter and hot and humid in summer. In 1927 it was a relatively sleepy town. Mixed-race people and Indians ran the market gardens situated in the hills above the city, and Zulus provided domestic help. On Sundays the Zulus would dance on the beach. For the whites, life was pleasant enough. Moreover, for the Klug boys there were lots of things to do – there was the beach and the bush full of gum trees, poisonous snakes, and troops of chattering vervet monkeys, all accessible on a bicycle[2].

The Klugs moved between houses in the Glenwood area, which lies above the town on the ridge known as Berea, then near the edge of the town. They lived a modest middle-class life with two Zulu domestics. The sound of singing emanated from the Zulu labourers building new roads all around their house. The young Klug brothers enjoyed their environment, which included the neighbouring Greyville Racecourse. They would crawl though the hedge to watch the races. Benjamin (Bennie), endowed with the excellent memory that was a characteristic

---

[2] I am indebted to Aaron and Liebe Klug for many conversations and hundreds of e-mails providing much essential information. Some information has been taken from an autobiographical account: Klug, A. (2010) *Ann. Rev. Biochem.* **79**, 1–35.

Figure 1.1 Bennie and Aaron (courtesy of the Klug family)

of both brothers, soon knew the names of all the horses and the winners of the Durban July Handicap for the past few years. A cycle ride away across the town, next to the beach, was the Jewish Club where Lazar liked to play chess.

Figure 1.2 Aaron, Bennie and their mother Bella (courtesy of the Klug family)

The annual visits of the Royal Navy South East Asia Squadron, fully beflagged, to Durban Harbour under the command of 'Evans of the Broke' (Admiral Lord Mountevans) were big days for the colony and interesting for the Klug schoolboys. The 25th Jubilee of George V was celebrated by a procession representing the diverse peoples of this great Commonwealth. In spite of the heat, poor Bennie went as a Newfoundland fisherman complete with heavy oilskins!

When Aaron was three and a half, Bella made an unusual discovery. In spite of the family switching from Yiddish to English when Aaron was two, he was now reading the English newspaper. What was to be done with such a precocious child? It was decided he should attend the local Penzance Road Primary School together with Bennie, who was nearly two years older. For Aaron, this had its positive and negative sides: Aaron was easily able to keep up with the curriculum, but because he was much younger and smaller than his classmates, he needed Bennie to protect him from bullies. Moreover, Penzance school was right on the edge of the bush, and frequently the Klug brothers had to defend their lunch bags from invading hoards of vervet monkeys.

In 1932, family tragedy intervened: Bella died in hospital of pneumonia following a minor operation. The trauma was somewhat ameliorated by the presence of Aunt Rose, who became Aaron and Bennie's surrogate mother. Later, Lazar married Rose and they had two children, Phillip and Ethel, who changed her name to Robin. Bennie, Aaron and Robin were bright, independent and socially competent, as Rose expected children to be. Phillip was different. He needed a lot of help even to keep abreast of day-to-day problems. His condition would nowadays probably be described as mildly autistic. Both Bennie and Aaron were very attached to Phillip and later gave him a lot of support. As an adult, Phillip left the family home in Durban to live with Aaron in Cambridge, England. This turned out to be too stressful, and Phillip finally moved in with Bennie in Johannesburg where he died in his 40s of a lung infection. Robin became a pharmacist and now lives in Israel.

Aaron early developed an interest in cinema, particularly Westerns. Their intrinsic morality, the predictable triumph of good over evil, appealed to him. Moreover, just across the town Durban offered its own speciality – Indian Cinema – in which the fat rich boy falls for the poor thin girl. This stereotyped art form also intrigued young Aaron.

At the age of eleven, Aaron transferred to the reputable Durban High School for Boys. He could have gone to the Glenwood Secondary School, but they did not teach Latin, and Latin was essential to get into medical school. Aaron was, again, the smallest and youngest member of the school. While Bennie was still able to offer protection from bullying, Aaron sometimes had to win freedom of action by helping his older but less talented school friends with their homework. He was always a good teacher.

Durban High School for Boys was established in 1866, modelled on an English public school. There were six houses, including one for boarders. Cricket culture percolated the school. Situated on the ridge of the Berea, the school had a slightly less oppressive climate than the town; nevertheless, in summer it was still hot. With scant regard for the pupils' comfort, school uniform (for the junior school) was short wool trousers, knee-high wool socks, a white shirt and a navy-blue woollen blazer. At temperatures in excess of 85 °F (30 °C) one was permitted to remove one's blazer. Aaron entered this microcosm of Englishness at eleven and stayed until he was fifteen. It had an enduring influence. Together with Bennie and Ralph Hirschowitz, who became his life-long friend, Aaron generally enjoyed his four years at the High School. The order in class settled into a pattern: Aaron at the top of the class, followed by Bennie and Ralph.

The ethos of the school, based on Rudyard Kipling, was to ignore those two imposters triumph and disaster. One did not make a fuss about things. Aaron had absolutely no difficulty in adopting this reserve and avoidance of overt emotionalism since it was germane to his own personality.

According to Aaron, the underlying philosophy of the school was quite simple – the bright boys specialised in Latin, the not so bright in science and the rest studied subjects such as geography and history. Aaron did Latin; Greek had been dropped to accommodate Afrikaans.

Bennie was a good athlete, renowned for his prowess at long jump. Furthermore, he was a cricketer, an opening bat for the house eleven and left-handed: a left-handed batsman is useful for confusing the opposing bowlers. Very much to the regret of the Klug brothers, their father Lazar would not allow them to play games on the Sabbath, which cramped Bennie's style and prevented him from progressing to the school first eleven.

By today's standards, the school offered few challenges other than Advanced Latin Prose Composition in the Sixth Form. Aaron was very good at Latin, which he positively enjoyed. On account of an optic atrophy in his right eye that affected his peripheral vision, the result of an accidental injury sustained as an infant, Aaron was excused some sports and spent much of his time reading books. The school library was very good but was kept locked; those who applied for the key were 'swots', which attracted disapprobation in a school devoted to cricket. Fortunately, there was also a good public library in Durban, which the Klug brothers used to visit after the required attendance at the synagogue on Saturday mornings. Aaron was a very fast reader and was equipped with a phenomenal memory. He was also endowed with unbridled curiosity, more so than his older brother. After studying Leopold von Ranke's treatise on the machinations of the renaissance Popes, Aaron developed a particular fascination for the Roman pontiffs. Furthermore, from his knowledge of the Hebrew alphabet, which is based on the Phoenician alphabet with roots in Egyptian characters, Aaron became interested in hieroglyphics. He was fast becoming a polymath.

Two teachers in the school, Charlie Evans in History and Neville Nuttall in English, made a memorable impact. Under Nuttall's tutelage Aaron developed a lively and sustained interest in English literature and poetry. Evans pointed out that tensions over the 'Polish Corridor' – the narrow strip of land connecting Poland to the Baltic Sea, and dividing East Prussia from the main body of the German Reich – could potentially cost some of the students their lives. Indeed, a few months later it led to the outbreak of the Second World War. Finally, Aaron began to take an interest in science. In particular, Paul de Kruif's *Microbe Hunters*, which has been an inspiration for many aspiring physicians and scientists, had a lasting influence. This book determined Aaron's schoolboy life plan, namely the study of microbiology.

On leaving school, Aaron and Bennie split up. Aaron went to Johannesburg; Bennie stayed in Durban and went to the Natal University to study engineering. Later, Bennie was to work for the structural consulting engineers Arup Associates in Johannesburg, Israel and London. For some time, Bennie was the quiz champion on South Africa Radio – until he was requested to resign, to give others a chance.

In 1941, at the age of 15, armed with a scholarship that paid for tuition but nothing else, Aaron entered medical school at Witwatersrand University in Johannesburg. He was not intending to become a medic, but this was apparently the only way into microbiology. Uncle Isaac (Itzic) could offer a *pied-à-terre*. He lived in Booysens, an industrial suburb to the south of Johannesburg where there were big mine dumps. This was far from a convenient arrangement, since Isaac's home and the University were 8 km and two tram-rides apart. Aaron was still very young, and he palled up with his younger cousin Bommie (Benjamin) to run up and down the mine dumps for fun. Later, they tried out a local boxing club, but interest quickly flagged.

The poetically named University of Witwatersrand ('ridge of white waters'), prosaically known as Wits, lies to the northwest of the centre of Johannesburg on the ridge after which it is named. This ridge, which runs east–west through Johannesburg, extends over a hundred miles. It is remarkable for having yielded nearly half of all the gold ever mined in the world. To the northeast, separated by a mile from the main campus, lies the Parktown Campus, the site of the Medical School and the Department of Anatomy. Between the two campuses, on the brow of the hill, there was a fort then used as a prison.

For his first year Aaron elected to do the preliminary course for medicine, the *prima* (similar to British A levels): botany, zoology, chemistry and physics. The second year was in the medical school and consisted of physiology, biochemistry (then still called physiological chemistry) and anatomy. Aaron was accompanied in this venture by his school friend Ralph Hirschowitz, who also wished to study medicine. Another student studying medicine at this time was Sydney Brenner, who would later be a colleague of Aaron's in Cambridge.

The commanding personality in the Anatomy School was Raymond A. Dart, a native of Queensland, Australia, who in 1922 was made head of the newly established Department of Anatomy. In 1924, a limestone quarry owner at Taung in northwestern South Africa shipped Dart a box of rock in which Dart discovered a fossil skull of a child. This, he declared, was an example of *Australopithecus africanus* (a missing link between humans and apes). Dart lectured in physical anthropology, giving demonstrations of how reptiles walked and how the articulation of the limbs differed from that of mammals. One of the highlights of Dart's lectures was his proudly producing the Taung child fossil skull. Another strong influence in the Anatomy School was Lawrence Herbert Wells, who was single-minded about dissecting cadavers. In spite of Dart's skull, after some months of dissection Aaron found the going tedious and shifted his focus back to the physics and chemistry schools on the main campus.

Chemistry and biochemistry seemed to offer more hope of understanding microbiology than dissecting cadavers. Moreover, as a result of a public lecture on modern physics Aaron discovered the Schrödinger wave equation, which opened new vistas for him. During this time Aaron added a new dimension to his life, namely mathematics. He discovered that he had an innate ease with mathematical formalism. Mathematics allowed him to appreciate the intellectual elegance of the Schrödinger wave equation and quantum mechanics. He also became fascinated by functions of a complex variable. However, although he turned out to be very competent in this area he never saw mathematics as an end in itself but rather as a means to an end: he was never a pure mathematician. For the next three years, after savouring many disciplines, Aaron devised his own course of study in physics, chemistry and mathematics, treating the University as a kind of intellectual supermarket. He graduated in science with three majors rather than the normal two. Fortunately, the Dean, Dr Van der Horst, arranged for Aaron to get exemption from University regulations when necessary. The Dean was understanding of Aaron's apparent dilettantism, especially since Aaron finished with firsts in all three subjects.

Aaron was still an enthusiastic cinemagoer. A movie was then called 'a bioscope' to be seen at a 'Café Bio'. The Café Bios ran continuously. Moreover, one got a cup of coffee or an ice cream with the tickets, which were quite cheap. Aaron went most days and became something of an expert in Westerns. He maintained this interest: later during his time in Cambridge, he was known to slip off to the Arts Cinema for a choice Western.

Since it was wartime, Aaron perforce joined the Officer Training Corps. In fact, if the war had continued, at the end of his studies he would have been enlisted. For six weeks in the summer and then two weeks at Easter they went to camps. Since he was far younger than most, Aaron was never required to do very much. This was not really Aaron's métier. In contrast, one of the profound influences on Aaron at this time was the Hashomer Hatzair[3], which Ralph Hirschowitz urged Aaron to join. The Hashomer Hatzair was a romantically idealistic Zionist youth movement that resulted from the merger of the Hashomer ('The Guard') and Ze'irei Zion ('The Youth of Zion'). While in Durban, Ralph and Aaron had been members of the Habonim, a Jewish scouting movement based on the German Wandervogel and Baden Powell's scouts. However, the Hashomer Hatzair offered an ideological and cultural depth that the Habonim, a mass movement, was not able to provide. Believing in the equality of man, in Marxism, and in Zionism, the movement was also influenced by the German Wandervogel movement and by the ideas of Ber Borochov, the father of socialist Zionism. Labour Zionists believed that a Jewish state could only come about by the Jewish working class settling in Palestine and constructing a state through the creation of a progressive Jewish society with rural kibbutzim and an urban Jewish proletariat. The Wandervogel movement sought to liberate and educate youth by returning to Nature whereas

---

[3] Much of the information about Hashomer Hatzair came from *Shomrim in the Land of Apartheid: The Story of Hashomer Hatzair in South Africa 1935–1970*, edited by Chaim Shur, and published by members of Hashomer Hatzair South Africa and Havazelet in conjunction with Yad Yaari (1998). I am grateful to Ralph Hirschowitz for introducing me to this book.

Hashomer Hatzair thought that the liberation of Jewish youth could be accomplished by living in kibbutzim, which was perhaps not very different. They addressed each other as 'shomrim' in the sense of 'comrades'[4]. The Hashomer Hatzair, which is secular, organises activities and machanot (camps), which are used to educate and also to propagate the aims of the movement. In the Palestinian Mandate the movement also formed a political party, which true to its socialist convictions advocated a bi-national Palestine with equality between the Arab and Jewish workers – united by their class struggle. By 1939 Hashomer Hatzair had 70,000 members worldwide, the majority in Eastern Europe. During World War II and the Holocaust, thousands of members of Hashomer Hatzair died resisting the Nazis. Together with the Habonim they played a major role in the Warsaw Ghetto Uprising in 1943. The local groups of the Hashomer Hatzair actively supported immigration into Palestine, both legal and illegal, to the kibbutzim to which they were attached. This act of 'Aliyah' or 'ascent' was of central importance to the movement. The rules of Hashomer Hatzair required that a person acquiring majority should fulfil the goal of self-realisation by going to live on a kibbutz.

However, in South Africa in the 1940s there were also other political forces at play, in particular those supporting Trotskyism and world revolution. Here Zionism played a secondary role. Members of the Hashomer Hatzair were also anti-apartheid, which led later to the enforced closure of the South African branch. One of the important figures was Baruch Hirson, later a friend of Aaron's, who turned from Zionism to working with the African National Congress. For his apparent association with a terrorist outrage Hirson later served a devastating 9½ year prison sentence. His departure from the Johannesburg Ken (literally 'nest') of the Hashomer Hatzair coincided with the emigration to Palestine of a number of influential leaders, which led to a power vacuum in Johannesburg. At this moment, Aaron and Ralph appeared on the scene. Ralph, good-looking and charismatic, was a natural leader and he and Aaron created a strongly academic and influential group within the Johannesburg Ken. By attempting to fuse radical Marxism with psychoanalysis Ralph was led to the works of Wilhelm Reich,

---

[4] The Hashomer Hatzair motto is: 'We are the last generation of slaves and the first of free men.'

who soon became a dominant influence both in Johannesburg and Cape Town. The Vanguard bookshop in President Street carried much of the left-wing literature and the works of Wilhelm Reich. These were sought out and avidly read by the student members of the Hashomer Hatzair. According to Reich, life's difficulties and neuroses are the result of a patriarchal authoritarian education with its sexual suppression. Fulfilment could be obtained via the release of frustrated sexual energy: a 'Sexual Revolution' rather than a proletariat revolution. Lively discussions of 'Orgone Energy' (freedom through orgasm) and the understanding of 'Character Armour' quickly usurped the striving for fulfilment through 'Aliyah' and 'Kibbutz'. The contrast between Reich's invitation to sensual behaviour and the strictly puritanical rules of the Hashomer Hatzair often led to discussions in which Aaron poked fun at Reich's theories. Furthermore, communism did not interest these young academics: for their situation it just seemed irrelevant.

Besides providing Aaron with a forum for youthful utopian discussions and a network of friends whose various abodes turned out to be useful alternatives to the nightly trek back to uncle Isaac, these heady days with the Johannesburg Hashomer Hatzair were to have a profound influence on Aaron's life. Amongst other things, they led to his meeting with members of the Cape Town Ken, including Vivian Rakoff, who was to become a life-long friend. Vivian Rakoff attended a machaneh (summer camp) in 1944 at Magaliesburg, a holiday venue near Johannesburg renowned for its bucolic beauty. He recounts that an afternoon walk with Aaron and Ralph was worth a year at university. In 1945, the Johannesburg Ken set up a large machaneh in an idyllic venue near Newcastle in the Drakensberg, the line of mountains that marks the border of Natal. This camp lasted six weeks – long enough to build a symbolic watchtower. However, everything one needed (including lime for the latrines, which became Aaron's special responsibility) had to be carried in. It turned out that the camp had been pitched with insufficient regard for the logistic needs of nearly 200 young people over an extended period. Stress included an outbreak of scarlet fever. Somewhat in desperation at the shortage of supplies, Aaron approached the local farmer and purchased some sheep – on the hoof. Confronted with the necessity of turning these furry behemoths into lamb chops, Aaron sought out a sharp knife. Luckily, a local Zulu farmhand intervened, saying: "This is no work for you, baas-sir." This incident illustrates one of Aaron's endearing qualities: the ability to get someone else to take over onerous practical work by indicating a hesitancy that might be construed as lack of competence.

South Africa is vast, but travel was cheap. The whites were a natural aristocracy with an informality arising from an innate self-confidence. So long as you were not rabidly anti-apartheid, everyone was your friend. Travel for students was cheap since it mostly involved hitchhiking or riding in the brake vans of goods trains. Thus young people from Cape Town, some 1500 km away, turned up at the camp in the Drakensberg, among them Liebe Bobrow. Ralph, who had previously visited Cape Town on a fact-finding mission on behalf of the Johannesburg Ken, introduced Aaron to Liebe. With his sober Lithuanian family and colonial English public school education, Aaron contrasted starkly with Liebe, a vivacious sixteen-year-old direct out of school, who came from a very different culture. Her father, Alexander Bobrow, was an analytical chemist, a scholar, and very left wing; Liebe was a musician and a dancer. Moreover, Cape Town is not Durban. During this camp they merely became acquainted. However, the following year the Johannesburg Ken of Hashomer Hatzair decreed that Aaron should go to Cape Town.

2

Cape Town

Alexander Bobrow, known to his friends as Alter, grew up in Pinsk, now part of Belarus near the Ukrainian border. He attended the same high school as Chaim Weizmann, and went on to the University at Vilna to study chemistry. After University he took up work as a chemical engineer in a sugar factory.

In 1917 the towns of the Pale of Settlement, already poor, suffered dreadfully from the ebb and flow of the revolutionary and imperial armies. Returning to visit his family in Pinsk, Alter was confronted with appalling scenes of destruction: children could be seen crawling around in the wreckage that had been their homes. Many had watched their entire families butchered before their eyes. Typhus was rife. Alter resolved that these children must be offered succour. He belonged to a Poale Zion Group (Workers of Zion), and on his initiative all the members of the Pinsk group left their jobs to help save the children. They walked from place to place collecting starving children from the rubble. The group commandeered a disused old-age home and converted it into the first Pinsk orphanage. Within two years there were three orphanages in Pinsk alone.

Emigration seemed the only way out of this mess. Fortunately, the influential Jewish Community in Cape Town established an orphanage for children from the Pale of Settlement. In 1921 they accepted 200 children, to include some of the children from Pinsk but with the proviso that Alter should come too. Thus in 1921, via Danzig,

London and Southampton, Alter sailed to Cape Town accompanied by a group of children from Pinsk.

In 1921, Cape Town was still a smallish city. The town is blessed with a congenial Mediterranean climate and one of the world's most dramatic settings. From the waterfront the skyline consists of the spire of Devil's Peak, the flat mesa of Table Mountain – 1000 m high – and the dome of Lion's Head and Signal Hill. The older residential part (referred to as the City Bowl) snuggles in this ring of mountains and looks out to the northeast over the remarkable Table Bay. The Jewish Orphanage was located in the suburb Oranjezicht on the slopes of Table Mountain, but still in the City Bowl. The harbour sits in Table Bay in the cradle of the mountains. The older suburbs snake around the northerly slopes of the Devil's Peak and turn south along the east side of Table Mountain to form a string that eventually reaches False Bay. This beautiful bay, on the eastern side of the peninsula where the sand is soft white, faces the Indian Ocean. Its name stems from the frustration felt by many mariners returning from the Far East who turned into the bay thinking that they had already rounded the Cape of Good Hope.

A spirited young lady, Hannah Gamsu (known as Annie), who originally came from Dvinsk in Latvia, taught sewing and dressmaking at the orphanage. Annie was decidedly left-wing and was the rebel of her family. She had great taste and style for visual things, whether for clothes or house decoration. Annie was introduced to Alter by one of the orphan girls that he had saved. There ensued a meeting and merging of souls, but the pair did not marry for another five years because Alter felt an obligation to send much of the money he earned back to his parents. At the time of their wedding in 1927, they lived in a flat in an old house in Jagersfontein Avenue. When Annie became pregnant, they moved to the beautiful suburb of Muizenberg on False Bay so that the child should benefit from the ozone-rich sea air. On 24th May 1929, Annie gave birth to Liebe Bobrow, who 19 years later would become Liebe Klug.

The 'Southern Line' railway links Muizenberg with the city centre, but the journey took about 45 minutes, so that after three years of living on the seashore the Bobrows decided it was time to move back into town. They purchased a house in the foothills of Table Mountain in a new suburb called Vredehoek, to the east of Oranjezicht. They named the house Espero (Esperanto for hope), embodying one of their dreams: if all were to speak one language then we should have peace. At the time their road, Rugley Road, was still the highest on the mountain. From the back window of the house, young Liebe was able to watch the cable car travelling up and down Table Mountain. The view from the front stoep (veranda) was the wonderful vista of Table Bay. As a child, she would watch the ships rounding the bluff into the bay to dock at the end of Adderley Street. When the wind blew from the southeast, as it sometimes did with great ferocity when bringing the hot summer weather, it covered the mountain with a white cloud known as the 'Table Cloth'.

Despite the Board of Governors' original insistence on his coming from Pinsk, Alter's position at the orphanage was not secure. Perversely, this may have come about because he was very popular with the children. There was envy among the staff. Furthermore, the Bobrows were not at all conventional in their views, being remarkably left-wing and politically active. Annie worked very hard for a Jewish Workers' organisation called 'Geserd'. This decreased Alter's popularity with the management. His contract was not renewed.

There ensued the Bobrows' 'Wanderjahre'. Being influenced by Buddhism and the Indian poet and mystic Rabindranath Tagore, Alter first tried meditating in a little shack by the sea in a deserted place called Bakhoven. Later he started teaching Hebrew privately and preparing boys for their Bar Mitzvahs. However, with a family to feed, Alter was obliged to seek more permanent employment, which he found in the Johannesburg area. The Bobrows sold their wonderful house and wandered into an economically imposed exile from beautiful Cape Town. Their banishment ended only in 1944 when Annie and Alter were appointed Matron and Principal of Herzlia, a boarding facility (and later school) in Cape Town that enabled children from smaller communities to come to the city and be properly educated in a Jewish environment. In the meantime, Liebe sampled schools in Muizenberg, Johannesburg, Johannesburg Springs and a convent in Benoni, each with its own curriculum. Thus Liebe's schooling became a wide-ranging pedagogic survey of South Africa beginning and ending at the Good Hope Seminary for young ladies.

Liebe played the piano rather well, and after completing school she started a BMus degree at the College of Music in Cape Town. She also loved to dance. Sometimes Alter would play his mandolin to accompany her. In addition, Alter introduced her to the ideas of modern dance (he somewhat disapproved of classical ballet as being the plaything of the Russian aristocracy). He also encouraged Liebe to improvise. After a year at the College of Music, the siren call of dance became too strong, and Liebe changed to an unconventional dance studio called 'The School of Charm'. The studio earned money by running classes in energetic dancing ('dancercise'). The serious dance students had the freedom to concentrate on developing their own kind of modern dance. It was way outside the South African style, more akin to Isadora Duncan and Martha Graham. Like everything else the Bobrows were involved in, it seemed to be about 30–40 years before its time. Alter was rather proud that Liebe was becoming a dancer.

During 1947, his second year in Cape Town, Aaron Klug went to a performance at which Liebe danced to the music of Vaughan Williams' setting of 'Greensleeves', and he fell in love. Thus the fate of the Bobrows and the Klugs became entwined.

But how had Aaron come to be in Cape Town? At the end of 1945, after four years at Wits, Aaron knew that now he really wanted to study physics. Physics as a discipline was in turmoil. Two atomic bombs had demonstrated, devastatingly, the correctness of Einstein's formula equating mass and energy. This relationship came out of special relativity: there was no 'ether' as had previously been thought, only space. In 1919 Eddington had exploited a total eclipse of the sun to measure the displacement of star images near the sun. The mass of the sun bent the starlight exactly as predicted by Einstein's general theory of relativity: even space could be bent. This was the death knell of Newtonian physics. Perhaps even worse for classical physics was the notion that energy came in packets or 'quanta' of determined size. All this seemed to give modern physics the attributes of Alice's sojourn in Wonderland. Nevertheless, experimental data fitted the new physics with wonderful accuracy. Moreover, quantum mechanics was very good at accounting for the behaviour of electrons. It predicted atomic and even simple molecular structures, thereby turning chemistry into an exact science. The downside was that you needed to understand a lot of mathematics to come to grips with quantum mechanics. Most undergraduate physics courses now had to teach more applied mathematics simply in order to make modern physics accessible. Aaron was a special case. He had taught himself enough of the appropriate mathematical language to wonder at the power and subtleties of the new physics, and he wanted to know more.

After obtaining his degree in November 1945, Aaron was offered a job at the Department of Scientific and Industrial Research (DSIR) in Johannesburg. It turned out that this was not the new physics he was looking for; rather, it was applied electronics. After six weeks, Aaron resigned.

In a further thread in the story of Aaron's future, in 1943 Baruch Hirson had come to Cape Town to try to convince the Hashomer Hatzair leaders to leave the organisation in order to promote the wider aims of world revolution. In this he was broadly successful: the Cape Town Ken was splintered by the same defections and doctrinaire disputes that had beset the Johannesburg Ken. Thus in 1945 the Johannesburg Ken, noting the disarray of the Cape Town Ken, arranged that a shaliach (youth leader and emissary) should be sent to reorganise the remnants. Blue shirts and neck scarves were reinstated. Study groups were instituted. In addition, the group became renowned for its enthusiastic singing and dancing. The venue for these noisy activities was the Maon or meeting place in rooms above a dentist's practice in Long Street. The long-suffering dentist must have regretted ever having signed a rental agreement with this lively bunch. A great deal of earnest discussion happened both there and in 6 Annadale Street, a student lodging-house belonging to the Bobrows, which turned out to be a second Maon. The courtyard in Annadale Street was a social hub; one could nearly always find someone there. Furthermore, Peter the Greek, who ran a small 'tea room' on the other side of the street, offered welcome opportunities for eating, socializing and dating in an affordable price range. Naturally, the young Hashomer Hatzair members studied the official texts, but they also organised seminars on Marx, Freud and Reich. Poetry readings were interspersed with Israeli dances such as the Hora. They learnt and sang the Hebrew songs of the future state. Although the success rate in terms of promoting Aliyah was modest, this romantic brain-washing brought together a group of young people at an impressionable age: they connected and remained a cohesive group of friends for the rest of their lives.

In order to strengthen the Cape Town Ken further, the Johannesburg Ken of the Hashomer Hatzair decided that Aaron should be posted to Cape Town as second in command to the shaliach. The idealism of the Hashomer Hatzair was not particularly concerned with mundane factors such as earning a living – but then Aaron saw an advertisement offering £20 a month to demonstrators in physics at the University of Cape Town (UCT) coupled with the possibility of doing a MSc in physics. It transpired that the UCT had the best physics department in South Africa. Aaron applied and was accepted.

Thus, in January 1946, Aaron moved to Cape Town. Unlike Durban and Johannesburg, Cape Town is an old city with a distinctive history and culture. As an impressionable 19-year-old, Aaron thought it was marvellous. He found lodgings in 6 Annadale Street. In fact, Aaron shared a room with the shaliach Mike Levine, who was originally from Johannesburg and had gone off to Israel in the general exodus, living in the kibbutz Shoval in the Negev. However, Levine had soon returned to South Africa to study medicine and help with organising the Hashomir Hatzair. The Annadale Street house was built around a courtyard, as are many houses in Cape Town. At the back of the courtyard were four rooms in a building that had been the servants' or slaves' quarters. A slave's room was to be Aaron's home. There was a small kitchen, bathroom and shower in the courtyard. A mature Malay woman called Annie ran the place. Very much in the spirit of a Cambridge bedder or an Oxford scout she cleaned the rooms, provided linen and brought everyone an early morning tea, made with sweet condensed milk.

Aaron was befriended by Vivian Rakoff, who lived but two blocks away. An erudite and witty person, Vivian had played a pivotal role in re-establishing the Ken in Cape Town. In the Maon above the dentist's practice, Vivian had painted a mural showing a group dancing the Hora. He was a man of many talents: he could design, paint, write poetry, sing and act. Having recently acted the part of Macbeth in a school play, he was brimming with enthusiasm for the theatre and often in demand in the amateur productions in the town. After studying English at the University of Cape Town he planned to go to Downing College Cambridge to read English with Frank Raymond ('F. R.') Leavis, and then perhaps become an actor. In fact, he was to become a well-known psychiatrist, head of the Clarke Institute and chairman of the Department of Psychiatry at the University of Toronto. Vivian's early literary aspirations overlapped with Aaron's interests, which helped to foster the deep and lasting friendship that developed between them. Moreover, in fulfilment of their support for the Hashomer Hatzair, Aaron and Vivian became involved in a number of slightly risky attempts to help potential candidates for Aliyah out of the Cape Town Docks and on their way to the Promised Land. On one occasion, an Argentinean member of Hashomer Hatzair was to arrive on a training vessel. The plan was that he would jump ship and should be met and helped on his way to Johannesburg from where he would be translated to Palestine. On the appointed evening, Aaron and Vivian, wearing Hashomer Hatzair blue shirts and carrying Jewish newspapers, proceeded to the docks to await the sailor. They were duly recognised by the candidate, and the group swiftly repaired to Annandale Street. In something of an anticlimax to the event, the arrangements in Johannesburg had not worked out properly, with the effect that the sailor promptly left and returned to his ship. Despite such extramural activities, Aaron's academic career proceeded apace.

The University of Cape Town lies to the east of Devil's Peak, shielded from the setting sun by the mass of the mountain. It was founded in 1829 as the South African College, a hundred years before Wits, and is the oldest university in South Africa. The main teaching campus, known as the Upper Campus, is located on the slopes of the Devil's Peak. It includes the faculties of Science, Engineering, Commerce, and most of the faculty of Humanities. The buildings and layout of Upper Campus were established around 1930. Since that time, many more buildings have been added as the university has grown and the site has extended into the lower or 'Rondebosch' Campus. The journey from Annadale Street to the university entailed walking to the main station in Adderley Street and taking the Southern Line to the sixth stop, Rondebosch. The university could then be reached by an uphill walk through the grounds of the College of Music, which was ensconced in a lovely old Cape House, passing the statue of Cecil Rhodes pointing north, saying 'Your hinterland is there.' Aaron soon discovered a more congenial mode of transport: a lecturer in the physics department, John Walter Faure Juritz, who lived nearby, could pick up Aaron in the mornings in his 'Traction Avant' Citroën car. Juritz too was something of a polymath. He flew small planes, was an accomplished pianist and played the organ in the neighbouring church[1]. He also played the bassoon well enough to perform in the Cape Town Orchestra when an extra bassoon was required. He remained a firm friend of Aaron and Liebe until his death in 2007.

---

[1] 6 Annadale Street has since been replaced by a community hall attached to the neighbouring Presbyterian Church.

Another friend Aaron made, this time in the English department, was Bernard John Krikler, known as Bunny. Bunny was a couple of years older than Aaron and had served in Burma towards the end of the war. On discharge from the army he was given UCT quarters that were constructed largely out of corrugated iron. These were rather insensitively referred to as 'Belsen'. Later, while the Klugs were in Cambridge, Bunny Krikler and his wife Berenice came to live in London. They also remained life-long friends of the Klugs.

In addition to the MSc course lectures, which were given by Reginald William (R. W.) James, Aaron attended lectures on mathematics, philosophy and poetry. The senior lecturer of ethics and philosophy, Martin Versfeld, was an eccentric Afrikaans intellectual who had converted to Catholicism and was keen to proselytise, particularly while climbing Table Mountain. The brother of one of Aaron's Hashomer Hatzair friends fell in love but was rejected. As a result, he tried to hang himself. Martin Versfeld not only discovered him and cut him down but also converted him to Catholicism. Subsequently the man became Brother Elias (Elijah), a Carmelite monk, and lived the rest of his life on Mount Carmel in Israel under the protection of his biblical namesake. Aaron found himself being urged to turn from agnostic Judaism to Catholicism. Aaron, who is not devoid of spirituality, explored the situation in some detail but decided to remain as he was. Nevertheless, in so doing he learnt more about the early Christian Church than is known to most clerics. He became something of an authority on Origen, an early Christian philosopher from Alexandria who wrote commentaries on the Bible and interpreted the scriptures allegorically. Despite delving into this fascinating topic, Aaron still managed to complete the two-year physics MSc course in one year and to pass with first-class honours. Aaron is a phenomenally fast reader, but even so, one wonders at his ability to come to terms with so much material. Moreover, at the same time, for his MSc under the tutelage of R. W. James, he generated a research paper on the structure of a halogenated benzene molecule by X-ray diffraction. He was still only 19. The next year, this work was published as a short letter to the research journal *Nature*, forming Aaron's first research publication[2].

---

[2] Klug, A. (1947) *Nature* **160**, 570.

For his second year in Cape Town, Aaron readily accepted a position as junior lecturer that carried the princely remuneration of £30 per month. His main task was teaching physics to trainee nurses, who were required to know about medical physics including X-rays and dosage. For some reason they were also required to know about negative numbers, which gave some trouble. Aaron was grateful that they did not need to know about imaginary numbers. Nevertheless, Aaron had plenty of time to continue with research, and he became enthusiastic about working with James.

R. W. James came from Paddington and was a Londoner born and bred. He won a scholarship to St John's College Cambridge and after obtaining a double first in Natural Sciences started research at the Cavendish laboratory under J. J. Thomson. Conditions were not good: J. J. Thomson had a mass of students and no funds. Therefore, in 1914 James responded with some alacrity to an invitation to join Ernest Shackleton's Antarctic expedition as the expedition's physicist – with J. Wordie, later to be Master of St John's, as geologist. Shackleton was known as 'the Boss'. One of the questions the Boss put to James was 'Can you sing?', this being considered an important recreational activity in the long polar nights. Until this point, apart from fell walking, James (nicknamed Jimmy by his new colleagues) had led a rather quiet life that contrasted starkly with the rough seafaring manners of his companions. To them, Jimmy was clearly just a landsman but was tolerated. Unfortunately, after a propitious start, the expedition developed into a catastrophe. Their ship, the *Endurance*, became locked in Antarctic ice and crushed. The party escaped in three small boats. Soon the chronometers became unreliable, and they did not know where they were. Here, the unassuming landsman Jimmy made a very important contribution to saving the expedition: he was able to register the lunar occultations of the planets and, with the aid of a nautical almanac, use these observations to calculate the party's longitude. Thus after 400 days at sea, Elephant Island was found, and the party of 30 men had solid ground under their feet. However, Elephant Island was far from any shipping routes, so Shackleton with five colleagues decided to risk a 1200-km open-boat journey to the South Georgia whaling stations to seek help. Shackleton's heroic journey was successful; and somewhat miraculously, in August 1916 all the members of the expedition were rescued and were able to return to England. In January 1917, James was commissioned into the Royal Engineers and went to work with William Lawrence Bragg on sound ranging methods for detecting gun positions near Ypres. The importance of this meeting went beyond their ensuing accurate location of enemy guns: in 1919 Bragg took up the chair of physics at Manchester University, and James joined him there.

William Lawrence Bragg was born in Adelaide but moved to England when his father was appointed professor in Leeds. Lawrence Bragg entered Trinity College Cambridge at the age of 16. Subsequently he started research under J. J. Thomson at the Cavendish Laboratory. By then, his father had awakened his interest in German physicist Max von Laue's work on the diffraction of X-rays by crystals. Von Laue had shone a beam of X-rays on a crystal of zinc sulphide and recorded the diffraction pattern on a photographic film. X-rays have a wavelength close to the distance between atoms and therefore are strongly diffracted – scattered and diverted into specific directions – by the regular atomic arrays within crystals. On an X-ray film the diffraction pattern consisted of regularly arranged sets of spots, which are known as X-ray 'reflections'. The spots happen as a consequence of the diffraction from a regular crystal lattice. Von Laue produced a set of equations that allowed one to estimate the strength of diffraction in any particular reflection from a crystal of known structure. These equations were not especially intuitive and did not help much with the more interesting problem of working out the crystal structure from the diffraction patterns. Lawrence Bragg's studies of von Laue's diffraction patterns obtained from zinc sulphide crystals led him to postulate that the structure of zinc sulphide crystals was based on a three-dimensional (3D) pattern known as a face-centred cubic lattice, an amazing piece of insight. It also led to the famous Bragg's Law of crystal diffraction. In essence, this says that a beam of X-rays is reflected from planes of atoms in a crystal at an angle that depends inversely on the spacing between the planes: small spacing – large angle; large spacing – small angle. Intuitively much simpler than the von Laue equations, Bragg's law also allows one by inspection of simple crystals to estimate how strong a particular X-ray reflection should be. After demonstrating the power of his law in the analysis of the structure of diamond, Bragg was appointed a Fellow of Trinity College just before the outbreak of the First World War. From 1912 to 1914 he had been working with his father, and the results of their work were published in 1915. It was this work that earned them jointly the Nobel Prize in Physics in the same year, making William Lawrence Bragg at 25 the youngest-ever Nobel Laureate in science.

When Lawrence Bragg took over the Chair at Manchester University he was neither a skilled lecturer nor a good administrator. It was James who kept the place running and contributed very substantially to the success of the laboratory. On balance, this was a fantastic time: the silicate structures were identified and the optical theory of the diffraction of X-rays was developed. The lab was abuzz with famous visitors. Father Bragg was by then Director of the Royal Institution in London where he had appointed Kathleen Lonsdale and John Desmond Bernal as co-workers. Together they worked out how to do X-ray structure analysis of complex organic molecules. This introduced a whole new world to chemistry: instead of guessing the structure of a molecule on the basis of often tedious chemical synthesis, one could visualise the structure as worked out from X-ray crystallography.

James stayed at Manchester for another 18 years, where his important research includes the demonstration that atoms do not stop vibrating at a temperature of absolute zero (which seems intuitively wrong but is predicted by quantum mechanics). He became Reader in Experimental Physics. In 1936, when everyone had assumed he was a confirmed bachelor, James married. He now had a family to feed. After an unsuccessful application for a chair in Aberdeen he accepted the chair of physics at the University of Cape Town. He moved to Cape Town in 1937 and spent the rest of his life there. He was a fine teacher and founded a successful school. Posthumously, he achieved a distinction earned by few scientists: two of his ex-students from UCT were awarded Nobel Prizes (Allan Cormack, mentioned below, and Aaron Klug).

In the 1930s, James started writing a book entitled *The Optical Principles of the Diffraction of X-rays*. This was an encyclopaedic tome containing all that was known about the diffraction of X-rays. It was finished in 1946, and Aaron was given the spare-time job of checking the page proofs. Since the book was full of equations, this was a demanding task. Because Aaron always remembered everything he had ever read, he ended up knowing a great deal about the diffraction of X-rays. In particular, he came across James's account of Fourier series, which are ways of expressing periodic functions (such as the electron density in a crystal) as the weighted sum of all the sine and cosine waves that will fit into the crystal. Fourier series are used to work out the electron density in a crystal from the measurements of the strengths of the X-ray reflections. Aaron realised that a Fourier series was a special case of a more general relationship, the Fourier transform, that will also work for non-periodic objects. He was to make much use of Fourier transforms in his subsequent research. Moreover, Fourier transforms have some powerful properties that Aaron later exploited in his Nobel Prize work. James did not know anything about Fourier transforms, but in the next edition of his book there was an appendix featuring them.

After completion of his master's degree, Aaron suggested to James that he might carry out an accurate structure determination of triphenylene by X-ray crystallography in order to measure the bond lengths of the various classes of carbon–carbon bond. This would form the basis of a possible doctoral thesis. Triphenylene is a ubiquitous combustion effluent that pollutes the environment. Its structure is actually predictable from its composition: it is a planar molecule with three benzene rings grouped round a central benzene ring, a kind of miniature graphite sheet. Crystals could be made according to a published recipe. Data were obtained on X-ray film using a so-called 'Weissenberg' X-ray camera. The film data (intensities of X-ray reflections) were turned into numbers with the help of a film densitometer that Aaron constructed based on an instrument used for measuring stars at the observatory. Moreover, the work earned the support from the South African Council for Scientific and Industrial Research – Aaron's first research grant.

Diffraction data are three-dimensional: as you rotate the crystal, every view gives a different set of X-ray reflections, and you really need to measure them all. In the days before computers, there was no way of dealing with all these data; instead, one concentrated on measuring the diffraction from two or three views, usually looking along a symmetry axis of the crystal. Feeding these data into a Fourier series yields a map of the electron density in the crystal but projected along the chosen axis (flattened into a two-dimensional view). With an object as symmetrical as triphenylene, two views at right angles give enough data to determine the positions of all the atoms, which show up as separate round objects in the projected electron density. Nevertheless, there is a catch. Diffraction measures intensities, whereas the numbers required for the analysis are the amplitudes, the square roots of the intensities (see Appendix B). One does not know whether these should be positive or negative (a problem Aaron's student nurses should have been familiar with). An initial model is often used to resolve the ambiguities. One calculates the expected diffraction from the model and takes over the signs. However, if the initial model is wrong, it is easy to be caught in a false solution. Aaron used a sophisticated and original method to obtain a starting model to solve this sign ambiguity. This involved calculating the diffraction pattern of a single molecule (this operation is actually a Fourier transform), and then moving the molecule around in the crystal so as to modify the calculated diffraction data to match the observed diffraction data. He then used the structure that gave the best fit to the observed diffraction data as a starting solution to give the signs to calculate the density from the observed data. These ideas became known by the term 'molecular structure factors', and the method established Aaron's early reputation.

Aaron carried through this analysis for triphenylene, but the answers displayed an unexpected result: some of the distances between the carbon atoms in the molecule itself deviated from the expected values by being rather long; furthermore, some of the distances between atoms in neighbouring molecules were remarkably short. This led Aaron into a ramified wild-goose chase involving calculations of bond lengths by the method of molecular orbitals, which is based on quantum mechanics. At this stage of his life, he was enamoured of quantum mechanics and was perhaps too ready to accept his X-ray crystallographic result as a *raison d'être* for indulging this interest.

A paper published six years later showed that Aaron had placed the molecule in the wrong position in the crystal unit cell. It turned out that a student who had been helping Aaron had made a serious error in copying out lists of intensities of Bragg reflections. With the correct data, the molecule came out in a different position in the unit cell and all the carbon–carbon distances came out in the expected range.

Later, a method very similar to Aaron's became standard in solving protein structures. If a protein has a close relative with known structure, this can be used to predict the signs or phases of the unknown structure. This is known as the method of molecular replacement. It appears that Aaron had published the first paper on molecular replacement. Experience has shown that it is fairly easy to get the correct orientation of the molecule in a crystal but also quite easy to get the position wrong. Here one has to be careful. This was exactly Aaron's problem. It is not always good to be a pioneer. Fortunately for Aaron's future, no one knew at this time that he was wrong. Aaron's two papers in *Acta Crystallographica* were well received and indeed determined his future.

A spare-time activity that appealed to Aaron was climbing. The first recorded ascent of Table Mountain was by the Portuguese explorer Admiral Antonio de Saldanha, who put into Table Bay in 1488 and then climbed to the top of the mountain in an attempt to find out where he was. Table Mountain offers superb climbing on compact sandstone and the views from the climbing routes are breathtaking, but getting to the climbing pitches is not all that easy. Thus Africa Face and Fountain Ledge are popular climbs because they are situated near the upper

Figure 2.1 Aaron climbing on Table Mountain (courtesy of the Klug family)

station of the cable car. Sometimes the mountain becomes enshrouded in the mist of the 'Table Cloth', which is dangerous unless you know the mountain well. Aaron went climbing at weekends, always riding to Table Mountain on his bicycle.

The future Nobel Laureate Allan Cormack, who was a lecturer in the physics department of UCT at that time, was a climbing nut and had a climb named after him. One weekend in the winter of 1947 he invited Aaron to climb. Aaron himself described this adventure[3]:

> I used to be a kind of Sunday climber doing mild climbs on Kloof Nek, B climbs, occasional C climbs, and Allan was an established climber. I didn't know that he had a route named after him, but he said he'd take me up something really interesting. I'd done some climbing with ropes, but only a bit, just for short stretches, so he took me up the Africa Face, which is a D or E climb[4]. About half way up he was leading, above me, just the two of us. Suddenly I slipped off the ledge and swung out into empty space and swung back again, hurting my arm. However, Allan held me, belayed the rope, pulling me back to safety and I was able to climb the rest of the way. I often think: what would have happened if he hadn't been able to hold me.

Two future Nobel Laureates – hanging from one nylon rope.

In the middle of 1947, Aaron suddenly became aware of Liebe. He had known her for a couple of years but somehow, what with the Hashomer Hatzair blue shirts and Girl Guide skirts, he had never really noticed what a remarkable person she was. Besides, Liebe was going steady with Denis Rutowitz, known as Rut, also a UCT student and a keen climber (and apparently very good looking, a bit like James Dean), who courted her enthusiastically. Nevertheless, as mentioned earlier, one evening Aaron saw Liebe dancing and was hooked. Now Aaron's tenacity came to the fore. He courted Liebe with stories, not of danger and moving accidents as Othello to Desdemona, but rather of life's richness and meaning. His suit was successful: despite his being rather unmusical and certainly no dancer, he was accepted by Liebe, the dancer and musician. Alter seems to have been delighted with his daughter's

---

[3] Recorded in *Imagining the Elephant: A Biography of Allan MacLeod Cormack* by Christopher L. Vaughn, UCT Press and Imperial College Press (2008).

[4] In modern notation, climbs on Africa Face vary between 20 and 32.

choice: his prospective son in law was a scholar knowledgeable in Hebrew and Yiddish. Rut eventually gave up the chase and retreated to Johannesburg to fly planes, later joining the Israeli Air Force.

In the summer on a Saturday evening, the young couple would take the bus and walk to a part of the beach called Saunders Rocks. There, a little out to sea, was a big flat rock warmed by the sun where they could sit and dream. After sunset, they would run to catch the last bus back to Cape Town, and then walk past the Parliament Building through the Great Synagogue Gardens to the corner of Annadale Street. At the parting of the ways they would sit on the pavement chatting, not wanting to separate.

Aliyah beckoned, but immigration to Israel, even after independence, was limited. Those in the older age group such as Ralph and Aaron had the opportunity to go to Israel first. Often pairs married, or married out of convenience, since couples could enter Israel on the same immigration voucher or 'ticket'. By the middle of 1948, Ralph and his first wife had already left. Vivian Rakoff went with them on the mail boat, intending to travel to Cambridge. Somewhat by chance, he finished up on a kibbutz on Mount Tabor, where he stayed for a year. Aaron and Liebe started making plans: he could seek out an appointment at the Weizmann Institute; she could set up a modern dance school in a nearby kibbutz. Aaron had a ticket and by then a passport, but they wanted to stay together. A lively debate ensued over whether they should marry. Liebe felt the case was clear, but Aaron thought that they were too young. He was 22 and Liebe was just 19. Nevertheless, in June 1948 Aaron and Liebe suddenly decided to get married on 8th July, which was just three weeks away. They wanted a very quiet wedding and reasoned that with only three weeks' warning one couldn't generate too much fuss. Also the time chosen, a Thursday at 5 p.m., was selected to minimise interest. They did not reckon with the fact that Liebe's parents were well known figures in the Jewish community. So they had to go and see the Chief Rabbi, who, given the apparently precipitous timing, assumed the worst about Liebe's status but in any case read the banns. Aaron's parents were shocked when he phoned and announced, hesitantly, "Um. . . I'm getting married in three weeks' time!" Their brilliant son! What floozy had got hold of him? And a dancer! Nonetheless, Aaron and Liebe were duly married in the Great Synagogue in the garden at the end of the Avenue. The Synagogue is a grand and beautiful building, perhaps the most beautiful in Cape Town. Liebe's parents and most of her cousins had been married there. The choirmaster was an old friend of the family and had known Liebe since she was a baby, so there was a full choir. So much for a quiet wedding!

Lazar, Rose and Aaron's half-siblings came over from Durban. Aaron was taken to buy his first suit. As a reception, Annie had arranged a simple late afternoon cocktail party and invited as few people as possible. There was an announcement in the paper, and, since the Herzlia School, where the Bobrows lived, was just a stone's throw from town, the wedding party was well attended. Aaron's brother Bennie was best man, but was not enthusiastic at having been dragged all the way from Johannesburg in such a hurry. Liebe's dress was made by a parent of one of the children at the Jewish boarding school. The dress was ice blue in heavy taffeta, and the style, appropriately, was Christian Dior's New Look. Liebe herself designed a silver hat with a pale blue and silver threaded veil. With her beautiful dark hair and lovely dress, the bride looked wonderful. Nevertheless, it seemed that many thought this was an ill-assorted match, including Vivian, who wrote a letter to Aaron explaining why it shouldn't happen, and Ralph, who expressed rather muted enthusiasm. Nor were any of the Hashomer Hatzair in evidence. Aaron's family stayed for a few days, and were warmly entertained by the Bobrows. However, when Aaron and Liebe went to Durban a few weeks later the family reception remained cool: Liebe was not wealthy, did not have a degree, and went on the stage.

There was no honeymoon, nor did the Klugs have a married home in any formal sense. Liebe just moved in with Aaron in his room in Annadale Street, and everything continued as before. Each morning, Aaron was fetched by John Juritz and they drove together to the University. Liebe taught at a local nursery school.

One morning in early 1949, Aaron, in a rather confused state, phoned Liebe from the University. He had applied for an 1851 British Empire Fellowship and for a Trinity College Rouse Ball Scholarship. On account of his *Acta Crystallographica* papers, he had been awarded both. The prospect of doing his PhD in Cambridge, particularly in Lawrence Bragg's laboratory, looked too good to turn down, but their life's plan was to go to Israel. After some debate, it was decided that Aliyah had to take a rain check until Aaron had a Cambridge PhD. Liebe could use the time to study modern dance at the Contemporary Dance School in

Figure 2.2 Aaron and Liebe, 8th July 1948 (courtesy of the Klug family)

London with Sigurd Leeder. Liebe's mother packed their voluminous luggage, cabin trunks, suitcases and two enormous crates, with linen, blankets, pots, pans, crockery – all the basics to start a home – and of course Liebe's clothes for three years and all seasons, plus a large square box just for Liebe's hats.

On 26th August 1949 Aaron and Liebe set sail on the S.S. *Edinburgh Castle* for England. Leaving on the mail ship was always a big occasion.

The Klugs' large circle of friends and relatives sent flowers, chocolates and fruit that were duly delivered to their cabin. Neither felt sad at leaving; it was the beginning of a huge adventure. Naturally, Liebe's parents were downcast at losing their only daughter, but they kept their feelings under control, not wanting to upset the young pair.

Getting out of Cape Town Harbour takes time. It takes a while for the big ship to be slowly towed out to sea. The Klugs stood on the deck watching the figures on the quayside slowly getting smaller. Once the ship reached the open sea, it was hit by the infamous Cape rollers and pitched and rolled to the sound of creaking timbers. The 1851 Fellowship entitled Aaron to a first-class ticket, but Liebe's parents could only afford a cabin-class ticket for Liebe, and Aaron had to downgrade. Nevertheless, they had a big cabin with a porthole. The stewards drew one's bath – salt water and very hot – and brought tea to the cabin first thing in the morning. The routine of the ship was much like that in the movies: hot beef tea on deck at 11.00 in the morning; deck games; officers in dark Navy uniforms. At 23° S, the officers changed into white uniform. That was when Aaron and Liebe saw flying fish and dolphins following the ship. The 'crossing the line' ceremony was done in style, with Father Neptune coming aboard and passengers being dunked in the pool. On the voyage they were befriended by Max Gordon, a well-known left-wing activist, who led the first strike of black workers in Soweto. He was an experienced traveller and knew the ropes.

At length, after 11 days, the ship stopped at Madeira to give Aaron and Liebe their first experience of continental ambience and food. The passengers disembarked into small ships to get to the harbour. Gordon showed Aaron and Liebe around and took them to a typical restaurant: fish Spanish style in olive oil. After a few hours in Madeira, it was back to the ship and on to England.

3

Cambridge

Aaron and Liebe's first glimpses of England were the lights of Weymouth as they approached the English coast at night through the Needles channel. They docked at dawn. To the young couple impatient for the new life, mooring the ship at Southampton seemed interminable. Even more time was consumed in getting through customs and immigration. Finally, they were on the train to London. Near London, the train ride to Waterloo Station exposed miles of monotonous grey back-to-back houses. Moreover, it was already the equinox, when the sun struggles up to 38° above the horizon to yield a pale autumn light. It was not Cape Town.

The Klugs arrived at London around midday. They were met by Elana, an old friend from the Hashomer Hatzair, who had booked them into a hotel in Bayswater. With her help they collected the smaller luggage; the large crates and trunks had been sent on to Cambridge. After checking in at their hotel, they set out on a walk of discovery through Hyde Park to Piccadilly, past Buckingham Palace, the Mall, Trafalgar Square, the Strand, and then back to Piccadilly, essentially taking a stroll through the top row of the Monopoly board. London was very grey, with bomb-craters everywhere. The people looked depressed. Food and fuel were rationed. The streets were still lit by gaslight. However, it all seemed somehow familiar: over the years one had read and heard so much about London and seen it all, courtesy of Pathé News. Breakfast at the hotel was interesting: Aaron recalled that it was the first time he had eaten something called 'bloaters' (smoked herrings) for breakfast, although reading *Comic Cuts* had prepared him for the dish.

They met up with old friends from Cape Town, Bennie Kaminer and his wife Freda. As befits a cinema addict, on their first Saturday night in London Aaron took Liebe and the Kaminers to the cinema in Leicester Square. Here was a taste of cultural differences, queuing for the cinema. Nevertheless, the movie was memorable: Orson Welles in 'The Third Man'. Aaron possessed an up-to-date *Bacon's Guide*, which he proudly used to navigate on buses across north London to visit the sister of their old Cape Town friend Bernard John Krikler. Bernard and his wife Berenice would later move to London.

A couple of days later, on 17th September 1949, the Klugs proceeded by train from Liverpool Street Station to Cambridge, happily not scarred by bombing, arriving at one of the longest railway platforms in England, and duly sought refuge in the Blue Boar Hotel. The Blue Boar was across the street from Trinity College where Aaron held his research scholarship. Many inns, in spite of their dating from the fifteenth century, when Cambridge was still a busy port, nevertheless vanished with little trace as they were absorbed into the expanding colleges. Thus the Blue Boar Hotel, which was at one time a coaching inn – albeit a relative newcomer from the seventeenth century – is now an annex of Trinity College that has spread across Trinity Street to take up all available space as far as Sidney Street. However, in 1949 the Blue Boar was still an upmarket hotel where the better-heeled could stay while visiting their respected offspring.

The Blue Boar Hotel was not affordable for more than a few days. Where should they live? An enquiry with the omnipotent Trinity College porters indicated that the Klugs might enquire of Mrs Heffer of Heffers Bookshop next to the Red Lion Inn (long since demolished) in the small street known as Petty Cury. Mrs Heffer was associated with the Commonwealth League, founded to foster friendship between the peoples and students of the Commonwealth; she duly introduced them to Mrs Fuchs, whose husband Vivian Fuchs was away on his first expedition to the Antarctic. Bad weather on a massive scale marooned him there for three years. In the meantime, in her generous Barton Road house on the western perimeter of Cambridge, Mrs Fuchs indeed had space. She could offer the Klugs a room with its own bathroom and central heating. *Pro tem*, Aaron could use Vivian Fuchs's study. So the Klugs moved into a Georgian house with an immense garden and woodland – all for two guineas a week – although transport from this remote boundary of Cambridge necessitated owning two bicycles. The rent included the use of the kitchen, but as Liebe's family had always managed to have a cook, Liebe was without any real culinary experience. Nor was the very old-fashioned gas stove of any great help. Mrs Fuchs employed a live-in housekeeper; initially Liebe was rather messy and the housekeeper became most unenthusiastic about having her in the kitchen. However, things settled down, and the housekeeper did manage to teach Liebe how to make lemon meringue pie and a few other dishes.

Joyce Fuchs – formerly Joyce Connell, and Vivian Fuchs's cousin – had married Vivian in 1933. She was a skilful climber and outdoor enthusiast, who accompanied Fuchs on his expedition to East Africa in 1934. The findings from this expedition, in which two of their companions died, gained Fuchs his Cambridge PhD. However, his great achievement was the crossing of the Antarctic. On 2nd March 1958 he conducted his party into Scott Base on Ross Island, having made the 2,158-mile journey in 99 days. He was greeted by congratulatory messages from all over the world and was awarded a knighthood.

Although at this time Joyce Fuchs was living alone (her children were away at boarding school, and Vivian did not return until April 1950) she kept up her standards: she changed for dinner every evening. Moreover, lunch was served at the dining room table, with the serving dishes appearing through the hatch. Even food rationing was not allowed to have an impact on this genteel tradition. On Sundays, a friend came to lunch. To preserve the illusion that she was a guest even in times of rationing, the friend would come round to the kitchen door on Saturday afternoon and pop her chop into the fridge, ready for the housekeeper to prepare for Sunday lunch with Mrs Fuchs's chop.

Now that they had a *pied-à-terre*, Aaron needed to organise his life. He held the prestigious 1851 Exhibition fellowship, which was endowed with £350 per annum. The Exhibition of 1851 had been an enormous success and had made a substantial profit. With the proceeds the Royal Commission purchased 86 acres of land in South Kensington and established three famous museums, the Royal Albert Hall, Imperial College, and the Royal Colleges of Art and Music. When this huge undertaking was largely complete, there still remained sufficient funds for the Commission to set up an educational trust giving fellowships and grants for research in science and engineering. Aaron was paid from this trust.

Trinity is the largest college in Cambridge. Over the great Tudor gateway presides its founder Henry VIII; following some long-ago prank, he holds a chair leg rather than the more appropriate sceptre. On the right, embedded in the gate, is the Porters' Lodge. Above the Porters' Lodge are rooms at one time occupied by Isaac Newton. Through the gate is the Great Court created at the beginning of the seventeenth century by Richard Neville who razed a number of existing buildings to clear the necessary space. On the north side of the court is the famous clock that chimes the hour twice, once for Trinity, and once, a fifth higher, for Trinity's neighbour, St Johns. At midday this gives rise to 44 seconds of chimes, enough time for an Olympic athlete to run round the Great Court –although Aaron was not a great runner and had no intention of trying this feat.

As Aaron held a Rouse Ball scholarship from Trinity, the activation of his scholarship entailed seeking out the Junior Bursar. The Senior Bursar is responsible for managing the endowment of the College, a position of great financial responsibility, whereas the Junior Bursar tended to be entrusted with more mundane functions such as organising Aaron's scholarship. Aaron went to Trinity Great Court to see Charles Kemball, the Junior Bursar, who was a surface chemist. The Bursary is housed in a Tudor staircase on the east side of the Great Court to the south of the Great Gate. Kemball, a large redheaded Scott, held a half-blue for judo and was member of the Hawks' Club[1]. He announced to Aaron that, in view of the fact that he already held an 1851 Exhibition, his Rouse Ball scholarship would be a token £50 per annum. Although he was scarcely three years older than Aaron, on hearing that Aaron was married, Kemball became alarmed and delivered the riposte, "I suppose we'll be having perambulators on the lawns soon!"

While Aaron was in college, Liebe stood up against a college wall seeking warmth in the weak September sunlight. It was before the start of term and the town was empty. She felt lonely and out of place.

[1] Membership of the Hawks' Club is limited to male sportsmen in Cambridge who have earned a University Blue or Half-Blue, i.e. have taken part in the match against Oxford in a particular sport.

However, the native East Anglians are nice enough folk, if you can understand them, and Liebe was offered friendship by a porter in his bowler hat. Nevertheless, on Sunday evening when the bells of Great St Mary tolled for Evensong she knew that she could never really belong. How could an outsider, particularly a woman, relate to this exclusive male academic enclave? Liebe spent much of her life trying to solve this conundrum.

Aaron next visited his tutor, Walter Hamilton, who had translated Plato's *Symposium* into English for the Penguin Classics edition. Not that Aaron was expected to benefit from Dr Hamilton's academic skills: in the tradition of Cambridge Colleges, a tutor was responsible for general supervision. As a research student Aaron was *in statu pupillari*. Students mostly lived in college. However, since the colleges never seemed to be large enough, some lived in licensed digs (lodgings). In digs the landlady was in the first instance held responsible for the student's behaviour. Landladies were known to take their responsibilities quite seriously, even to measuring the depth of water in the bath so as to avoid squandering national resources (the war-time government had decreed that baths should be no more than 5 inches deep). Students were expected to eat in the college hall, for which purpose they would surrender their food ration books to the college. Moreover, in 1949 *in statu pupillari* certainly did not embrace the concept that a student could be married. Thus Aaron's case was not simple and challenged Hamilton's creativity. In the end, the Klugs' lodging in Barton Road was deemed to be licensed – and Liebe became Aaron's official landlady. She was required to submit and sign a list of 'kept nights' – that is, nights that Aaron had spent in Cambridge.

So on to the Cavendish Laboratory, at that time the fiefdom of James's one-time mentor, Sir Lawrence Bragg. The entrance to the original Cavendish from Free School Lane is through a solid mediaeval gateway. The stone-faced building framing the gateway housed the teaching labs, the notoriously uncomfortable lecture room, and the Cockcroft–Walton accelerator, built in the 1930s and capable of accelerating protons to huge energies by applying up to half a million volts. John Cockcroft and Ernest Walton used this apparatus to split a lithium atom in two and were duly awarded the Nobel Prize for Physics in 1951. Opposite the gateway in a large courtyard stood the Austin Wing, a rather functional four-storey brick building built in 1930, containing

Figure 3.1 The gateway to the old Cavendish Laboratory (from Wikipedia; photo by William M. Connolley. Licence: CC-BY-SA 3.0)

the Cavendish Professor's office, and numerous departments of the Cavendish. Adjoining the Austin Wing was the Royal Society Mond Laboratory for low-temperature research.

On the basis of R. W. James' recommendations, Aaron had already been accepted by Bragg to do his PhD in the Cavendish. On the first floor was the nascent protein crystallography group founded by Max Perutz and now funded by the Medical Research Council (MRC). James was in regular correspondence with Lawrence Bragg. Bragg told him about the work of Max Perutz on the structure of the protein haemoglobin and how Perutz had been joined by the dynamic young John

Kendrew, who was working on myoglobin, haemoglobin's smaller cousin. In Cape Town, James had read some of these letters to Aaron. Aaron had thought of working on something crystallographic but unorthodox: protein structure sounded about right. Therefore, he was somewhat taken aback when Bragg calmly announced that Perutz's MRC group was full. Aaron, who knew he was a clever fellow, contemplated long on what might have gone wrong. His analysis was that the previous year James had sent a student who had given up after one year, which would have put a negative spin on Bragg's appreciation of James's judgement. Be that as it may, Aaron arrived just after Perutz had recruited Francis Crick, who with his continuous loud voice and dominant personality may well have convinced Lawrence Bragg that Perutz's group was full to overflowing.

Substances as diverse as sand, amethyst, mica, cement, glass and asbestos all consist basically of silicon and oxygen. How come there are so many different minerals? One of Bragg's achievements in Manchester was to work out many of the silicate structures. On the basis of these and his own results, Linus Pauling was able to formulate the basic rules for making the silicate structures. They are all built from the same unit: a silicon atom surrounded by four oxygens to form a tetrahedron. These tetrahedra can then combine with each other with various symmetries, some crystalline, some fibrous, and with various metal ions and water to form the host of silicate structures that make up most of the solid earth we live on. Bragg offered Aaron the chance of working on the phenomenon of disorder in silicate structures, a subject he had initiated. The problem is actually rather interesting, but Aaron was unimpressed and decided to look further afield. Since Aaron felt it necessary to consult James on every decision, and since his advice could only be sought by mail, it was a couple of months before Aaron finally settled on a PhD project.

Aaron next consulted John Lennard-Jones, the Plummer Professor of Theoretical Chemistry. Lennard-Jones was famous for working out the force between atoms of a noble gas (such as helium), which is known as the Lennard-Jones potential. Now one great success of quantum mechanics was to allow one to calculate the distribution of electrons around the nucleus of a hydrogen atom. This could be derived exactly from the Schrödinger equation. The background to this equation was the discovery that particles (such as electrons) sometimes behave like waves.

Louis de Broglie had proposed in 1923 that there is a wavelength associated with any particle that depends inversely on its momentum (speed times mass). Schrödinger's equation was an extension of Newtonian mechanics that takes account of the wave nature of particles. Since electrons are very light, their small mass gives rise to wavelengths about the size of an atom. Thus the wave-like properties are very important when trying to calculate how electrons move around, especially in atoms. However, the Schrödinger equation can be applied to any mechanical problem, even a game of cricket, but here the wavelength associated with the cricket ball is so incredibly tiny that it's best to forget about it and just keep your eye on the ball. It turns out that the possible solutions to the equation applied to the hydrogen atom (one proton plus one electron) are grouped in shells with different symmetries and characteristic energies. These are known as atomic orbitals. Electrons, it transpired, could not just be in any old place but had to reside in one of these orbitals. If they jumped from one orbital to another they had to give up energy (or absorb energy) as light of a well-defined colour. For atoms with many electrons (that is, all atoms except hydrogen), the formalism becomes very complicated, and solutions to the Schrödinger equation can only be obtained by numerical methods. Nevertheless, the symmetries of the orbitals from the hydrogen atom can be carried over to atoms with many electrons. Metaphorically, as you add more electrons you fill up orbitals with ever more complicated symmetry. With some wonder, it was noted that the enumeration of the possible orbitals provided the theoretical basis for Dmitri Mendeleev's periodic table of the elements. Furthermore, it became possible to understand what chemical bonds were: they were electrons sharing orbitals between two atoms. Suddenly chemistry became a numerical science like physics.

In the 1920s, Lennard-Jones had been Professor of Theoretical Physics at Bristol where he produced a very important method for computing molecular orbitals (the distribution of electrons in a molecule) out of mixtures of atomic orbitals. One of Lennard-Jones' pupils was Charles Coulson, Professor of Theoretical Physics at King's College London, whose work on the electronic structure of organic molecules had so excited Aaron while he was working on triphenylene in Cape Town. This looked like Aaron's métier. Nevertheless, despite his early enthusiasm, after his discussion with Lennard-Jones, Aaron decided he did not wish to become just another theoretical chemist.

One of the effects of Bragg having come to Cambridge from Manchester was that it encouraged a co-migration of colleagues from Manchester, all of whom were friends of James and therefore available to Aaron. Douglas R. Hartree, Plummer Professor of Mathematical Physics in Cambridge, had been in Manchester until 1946. Hartree was very good at sums, or to put it more formally, was an expert in numerical analysis. In the 1920s, he was thrilled by the Bohr theory of the atom, the precursor of quantum mechanics, and his PhD was on numerical solutions to Bohr's equations. As he finished it, Schrödinger's equation was published. Hartree immediately set to work to obtain numerical solutions.

Schrödinger's equation is a differential equation. Such equations tell you how much one property of some problem alters when you make a small change in another property. This provides a method for getting numerical solutions: you make a small change in a basic property of the system (such as time) and work out from the differential equation what this does to the output (such as distance). Then by adding up (integrating) the outputs from different terms in the differential equation, you end up with, for example, the trajectory of a cricket ball. However, after a few times of going through this procedure with a pencil and paper, boredom sets in and one begins to devise machines to take out the drudgery. The use of mechanical devices to solve equations started in the nineteenth century with Charles Babbage's 'difference engines'. Babbage's machines manipulated numbers and were basically mechanical computers. The results of a calculation were printed out as numbers. Thus Lord Byron's daughter, Ada Lovelace, who worked with Babbage, could claim to be the world's first computer programmer. Another approach was the 'differential analyser', which was invented in 1876 by James Thomson, brother of Lord Kelvin. This was an analogue device. It did not deal in numbers but rather used cams and gears to produce movements responding to some input, such as time. The crux of the system was the integrator: this produced a final movement that was the sum of all the movements arising from the input movement, for example by having cams riding on cams, or summing torques by twisting a rod. Some astronomical clocks are built round integrators (the Strasbourg Astronomical Clock from 1842 in the Cathedral has three cam-type integrators for summing Fourier series: see Figure 3.2). The output is a displacement (for instance a pointer on a dial). About

Figure 3.2 The Strasbourg Astronomical Clock contains three mechanical integrators for summing Fourier series that calculate the positions of the sun and the moon. The series is summed mechanically by using cams riding on cams. The cams have a cosine wave profile that repeats once, twice, three or more times per revolution. The size of the profile determines the strength of contribution to the summation. The left integrator has two terms in the series, the centre six terms and the right integrator just one. (From Wikipedia; photo by David Iliff. Licence: CC-BY-SA 3.0)

1931, a practical version of Thomson's differential analyser was constructed by Henry Nieman and Vannevar Bush at the Massachusetts Institute of Technology (MIT) in Cambridge, Massachusetts.

Douglas Hartree constructed a version of the MIT machine at Manchester and became an enthusiastic user. During the Second World War, Hartree and his differential analyser became an essential part of the national war effort, doing many of the design calculations for the magnetron that was the mainstay of British radar. One of the problems with which he and his analyser were involved was how to make steel more rapidly. During the war, the usual practice in the steel industry was to take a white-hot iron ingot, quench it in water and wait for nine hours. Could one do better? It turns out that this is a difficult problem to simulate. Nevertheless, using an approximate theory, Hartree did manage to show that you could produce complete transformation to steel in four hours, which was a huge saving.

Through the Manchester connection, Aaron went to see Hartree, who suggested that for his PhD he might sort out the steel problem properly. The iron used for making steel contains carbon dissolved in the liquid iron. When steel solidifies it initially forms a 'solid solution', with the carbon remaining dissolved throughout the iron to create an iron–carbon phase called austenite. As the temperature falls, grains of ferrite, which is pure iron, start to form out of the austenite. As more grains of ferrite form, the remaining austenite becomes richer in carbon. At about 723 °C, the austenite, which now contains 0.8% carbon, changes to pearlite. The structure that results is a mixture consisting of white grains of ferrite intermingled with darker grains of pearlite. Each of these transitions gives rise to a burst of heat. The properties of a steel depend on the details of the final mix. To understand all the different crystal forms, you need a knowledge of crystallography; to work out how the heat gets out requires a good head for calculations. Aaron seemed to have the requisite talents. However, it turned out that the problem needed more: it needed the concept of the speed of nucleation of the different crystal forms of steel and how their growth continually modified the properties of the bulk ingot of steel. Moreover, the rate of formation of the nuclei depends upon the rate of cooling. These ideas came from Aaron: he managed to develop a mathematical model of the austenite to pearlite phase transition, realizing there were two problems, nucleation of a new crystal and subsequent growth. He derived a differential equation that introduced a new time scale for each nucleation. By numerical analysis of his equation, he was able to calculate the rate of growth of new phases and got a very good fit with the available experimental data. From the mass of technical literature, Aaron was able to deduce the dependencies of nucleation and growth on the ambient temperature. He also drew on results from simple experiments with cylindrical bars that he had set up with Hartree at Baldwin's steel mill in Sheffield.

The problem appealed to Aaron because its solution was useful to society. Nevertheless, Aaron's thesis was never published, which accounts for the fact that in the ensuing 50 years his results have been rediscovered in various forms without any reference to his pioneering work. In later life he was always somewhat dismissive of his steel work, proudly remembering only the ideas of nucleation and growth as the concepts necessary to explain how and why the phase changes happened during cooling.

Aaron needed to solve a partial differential equation for his work. He started by hand using a Brunswick calculator. Very soon, however, better methods became available. A motor-driven Marchant calculator could multiply and divide numbers through the offices of a fearsome collection of gears. Better than gears, electrical circuits and relays can be arranged to store numbers; even better are vacuum tubes (in UK English, valves) used as relays. To do anything useful you also need circuits that can add the numbers together. Then, by repetition of a set of instructions, one can calculate anything. The first important electronic calculator of this type was ENIAC (Electronic Numerical Integrator and Computer) built in the Moore School of Engineering in Philadelphia. ENIAC was designed to calculate artillery firing tables for the US Army, but its first serious use was in calculations for the hydrogen bomb. For a particular problem, the list of instructions that had to be carried out was set up by connecting cables and rows of toggle switches. ENIAC could not store numbers (or not many). It read in and put out answers on IBM punched cards. Hartree became an ENIAC enthusiast. But ENIAC was not quite a computer. John von Neumann and the group at the Moore School were thinking ahead and produced the influential report entitled 'Draft report on the EDVAC' (Electronic Discrete Variable Automatic Computer). EDVAC was a stored program computer with the same store for numbers and instructions, and with instructions executed serially.

But what constitutes a store? Nowadays numbers (and instructions) are mostly stored as magnetic blips on a thin coating of a magnetic oxide of iron that is somehow moved past a coil of wire to sense the magnetism. However, the technology of magnetic drums and tapes was not worked out until the 1950s. In the late 1940s, in a technology taken over from radar, numbers were stored in mercury delay lines: a loudspeaker makes a noise into one end of a long tube filled with mercury, and a few tens of milliseconds later a microphone can hear the sound arrive at the other end. Then you feed it back in again. Hence, circulating pulses of sound can be used to represent numbers.

In 1937, John Lennard-Jones founded the Cambridge University Mathematical Laboratory. Maurice Wilkes had spent the war working on radar and joined the Mathematical Laboratory after the war. In 1946, he visited the Moore School and was so convinced by von Neumann's plans for EDVAC that he went home and built one. EDSAC (the Electronic Delay Storage Automatic Calculator) was built with a small budget and with only two assistants, but even so it took up two floors in the old Anatomy School adjoining the Cavendish. EDSAC executed its first program on 6th May 1949. This was the first full-scale electronic computer to implement the stored-program principle. EDSAC became Aaron's differential analyser.

Programming in those days was a bit more strenuous than using a personal computer today. Instructions to the machine were prepared on punched paper tape that was read into the computer at the beginning of a session. One had to tell the machine in great detail what it should do. Numbers had to be fetched from the mercury delay lines and put into local stores (accumulators, made from vacuum tubes). Then the current instruction said how the numbers in these stores should be manipulated (added, subtracted, multiplied or divided) and where they should be put back into the delay line memory. Then the next instruction had to be found from the delay line and be worked on. This went on and on until the procedure is finished. The last instruction told the computer to put the answers out on punched tape. Just to keep you on your toes, there was a good chance of one of these 10,000 vacuum tubes (all ex-radar stock after the war, much of it German) packing up while you were working, which put you back to square one.

Alongside Aaron were John Kendrew and John Bennett, working out the world's first program to sum up a Fourier series on a computer (a digital version of the Strasbourg clock). During the long nights, John Kendrew's wife Elizabeth used to bring them coffee and sustenance (John's home was nearby in Tennis Court Road). With the energy of youth, Aaron finally got all his programs to work. However, in later life he was always wary of going too close to computers and avoided ever having to send an e-mail himself.

Aaron worked in the theoreticians' room at the Cavendish, a huge room with big windows down to the ground. His fellow research workers included Oliver Penrose, Charles Kuper and Stefan Machlup. Machlup had been a graduate student at Yale and was now a postdoctoral researcher in Cambridge. The impetus for his visit was provided by the sabbatical of his PhD supervisor, the Nobel Laureate Lars Onsager, who came as a Fulbright Scholar to work with David Schoenberg at the Mond Laboratory. Machlup was born in Vienna in 1927 and emigrated as a child to the USA. He shared a Jewish background with Aaron but unlike Aaron was a very bouncy fellow. He was fluent in five languages and an accomplished cellist. Arising out of his time in Cambridge, Machlup published with Onsager two frequently cited papers on the Onsager–Machlup–Laplace approximation. The theory addresses losses in thermodynamic systems not in equilibrium and applies to a wide variety of constituents and forces: nuclei in magnetic fields, atoms in a laser, molecules in chemical reactions, and even ions passing through biological membranes.

Onsager's original theory, for which he was awarded the Nobel Prize in Chemistry in 1968, applied to the coupling of forces and flows in changing systems near equilibrium (for instance a slowly flowing river). Pressure differences will lead matter (water, in our example) to flow from high-pressure to low-pressure regions, and temperature differences will lead to heat flow from the warmer to the colder parts of the system. When both pressure and temperature vary, it can be observed that pressure differences can cause heat flow, and temperature differences can cause matter flow. Even more surprisingly, the heat flow per unit of pressure difference and the matter (water) flow per unit of temperature difference are equal. Onsager showed that this symmetry was a necessary consequence of the statistical behaviour of molecules moving around. Aaron was fascinated by the Onsager relationships and incorporated his own extensions of the theory into his PhD thesis.

Cambridge is a great place for a polymath, and with Liebe often away in London, Aaron had time to spare. He sat in on the Part III Mathematics Tripos lectures, particularly on group theory and Dirac's quantum electrodynamics that had predicted the existence of the positron (a positively charged electron). The ability even to understand Dirac's esoteric formalism is only granted to a few gifted initiates. Naturally, it delighted Aaron.

Vivian Rakoff, Aaron's close friend from the Cape Town days, had whetted Aaron's appetite for the controversial English literature specialist F. R. Leavis, and Aaron attended his lectures. Leavis applied rigorous intellectual standards and attacked dilettante elitism, which for example he thought characterised the Bloomsbury group. In 1948, Leavis made a general statement about the English novel in *The Great Tradition* where he traced a path through Jane Austen, George Eliot, Henry James and Joseph Conrad to D. H. Lawrence. Leavis's main tenet was that great novelists showed an intense moral interest in life, which determined the form of their works. This echoed Aaron's humanist canon, which had been formed while observing the dreadful inequalities of apartheid. In the Leavis lectures Aaron made friends with the future novelist and literary critic Dan Jacobson and his wife Margaret.

Later, Leavis vigorously attacked C. P. Snow's suggestion that practitioners of the scientific and humanistic disciplines should have some significant understanding of each other, and that a lack of knowledge of twentieth-century physics was comparable to an ignorance of Shakespeare. Leavis's angry rejection of Snow and the 'two cultures' would have found little resonance with Aaron; but fortunately for his peace of mind, the confrontation between Leavis and Snow would not happen for another ten years.

In Cambridge, Aaron met his Cape Town colleague Allan Cormack, with whom he had had the alarming climbing experience on Table Mountain two years before. East Anglia is not renowned for its mountains, and apart from the risky nocturnal ascent of King's College Chapel there was not much climbing around Cambridge. Cormack had been trying to do his PhD in Cambridge; he was attached to St John's College, and his PhD Supervisor was Otto Frisch. In Birmingham in the summer of 1939, Otto Frisch and Rudolf Peierls had demonstrated that the splitting of uranium-235 would create a chain reaction that could be used to develop an extremely destructive weapon. After some delay, Otto Frisch was granted British citizenship and allowed to participate in the Manhattan project. In 1947 he was appointed Jacksonian Professor of Natural Philosophy at the Cavendish. Unfortunately, although he was a much-liked lecturer, his skill at running a research department left much to be desired. After two years of frustration and no results, Cormack got the offer of a lectureship from James back in Cape Town. Since Cormack, like Aaron, was a married PhD student the offer of a secure post seemed more important than a PhD. Soon after Aaron arrived in Cambridge, Cormack returned to Cape Town. He would never obtain a PhD; but shortly before Aaron, he was awarded a Nobel Prize.

Liebe and Aaron went to the Arts Cinema often: it was relatively cheap (1/6d – one shilling and six pence) and the programme changed twice a week. Nevertheless, while Aaron was accepted and integrated into the Cambridge scene, Liebe had neither friends or acquaintances in Cambridge, nor did she feel adequately fulfilled just by going to the Arts Cinema. She had hoped and expected that she could study with the Jooss–Leeder School of Modern Dance, which had been sited in Cambridge throughout the War. Unfortunately, in the autumn of 1949 the Jooss–Leeder School was no more.

At the beginning of the twentieth century, Isadora Duncan had developed a new form of dance based on classical Greek models, untrammelled either by the rigid rules of classical ballet or by their tight-fitting shoes. At the time, Isadora Duncan found more resonance in Europe than in her native USA so that, backed by the wealth of her lover's Singer Sewing Machine Empire, she opened an influential dance school in Berlin. During the Weimar Republic there was an explosion of dance activity – 'Ausdruckstanz' or expressionistic dance – in which Rudolf von Laban played a leading role. In 1925, Kurt Jooss and Sigurd Leeder, both students of von Laban, opened a new dance school, the 'Westfälische Akademie für Bewegung, Sprache und Musik', in Munster. In 1927, Jooss moved the Westfälische Akademie to Essen, and it became the 'Folkwangschule'[2]. Jooss's most important choreographic work, 'The Green Table', won first prize at an international competition for new choreography in Paris in 1932. It was a strong anti-war statement. On account of their ballet having Jewish members, in 1933 Jooss and his company had to flee Germany. They eventually found support from Dorothy and Leonard Elmhirst, the philanthropic owners of Dartington Hall in Devon. Dorothy's interest in dance arose from observing the development of Martha Graham's Modern Dance in

---

[2] Information about Leeder–Jooss has been obtained from *Dancing in Utopia: Dartington Hall and its Dancers* by Larraine Nicholas (Dance Books, 2007).

New York; Leonard had grown up in the British Raj and was friendly with Rabindranath Tagore (revered by Liebe's father) who enthusiastically furthered Indian dance. Thus the Elmhirsts welcomed the opportunity of establishing a world class school of modern dance. The Jooss–Leeder School of Dance was founded at Dartington Hall in 1934. There was also a companion touring company, the Ballets Jooss.

But after six successful years, this important experiment in establishing modern dance in England was terminated in 1940 because both Jooss and Leeder were interned as enemy aliens on the Isle of Man. They were released in 1941 and reformed the Ballets Jooss and the School of Dance, but the Dartington Hall connection was no longer available. Instead they finally found a home in Cambridge. They were put up in a large house on the western perimeter of Cambridge belonging to the Roughtons, 9 Adams Road. F. J. W. (Jack) Roughton was professor of colloid science. Alice Roughton, who was a family doctor, helped look after the dancers, and even installed a dance floor in the large living room. The Jooss School was supported by CEMA (the precursor of the British Arts Council), which was chaired by John Maynard Keynes. He, much to the dismay of his associates in the Bloomsbury group, had married a famous ballet dancer, Lydia Lopokova. Jooss added new works to his repertoire, including 'Pandora' (1944), containing disturbing images of human disaster and tragedy, which was later interpreted as foretelling the dropping of the atom bombs on Japan. Disappointingly, it was not a success[3]. Moreover, in spite of CEMA support, 9 Adams Road lacked the magic of Dartington Hall. Soon after the war, Kurt Jooss closed his ballet to take up an appointment in South America. In 1949 he returned to Essen where he re-established the Folkwangschule. In 1947, Sigurd Leeder opened his own school at Morley College, London, behind Waterloo Station, where he worked out his own teaching method, later known as the European Contemporary Dance Technique.

It became clear that Liebe's future lay in London, which could offer her both dance and friends. Thus on 5th December 1949, Liebe moved to digs in London on a weekend commuting basis. At the weekend, Liebe sometimes returned to Cambridge, sometimes Aaron went to London. Soon they were joined by Vivian Rakoff.

---

[3] In 'Dr. Alice Roughton', obituary by John Gregory. *The Independent* (8 July 1995).

Rakoff had travelled on the mail ship to England with Ralph Hirschowitz and his first wife Tiby (Thelma), intending to leave them in England to go to Cambridge while they went on to join an immigrant ship in the south of France. But somehow this never happened. Instead of leaving them, in fact Rakoff travelled with the young couple to Israel. Here he joined a kibbutz on Mount Tabor, where a group of break-away Hashomer Hatzairs had already settled. He stayed there for about a year, and was there when Aaron and Liebe married. Then Rakoff returned to Cape Town by the same circuitous route. He decided he no longer wanted to be an actor but would rather pursue a solid middle-class career, so he took a Master's degree in psychology. In December 1949, after completing his Masters, Rakoff came to London to study medicine at University College Hospital. In due course he completed his medical degree and married his first wife Judy, who also came from Cape Town. However, the marriage was not a success: they returned to Cape Town and divorced. Rakoff returned to London and took up psychiatry. For some years he himself was analysed by Willi Hoffer, a student of Freud. Later, he returned to Cape Town, married his present wife Gina and then emigrated to Montreal, where he did an internship in psychiatry at McGill University. Then the family, now including three children, moved on to Toronto where Rakoff became Chair of the Department of Psychiatry and had an Institute named after him. He was much sought after for radio and television interviews, where no doubt his acting experience came to the fore. He also wrote a number of plays for radio and television.

Liebe (and Aaron at weekends) decided to share a London flat with Rakoff (who had come to terms with Liebe's marriage to Aaron) and another couple, the Fanaroffs. They found a flat in Clanricarde Gardens off Bayswater Road, where nowadays a one-room flat fetches £500,000. In those days, Notting Hill was not so salubrious. Indeed, prostitutes used to line both sides of Bayswater Road. The flat was filthy and no chair stood upright. Liebe's companions called it Clan-rickety Gardens. Nevertheless, although they had no money and food was rationed, they had plenty of fun. Refugees from apartheid South Africa were always turning up. Liebe started attending Dance School with Sigurd Leeder in a drafty church hall in Loudon Road, Swiss Cottage. In the second year it moved to somewhere behind King's Cross Station. At the same time, Liebe was trying to teach dance at a school in Clerkenwell. After the

War, this part of London suffered an economic decline. The school turned out to be a hellhole: no other teacher lasted for more than one term, nevertheless Liebe survived for two.

Liebe attended the Leeder School from January 1950 until the middle of 1951. However, the Leeder method was very structured and technical. In Cape Town, Liebe had had a teacher called Lou van Eyck, who had danced with the Ballets Jooss. He came to London and watched a day of classes. He confirmed what Liebe already had been feeling: while the Leeder method was successful for some students, for her the classes were arid (as well as expensive). Leeder's rigid dance notation left no room for improvisation. Van Eyck felt that Liebe would be better off going to a selection of classical ballet, or indeed any other sort of dance.

In the summer of 1951, quite by chance, Liebe attended performances of a West Indian group – the Boscoe Holder Dance Company – in a little theatre off the Strand. Arthur Aldwyn Holder (known as Boscoe) was a dancer, choreographer and artist. Born in Trinidad in 1921, he showed early talent as painter and pianist. As he grew up he became fascinated by his island's culture. He researched and learned the local dances and songs, and by the late 1930s had a dance company depicting the music, songs and dances of Trinidad. In 1948, Holder married a member of his dance company, Sheila Clarke, and two years later with their son Christian they settled in London. Holder was befriended by the designer Oliver Messel, who introduced him to his Mayfair friends, including Noël Coward. In 1950, Boscoe Holder and his Caribbean Dancers, with Sheila Clarke in the lead, had their own show on BBC television, Bal Creole. His dance company performed at the Queen's coronation. Liebe joined the company and danced with them from the middle of 1951 until the Klugs departed for South Africa in the autumn of 1952. No doubt influenced by her South African childhood, Liebe showed great talent for dancing these ethnic forms. She performed a number of times with the troupe in a theatre off Leicester Square. Vivian Rakoff came to one of the performances and remarked acidly that Liebe stood out: firstly, she was technically much better than most of the troupe; secondly, in no way could she be construed as being West Indian. Aaron came when he was in London and encouraged Liebe to continue dancing with the Boscoe Holder Troupe.

In the summer of 1950, Liebe and Aaron had at last visited Israel (of which more in the next chapter). The Cambridge University Jewish Society organised a tour including eight weeks as members of a kibbutz. Since they would be away for three months, Liebe gave up her place in Clanricarde Gardens. When they came back, Liebe lived for some months in a bed-sit in Belsize Park Gardens, just north of Primrose Hill, next to the house where her son David and his family live now. When the money ran out she became an *au pair* with the Golbergs, a South African couple who lived in a flat in Ennismore Gardens in South Kensington. He was a chest physician with an expertise in tuberculosis. They had a severely disabled child, and Liebe's main job was to wash the nappies (diapers). However, Liebe blotted her copy-book by not turning up after one weekend in Cambridge, and was replaced. In spite of this hiccup the Klugs and Golbergs stayed on friendly terms. Later, when the Klugs were living in London, they traditionally went to the Golbergs in Richmond for lunch on Boxing Day.

So for a few months Liebe bunked down and lent a hand with her old dance teacher Jennifer Craig from the 'School of Charm' in Cape Town, where Liebe had first started to dance seriously. Then for a short time she stayed in a flat that Vivian was sharing in St John's Wood. Subsequently, while she was rehearsing and dancing with the Boscoe Holder Troupe, she came to rest in a flat-share on the Bayswater road, near where she had started out.

In the meantime, in Cambridge, Aaron's lodgings were no less transitory. By April 1950, Vivian Fuchs's ship had been released from the ice and he was coming home. The Klugs were asked to leave. Aaron moved to share a flat in Chesterton Road with Doris Krook, an acknowledged expert on the later writings of Henry James, a fellow of Newnham and ex-University of Cape Town. There was drama associated with Doris: back in Cape Town, it was her spurning of his advances that had moved the future brother Elias to attempt suicide. Aaron had a tiny room in Doris's flat and shared her kitchen and bathroom. In the garden of the house next door, Aaron used to see a pretty young woman hanging out the washing. Years later Aaron learned that this was Stella Porter, wife of George Porter. He, like Aaron, was a future President of the Royal Society, and perhaps just as important, the future PhD supervisor of their second son David. While head of the Royal Institution, Porter mentored David to become a successful scientist.

In July 1950, Aaron moved again, to share a flat in Green Street. This apartment had views over the housetops of the town and, more important, a back door direct into Rose Crescent, which is but a stone's throw from the Arts Cinema. Aaron (and Liebe at weekends) shared with Richard, who was a librarian at the University Library, and Cecily, a teacher. This was very much an "open house" – anyone who cared to could pop in. The kitchen utensils were washed in the bathroom as the kitchen had no running water. The lavatory was two floors below. Later, their final share from June to September 1952 was with Asher and Shirley Korner in Luard Road.

Since Liebe was away in London during the week, Aaron had plenty of opportunity to eat in College, but somehow it never seemed to fit with EDSAC schedules; so he often ate with friends or made do with fish and chips in King's Street. He did not enjoy living without Liebe. Aaron dreaded the weekly partings and sought consolation in poetry. In one of his midweek letters[4] to Liebe, he quotes from W.B. Yeats collection 'The Wild Swans at Coole':

> Said Solomon to Sheba,
> And kissed her Arab eyes,
> 'There's not a man or woman
> Born under the skies
> Dare match in learning with us two,
> And all day long we have found
> There's not a thing but love can make
> The world a narrow pound.'

While Cambridge winters are cold, grey and miserable, the months of May and June are a delight. 'The Backs', the college gardens along the river, become a kaleidoscope of colour. The fragrant flowerbeds along the 'Wedding Cake' of St John's merit international recognition. Moreover, the dusk is so long and the dawn is so early that the hours of darkness are just an intermezzo. The College May Balls (held in the first week of June) are conducted in this magical ambience. Like Aaron, Asher Korner was at Trinity, but the Trinity May Ball was too expensive for impecunious pre-docs. However, at 2.00 a.m. the back gates of the

---

[4] Churchill Archives Centre: Klug Papers.

colleges onto Queen's Road were opened to facilitate a mingling of the summer night revellers. Thus, at 2.00 a.m., Asher, his wife-to-be Shirley, Liebe and a few other friends gate-crashed the Trinity May Ball. Naturally, one had to look the part. Liebe had it easy since her mother, apprehensive of clothes rationing, had provided her with a complete wardrobe for three years. They simply strolled in from Queen's Road and mingled. It was a warm summer's night, and there was music and a general air of gaiety – but the event was not as exciting as they might have hoped. That was the summer of 1950, and life was still a bit on the drab side in Blighty.

The Backs have more to offer. In May Week, the Cambridge Madrigal Society sings a concert from a raft of punts moored next to Trinity bridge. As dusk approaches, lanterns are lit and the raft is allowed to drift languidly past St John's under the Bridge of Sighs. Traditionally, the last item is John Wilbye's melancholy six-part madrigal 'Draw on sweet night – to shades of darkness find some ease from paining'. This is an enchanting experience. The Klugs had made friends with another South African couple, Margaret and Murray Carlin, who had two children and lived in the Grape House, Grantchester; Murray was studying for the English Tripos. During May 1951, Aaron and Liebe were walking to see the Carlins. The mile walk to Grantchester through the water meadows of the Cam is a delight, and they stopped on the way to picnic near the river. Suddenly they were regaled by ethereal music such as bewitched Ferdinand in *The Tempest*. It turned out, more prosaically, that the Cambridge Madrigal Society was rehearsing its May Week Concert on the river near Grantchester.

A couple of weeks later, Aaron went to talk over some aspects of his thesis work with Neville Mott, the Professor of Physics at Bristol. Murray Carlin was staying in town to swot up for his finals. In a break one evening, he invited Liebe to have coffee with him in Petty Cury, where she was introduced to Norman Podhoretz, who was also studying for his finals. Podhoretz was a clever Jewish student from Columbia University in New York, the star student of the great literary critic Lionel Trilling. At this time Norman was engaged to Jacqueline Clarke, who had been Harold Laski's secretary at the London School of Economics (LSE); she later married Huw Wheldon of TV Monitor fame and became a successful author. Laski recognised her potential and organised a bursary for her to be a mature student at LSE. The May

Balls were looming after finals. Podhoretz had a ticket for the Clare College May Ball but also had an infected big toe, so that he could barely walk. Moreover, it transpired that Jackie Clarke did not want to interrupt her studies to come up for the May Ball, so to use his ticket he invited Liebe to join him. Out of sympathy for his plight she agreed (moreover, an invitation to a May Ball is not to be turned down lightly). Liebe's store of clothes was at last depleted so she borrowed a ball gown from an American friend. Unfortunately, the friend was nearly six feet tall, a height difference too great to be accommodated with heels, so Liebe spent a day taking up the hem. She went to the ball, and spent most of the evening at Clare College in a strictly non-romantic mode commiserating with Norman Podhoretz. However, Liebe felt that she had created a situation in need of resolution, so she invited Podhoretz to meet her husband in Luard Road. Aaron and Podhoretz disliked each other at first sight. Yet there was a spark of recognition between them. Over the years, Podhoretz and subsequently his wife, Midge Decter, became close friends of the Klugs. Through Podhoretz, they became acquainted with Steven Marcus, who had also studied at Columbia with Lionel Trilling. Aaron, Steven Marcus and Norman Podhoretz became firm buddies, so much so that Liebe subsequently referred to them as 'The Three Musketeers'.

In the autumn of 1952, at the end of the three years allocated for the PhD, Aaron and Liebe were keen to get back to Cape Town. Aaron's 1851 Exhibition Scholarship included a return first-class ticket. Liebe went to the Headquarters of Union Castle Line in the City to try to persuade them to change this into two cheaper tourist-class tickets, but they refused. Liebe had to buy another cabin-class ticket and Aaron had to downgrade. They had been away for three years, but now the return date was fixed, even though Aaron was now assimilated into the Cambridge world and some large part of him wanted to stay. Rather belatedly, he decided to try for a Junior Research Fellowship at Trinity College. These Fellowships are awarded on the basis of publications, often a PhD thesis. Aaron suddenly realised that his thesis had to be in College before they sailed, which left them scarcely two weeks to get the whole thing together – and six copies were required. The Klugs were now residing with Asher and Shirley Korner. Shirley and friends helped with the organisation of Aaron's thesis. There ensued a frantic typing with five carbon copies, writing in of mathematical formulae and sticking in of figures and getting it all to the binders. Finally, on the evening before the sailing, a copy of the thesis *The Kinetics of Phase Changes in Solids* was deposited at Trinity Porters' Lodge. Then at the crack of dawn, driven by Asher, they were off to Southampton with a picnic breakfast in Windsor Great Park to get the mail boat to Cape Town. The first Cambridge adventure was over.

4

The First Visit to Israel

Let us make a small digression here, before returning in the next chapter to the young couple's arrival in Cape Town. As mentioned in the previous chapter, in the summer of 1950, after their first year in Cambridge, Aaron and Liebe decided to visit Israel. Bennie Klug and his wife Lucy were already in Israel. Bennie had sent a long letter warning his kid brother that Israel would not suit him. Nevertheless, after all the Hashomer Hatzair indoctrination, one would have to have a look. Summer visits were organised by the Cambridge Jewish Society in conjunction with the Jewish Agency. The students would travel around Israel and then work as volunteers in a kibbutz for a few weeks. Thus, at the appointed time, the group from Cambridge proceeded by ferry and rail to Marseilles and then embarked in a hulk that plied between Morocco and Israel transporting displaced Sephardic Jews from North Africa to the new homeland. The trip in this creaking unseaworthy vessel was an experience better missed. It flew a Panamanian flag, and had a mixed crew and a Greek captain, who was apparently often drunk. Living arrangements were primitive and unpleasant. The students slept on straw palliasses on deck. This was preferable to accommodation in the hold of the ship, which was hot, crowded and smelly. The Moroccans tended to keep their women below deck.

Food, mostly unripe dates and fruit, was served on deck in buckets. The Moroccans were always hungry so that even securing this lousy food became a serious undertaking. Some tarpaulins had been rigged

up to offer protection from the sun. However, there were no arrangements for washing or showering. Salt-water showers existed but these
were reserved for the officers. Since it was the height of summer, the
passengers were fast becoming odiferous. The students rebelled and
managed to persuade the Captain to agree to a time-table for the use
of the showers, one which actually included the Moroccan women.

In the evenings the Moroccan men would come on deck. They would
chat to the students in French and play stringed instruments or sing.
On account of its limited seaworthiness the ship tended to hug the
coastline. One night about two days out, en route to the straits of
Messina, when the moon was full and the men were playing and
singing, they sailed close to the active Stromboli volcano where they
were transcendentally illuminated by Vulcan's fireworks. Someone suggested that they should dance. Thus in this magical ambience Liebe,
clad only in a pale-blue swimsuit and diaphanous scarves, danced
Isadora-Duncan-like to Sephardic music.

South of Crete, the sea became rough and Aaron was sea-sick; but
eventually they reached Haifa. On arrival, all were sprayed with DDT
powder. After fumigation and immigration, they spent the night in
the town. Strolling around in the evening, the Klugs met one of the
immigration officers (still correctly dressed in his cream-coloured suit)
who offered to drive them to the top of Mount Carmel. He also proudly
pointed out the first and only 'ramzor' (stoplight) in Israel.

Mount Carmel is a long ridge running northwest–southeast. From
the northwest end, which runs out into the sea, the Klugs were able to
enjoy the wonderful view down to the Mediterranean. From ancient
times Mount Carmel's luxuriant verdure, arising from the vicinity of the
sea and abundant dew, was regarded as singularly beautiful. Moreover,
Mt Carmel has long been regarded as sacred – once the residence of
Elijah. Aaron was reminded of the famous contest between Elijah,
representing Yahweh the true God, and the priests of Baal. In spite of
its being one against 450, the contest turned out to be an easy win for
Elijah. This contest took place on the southeast end of the ridge near
Jezreel where Jezebel was subsequently murdered.

The next day, the group proceeded by bus to Jerusalem, from Israel
to Judea, a distance of 150 km. In Jerusalem, they were housed in a
school to the north of the old town. Because more summer visitors were
arriving than expected, within a day or two the school could no longer

cope with the influx. The water ran out, and the fire brigade had to be called to provide emergency water. Aaron's brother Bennie was at that time living in Tel Aviv where he was working as a civil engineer building a reservoir near Nazareth. Lucy was working in the Palestinian refugee camps. Aaron and Liebe took the opportunity of visiting them, as Tel Aviv is only 60 km away. The brothers had not met since Aaron and Liebe's wedding 18 months earlier, so there was plenty to talk about.

At that time in Israel, both food and water were in short supply. Moreover, public works such as main sewage were still being put in place. By the end of their stay in Jerusalem, many of the group had contracted 'shilshul'(diarrhoea). Nevertheless, after a few days they were all back in the buses and heading on to Beer-Sheva and the Negev. If you travel south from Jerusalem along route 40 towards Eilat, you come suddenly over the brow of a hill and the Negev is there. If it happens to be early morning and a Thursday, there will be swarms of black figures in flowing robes, outlined against the rising sun, coming down from the hills with camels and flocks of sheep to converge on a point that turns out to be the Bedouin market on the south side of Beer-Sheva. This gateway to the Negev is a place that once was little more than a watering hole for Abraham's sheep. Today Beer-Sheva is a city of 130,000 and home to Ben Gurion University. Modern Beer-Sheva, essentially founded by the Turks in 1900 as a railhead to help control the nomadic Bedouins, has become the effective capital city of the Negev. However, in 1950, apart from the small old Turkish town, Beer-Sheva was more like a frontier shanty-town out of the Wild West, which appealed to Aaron. Furthermore, the town caught Liebe's imagination. Thus a relationship with Beer-Sheva started that still enthrals them both. At that time there were many soldiers in town, mostly sitting around in jeeps. Movies in the evening were projected in the open on the side of a building. Accommodation was primitive, but if you're young and healthy, who cares?

Although called a 'desert', the Negev is not covered with sand. Rather, it is a mix of brown, rocky mountains and deep craters interrupted by wadis that bloom briefly after rain. It contains sites of great natural beauty, such as Ein Avdat, a spectacularly verdant desert gorge. Nearby are the ruins of the ancient town of Avdat, once an important part of the Nabataean trade route. The students explored the Negev on the backs of trucks, which turned out to be rather hot. They went as far south as the remains of the city of Subeta. Subeta (or Shivta) is another Nabataean city on the old spice route. The Nabataeans, who were Romanised Arabs, were masters of water conservation and even managed to grow grapes in Shivta.

After the fun came the work. The group split up. Aaron and Liebe had been allocated a kibbutz in the far north where they were going to work for six weeks. Sde Nehemia (Nehemia Field) is a small kibbutz in Upper Galilee, near the northern extremity of biblical Israel. The Jordan river originates in the kibbutz's grounds at the point where the Banias and Hasbani rivers meet. Both rivers flow off Mount Hermon and are always cold.

Sde Nehemia was founded in 1940 by immigrants from Austria and Holland on land purchased from the Arab village of al-Dawwara. It is a small (500 inhabitants) kibbutz located in the Hula valley between the Golan Heights and Lebanon. The area is blessed with a Mediterranean climate and is very fertile. The first modern Jewish settlement in the Hula valley was in 1883, during the first Aliyah. However, malaria was a serious problem and no additional Jewish settlements were established in the valley until 1939. Initially, these settlements also suffered from malaria, but the use of pesticides such as DDT eventually conquered the problem. In 1948 there were 12 Jewish and 23 Arabic settlements in the Hula valley. After the establishment of the State of Israel and the ensuing 1948 War of Independence, the Arab inhabitants fled from the valley.

Geologically, the Hula valley is an extension of the African rift valley. To the south of Sde Nehemia there is a basaltic outcrop that partly dams the valley. This gave rise to the large shallow Lake Hula and accompanying swamps, which were important for migratory birds. During the Ottoman Empire there were some half-hearted attempts to drain the lake. Later, during the British Mandate the idea was resisted by the British officers because of the excellent duck-shooting the lake afforded. The newly arrived Dutch immigrants, on the other hand, saw drainage as a challenge, and during the early 1950s Lake Hula vanished. So did the ducks and over a hundred other indigenous species. Initially this was perceived as a great national achievement for the State of Israel, but it soon became evident that the agricultural benefits were limited. Thus, a small section of the former lake and swamp region was finally re-flooded in an attempt to prevent further soil deterioration.

Sde Nehemia was a solidly agricultural kibbutz. The newly arrived male volunteers were put to work in pairs to lift large bags of chick-peas onto trucks. The status of volunteer was often used as a trial period to see if a candidate member of the kibbutz was going to fit in. Because Israeli immigrants had the most diverse origins, the leader of the team, who was Dutch and held a doctorate, was sometimes confronted with almost insurmountable cultural differences. Thus one of the team was a volunteer from Baghdad. The team leader, noting the Baghdadi's lack of lifting activities, remonstrated with him about work in a mixture of Hebrew, Arabic and French, whereupon in a display of strength the volunteer lifted up a dozen sacks on his own, but then stopped. A further exchange elicited the indignant Arabic response that he had just worked! As the owner of a barber's shop in Baghdad he didn't see that he needed to do anymore.

The women attended to household services and cooking. The kibbutz was somewhat run down, as the women and children had been evacuated and most of the men were on army duty. The Klugs shared a small wooden hut with two American medical students, who were studying in Switzerland. The showers and toilets were communal as was the laundry. To avoid the heat everyone got up at 4 a.m. At 8 a.m. came the usual Israeli breakfast of yoghurt, cheeses, salads and olives, and then it was back to work until about noon. The rest of the day was free. Apart from sleeping, there were no other activities.

Liebe was sent to work in the kitchens, which entailed peeling huge mounds of potatoes. After that came the dishes. However, the girl in charge of the kitchen, who was Dutch, took a liking to Liebe and took her to visit the Banias river. They picked grapes growing wild in the nearby Arab village of al-Dawwara, abandoned in the 1948 War of Independence, and cooled the grapes in the stream. The significance of this empty village was not immediately appreciated by the young volunteers.

Nevertheless, the Arab presence could still make itself felt. The Syrians, sitting 500 metres up on the Golan Heights, would randomly lob mortar shells into the Hula valley. One of these hit a cow-shed, causing death and panic among some bovine members of the Kibbutz.

Since the kibbutz had had very little entertainment for a long while, the management requested that Liebe together with another young female visitor should give a dance recital. In preparation, they were given a week off work and allocated a very good pianist. The kibbutz had been given a grand piano that sat in the dining room. Between meals they rehearsed. The other dancer had classical ballet training, so she did her programme and Liebe did hers. There were no costumes. Liebe performed in a pair of shorts and a chulza (an embroidered white Cossack blouse), which had been given to her by Vivian when he returned from Israel. They danced barefoot, which was not much fun on a rough concrete floor.

On the night of the concert, the kibbutz invited a neighbouring kibbutz to join in. Thus there was a large crowd sitting on benches. No lights, no make up, no costumes, just a pianist. The programme, which included Gershwin's Rhapsody in Blue and a Brahms waltz, was well received. The success reinforced Liebe's teenage dream of starting a kibbutz dance company. The next day the Klugs left. The members of the kibbutz expressed their gratitude to Liebe in the tangible form of a dozen eggs, which was a real prize in those days of food shortages. The eggs were gratefully received by Bennie and Lucy, whom the Klugs then visited in Tel Aviv.

During their time in the kibbutz, it had become clear to Aaron that his vision for the future did not include the concept of lugging sacks of chick-peas around for 8 hours a day, nor was Liebe enamoured of peeling hundreds of kilos of potatoes. Conversely, the members of the kibbutz had no understanding of Aaron's wish to obtain a doctorate. The gulf between their young socialist idealism and the actuality of a small agricultural village was becoming apparent. In fact, for some time, Aaron had envisaged an alternative scenario whereby he would do science at the newly founded Weizmann Institute and Liebe would teach dance in a nearby kibbutz. His Cambridge PhD had still two years to run, therefore there was no need to hurry. Nevertheless, while Aaron was with his brother in Tel Aviv, just 30 km from the Weizmann Institute at Rehovot, it seemed like a good idea to enquire directly at the Weizmann if he might get employment there.

Chaim Weizmann had established the Daniel Sieff Research Institute in 1934 in Rehovot, on Israel's coastal plain. In 1949, by which time the Institute had grown to a multi-faculty research organisation, the Institute was renamed the Weizmann Institute. At this time Chaim Weizmann was President of Israel, and his architecturally famous Erich Mendelsohn house on the Institute Campus doubled as official residence for the Head of State. The running of the Institute was entrusted to Ernst Bergmann, a renowned chemist and long-time colleague of Weizmann. Bergmann was also scientific adviser to the Ministry of Defence and is remembered as the father of Israel's nuclear programme. Aaron was interviewed by Bergmann, who was apparently not impressed by the relevance of Aaron's thesis work to Israel's defence needs. Nothing came of it.

A couple of days later, the student group re-embarked in Haifa. This time the allocated ship was pleasantly unremarkable and the journey home uneventful for the Klugs.

5

Back to Cambridge

At the end of Chapter 3, we left the young Klugs embarking on the mail boat from Southampton. In the autumn of 1952, two weeks after leaving Cambridge, the Klugs docked in Cape Town for what was to prove a relatively short stay. They returned to Annadale Street, to the same room they had left three years earlier. After the long absence, they enthusiastically sought out family and friends. Liebe was thrilled to be back with her family; Aaron was at last able to talk things over with R. W. James. There was so much to tell. Later they visited Ralph Hirschowitz in Johannesburg, and then Aaron's family in Durban. Since they had but little money, they hitchhiked everywhere.

At this point, Aaron had no clear plans for his future. In the postdoctoral tradition, after writing a doctoral thesis, a scientist will go for a year or two to gather experience in another land. Aaron's paper in *Acta Crystallographica* on molecular transforms had caught the attention of David Sayre, a young American of theoretical bent who, in 1952, had written his doctoral thesis in the Chemical Crystallography laboratory in Oxford with Dorothy Hodgkin on a method of determining the (unknown) positive or negative signs of X-ray reflections. Aaron was invited by David Sayre to go for a postdoctoral year at the Johnson Foundation for Medical Physics in Philadelphia where Sayre himself had taken up an appointment after his sojourn in Oxford.

A recurrent theme in the first part of Aaron's biography is the crystallographic phase problem, already referred to in Chapter 2 (see also Appendix B). The basic problem of X-ray crystallography is that the X-ray diffraction pattern of a crystal only records half the information necessary to compute the electron density in a crystal. The amplitude of the scattered wave for each point in the diffraction pattern can be measured, but its positive or negative sign (or in general its phase) is not experimentally available. However, one can get help from the fact that crystals are composed of atoms. The profile of electron density of an atom looks like a Gauss function, and one of the wonderful properties of Gauss functions is that if you square them they look the same. Thus, if you square the electron density of an atomic structure made entirely of the same atoms (or similar atoms), the squared function looks the same. In early 1952, David Sayre published a much-cited paper describing relationships that must exist between the phases (in cases of favourable symmetry, these reduce to signs) of the Fourier components arising out of the fact that if you take the square of the electron density of an atomic structure it looks like itself. These relationships between the Fourier components are known as Sayre's equations. In many cases Sayre's equations do allow one to work out the missing signs. Then it is easy enough to compute the density of the atoms in the crystal and see where the atoms are.

Aaron was keen to join in the fun, but he ran into a serious problem: he could not get a visa for the United States. The year of 1952 was the height of McCarthyism. Aaron applied for a visa at the US consulate in Cape Town and was turned down on the grounds of having belonged to a student organisation that the South African government had deemed to be communist, and therefore being guilty of 'moral turpitude'. Nothing could be done. Aaron's father Lazar was very upset by the US immigration's accusing his son of being a communist, since he himself had risked his own life defending his brother from the Red Army. He wrote Aaron a long letter in English explaining how he had killed a Red Army soldier who was attempting to take his younger brother hostage. Subsequently, Lazar had hidden in a village near Ukmerge with a family he knew. He came back to Zelva to attend his brother's wedding, where he was recognised by one of the soldiers (he was of noticeably short stature, as mentioned earlier). He tried to make a run for it but was ridden down whilst attempting to cross a brook and thrown into jail. After a drumhead trial, the Red Army soldiers took him out to be shot. However, an officer rode past and proclaimed that Lazar must be tried by a proper court, and that this was no way for Bolsheviks to behave. At this moment, the tide of revolution turned and the Red Army detachment withdrew in a hurry, abandoning their prisoner. Thus Lazar survived to become Aaron's father[1].

Aaron needed a job. In Cambridge, Jack Roughton was Professor of Colloid Science. Between the wars, colloid science became a label for research based on the fashionable idea that understanding liquid crystals would explain the secret of life. Roughton's aims were in fact much more mundane. Roughton was a physiologist working on the kinetics of oxygen uptake by haemoglobin, the oxygen-carrying protein that makes blood red. He was a pioneer of kinetic measurements of enzyme reactions. A Fellow of Trinity and very much a part of the Cambridge scene, Roughton drove around in a magnificent old convertible Rolls Royce that doubled as his filing cabinet. Roughton's department was not extensive: it consisted basically of his technician Ken Edwards, and his secretary. Otherwise, in Free School Lane he presided over a mixed bag of other people's research. In 1925 he married Alice Hopkinson, the daughter of the Professor of Engineering, and in 1933 they inherited 9 Adams Road, a large house on the western edge of Cambridge, from Alice's mother, who was a Siemens. Dr Alice Roughton was a most unusual personality, and her regime at 9 Adams Road has become legendary. As we have recorded, during the Second World War she rescued the Ballets Jooss and took them into her home. She was renowned for her humanity and philanthropy but not for her cooking. From her obituary in *The Independent*[2]: 'The only safe meal to consume on the premises was breakfast: newly laid eggs and daily baked bread.' Alice was most hospitable and organised her house to be the home of postdoctoral fellows and students. Jack Roughton was an orderly and somewhat austere fellow, and to escape from the confusion he built himself an apartment over the garage[3]. A story circulated that a German visitor appeared in the department asking for the Professor. On being told that he was working at home he asked: 'Where lives he?' 9 Adams Road, came the answer. 'So is it not possible, I live at 9 Adams Rd.'

---

[1] This is a shortened version of the story as told to me by Aaron Klug.

[2] In 'Dr Alice Roughton', obituary by Teresa Deutsch. *The Independent* (29 June 1995).

[3] Q.H. Gibson. 'Francis John Worsley Roughton 1899–1972' *Biogr. Mems Fell. Roy. Soc.* (1973) **19**, 562–582.

Once, Aaron visited Jack Roughton in his office over the garage. Roughton looked down into the garden where his wife was holding court to an audience that included some farm animals and commented: 'She should have been the Dame of Sark.'

Roughton went to Hartree with a problem: how do you deal mathematically with the uptake of oxygen by red blood cells as the oxygen is diffusing through the red blood cells and interacting with haemoglobin? Here was a chemical reaction linked with diffusion, a problem very like the quenching of steel. Hartree said he had a chap who really knew how to do these things and proposed Aaron. Aaron quickly said yes, but because the post between the United Kingdom and South Africa was not fast, it took a couple of months to put the whole thing together.

In March 1953, Aaron returned to Cambridge to take the oral examination (viva) for his doctorate. His examiners were Hartree and Neville Mott from Bristol (the next year, Mott was to replace Bragg as Cavendish Professor). In a letter to Liebe, who was still in Cape Town, Aaron reported that the viva went badly. Later Mott reported that he remembered there were indeed long silences because he did not know what questions to ask. Nevertheless, Aaron was awarded his doctorate and started working with Roughton.

Aaron went to work in the Colloid Science Department in Cambridge to solve the reaction/diffusion equations Roughton had set up. He used the numerical method of finite differences. For this he could employ some of the programs he had written for EDSAC 1 for the steel work. He also went to Fribourg in Switzerland with Roughton to do some experiments with Ferdinand Kreuzer, who had developed an accurate method for measuring oxygen uptake by haemoglobin. Their measurements on the rate of uptake of oxygen by thin films of haemoglobin reawakened Aaron's interest in biological phenomena. He calculated the rate of oxygen uptake by haemoglobin in solution (without the diffusion problem). The haemoglobin molecule is a tetramer and contains four oxygen binding sites. The first three sites take up oxygen at random, but then something happens so that the fourth oxygen goes on much faster: the binding of oxygen is cooperative. Aaron calculated the time course of the reaction whilst sitting in the sun in a little library room high up in the laboratory, cranking a small Brunsviga calculator by hand for a couple of days. The resulting curve was a good approximation to the experimental curve, and so he determined the rates.

This work was published in the *Proceedings of the Royal Society*. Thus Aaron's PhD, or at least the numerical methods he developed for his PhD, was actually published in the context of haemoglobin physiology[4].

Aaron was not overworked in his postdoc job, and he spent much of his spare time hanging around with Norman Podhoretz and Steven Marcus. Liebe stayed on for a couple of months with her parents in Cape Town. In a letter to Liebe, Aaron paints a vivid picture of the discussions of 'the three musketeers' and points out the great differences between a childhood in South Africa and a childhood in Brooklyn:

1 Park Terrace Cambridge

Sunday 22 March 1953

Saturday morning, up early (8.30 am) worked at the lab 'till about 3.00 pm and then popped in at Norman's. We talked and read and made coffee until supper, which we had at the Kingswood. By then Steve had joined us and we went to see 'The Third Man'. It's a great film, better than 'Odd Man Out' I think, and [Carol] Reed is the greatest of British directors.

Afterwards we had coffee at Norman's place and there a discussion was launched that was only to end at 5.00 am, when the first light appeared. It's something I haven't done for years but this time it was a real discussion with my rationalism pitted (vainly and half-heartedly) against Steve's idealism (in the philosophical sense). Norman was quiet for the first four hours. From a political discussion on China (which is how it started off – this after the two had 'cracked' the film) it led to a discussion of Asia, Africa, backwardness, Imperialism, the 'needs' of an Asian peasant, the 'needs' of the American (colour television this year), the American Economy; what does a man need?, the primitive man, the fallacy of the noble savage, what does Lawrence mean? By about 4.00 am, where we should, by all rights, have been stooped in the gloom of nihilism we still hadn't lost our values (a tribute to Norman and Steve's excellent training – they supported me like swimmers a drowning man). But we began to talk of childhood and I found that they both hated their childhood and longed to be grown up. This is due to the physical and emotional crowdedness of their lives in Brooklyn (Norman once talked

---

[4] Klug, A., Kreuzer, F. and Roughton, F.J.W. (1956) *Proc. Roy. Soc. B* **145**, 452–472.

about it to you, I remember). I guess that S.A. saved us with its wide open
spaces and freedom of action for the young.

I can't repeat all we said, but coffee (I don't know how many cups) kept
us awake and I was quite awake as I walked through the deserted streets.
That was this morning and I slept soundly till about 12.30 pm when
I dashed over to Shirl's.[5]

Fifty years later, Norman Podhoretz and Aaron were totally estranged
by a difference of opinion over Israeli politics.

When Liebe returned to Cambridge in May 1953, they found an
apartment in Clare Road, Newnham. Here they first met Jim Watson
of DNA fame. He was introduced to Aaron and Liebe by Beverly
Calsoyas, who was a friend of Watson's sister. Watson may have been
especially interested in the Klugs since they had quickly built up a circle
of young friends, some of whom were female. Watson's work with
Francis Crick on the structure of DNA had just been published. The
Cavendish laboratory organised an open day that included a display of
the two-stranded DNA model. Aaron went along and found the model
immediately convincing. The chemistry was good, the physics looked
good, and the possible biological ramifications were fascinating.

The job with Roughton was only for six months. What next? In spite
of the early rebuff from Bragg concerning Perutz's protein structure
group, Aaron still wanted to work in X-ray crystallography. However,
he did not just want to solve standard crystal structures because this was
too boring. Aaron had a long-standing interest in the problem of
transitions between ordered and disordered states. He was fascinated
by what happened to the scattering between the Bragg reflection spots:
that is, the background diffuse scattering that arises either from thermal
diffuse scattering or from not all the molecules being in the same
orientation. Going back to his molecular transforms, he had worked
out, but never published, that if you had two orientations with molecular transforms F_1 and F_2, then diffuse scattering must be F_1 minus F_2 all
squared. Later he discovered that von Laue had worked out the same
result. However, on the basis of this finding Aaron took up contact with
Kathleen Lonsdale, then Professor of Chemistry at University College
London. He thought of working on diffuse scattering from crystalline

[5] The Klug Papers, Churchill Archive Centre, letter from A. Klug, 22nd March 1953.

organic acids and put forward a research proposal to work at University College. Despite Kathleen Lonsdale's enthusiastic support, the research proposal was turned down.

Then Aaron discovered that in the Physics Department of Birkbeck College London there was another group doing protein structure determination. John Desmond Bernal was Professor of Physics at Birkbeck. In 1927, Bernal had been appointed the first lecturer in Structural Crystallography in the University of Cambridge. His laboratory, housed in the old Cavendish, quickly became an international centre. His work on the structures of steroids was very important and, in retrospect, he narrowly missed being awarded a Nobel Prize. With his student Dorothy Crowfoot (Hodgkin) he showed that crystals of pepsin diffract to high resolution (that is, the Bragg reflections that they produced are clear and well-defined) if maintained wet. What this implies is that all the molecules in the crystal must be identical in structure. This simple observation made the then-current colloid theories of proteins untenable, but nobody took any notice. Bernal's student Max Perutz started work on the structure of haemoglobin: it took another 25 years to establish its structure, a demanding subject for a doctoral thesis. Bernal's vision was that to understand life you would need to know protein structures. These could be unravelled by X-ray crystallography. This was essentially the dawn of molecular biology. In a talk in 1939, Bernal had already suggested how the phase problem for proteins could be solved with heavy atoms. It took Perutz another 13 years to make the method work.

In 1936, Bernal was joined by Isidor Fankuchen, who was a visiting Fellow of the Schweinburg Foundation. Fankuchen had first worked in Manchester (1934–36) with Lawrence Bragg. In 1936, Bragg took the Cavendish Chair of Physics, and Fankuchen moved with him to the Crystallographic Laboratory in Cambridge. Here he started work with Bernal on the structure of simple plant viruses. In 1937, Bernal was appointed professor of Physics at Birkbeck College, London, and Fankuchen again accompanied him to London. Max Perutz stayed with Bragg in Cambridge.

The Second World War began shortly after Bernal's move to London. Bernal and his friend Solly Zuckerman were pressed into war service. Zuckerman was from Cape Town, medically trained and Professor of Anatomy in Oxford. It was said he had the *chutzpah* of the devil. Later he became chief scientific adviser to the British government. The problem that Bernal and Zuckerman addressed was to understand the effects of bombing. Together they were very effective and showed great personal bravery as they inspected unexploded bombs and wrecked buildings around London. This was Operational Research (OR): they put bombing on a quantitative basis. Later, Bernal, Patrick Blackett[6] and Zuckerman advised against the bombing of German cities because it was a sheer waste of money. In 1943, Lord Mountbatten was appointed Chief of Combined Operations and set about planning D-Day. Bernal and Zuckerman were seconded to Mountbatten's team. Bernal's great contribution was to chart the Normandy beaches in detail[7]. A strong friendship sprang up between Mountbatten and Bernal. Late in the war, Mountbatten became Supreme Allied Commander South East Asia. Bernal joined him in Ceylon on bomb trials for jungle clearance. He found himself working alongside John Kendrew (also an early recruit to OR) and they fell to talking about proteins and the structure of proteins. This conversation had considerable repercussions. Shortly thereafter, Kendrew turned up in Cambridge to work with Perutz on haemoglobin.

After the war, Bernal resumed his professorial duties in the Physics Department at Birkbeck College, soon to be housed in its new buildings in Malet Street. In two adjacent Georgian houses, 21 and 22 Torrington Square, Bernal set up the Biomolecular Research Laboratory that was opened in 1948. In addition to groups working on organic crystals and proteins, he had a group working on the structures of cements (buildings and building materials were a life-long interest) and on the structure of water. In 1953, Rosalind Franklin joined him to work on virus structure.

The Biomolecular Research Laboratory advertised a Nuffield Fellowship (the Nuffield Foundation was established by William Morris – Lord Nuffield – of Morris Motors) to work on the structure of the protein ribonuclease. Aaron applied and was called for interview. The panel

---

[6] Patrick Blackett, a famous physicist, also renowned for his leading role in Operational Research in the Second World War, was awarded the Nobel Prize in Physics (1948). He was President of the Royal Society (1965–1970) and was made a Peer in 1969. Blackett was Professor of Physics at Birkbeck College before Bernal took over from him in 1937.

[7] See Andrew Brown, *J.D. Bernal. The Sage of Science*. Oxford University Press (2005).

consisted of Lord Todd (known even then as the Almighty), the mercurial Janet Vaughan and folk from the Nuffield. Since they did not understand what Aaron was doing about phase transitions in steel, which seems to have struck them as simply boring, nor did they know what a partial differential equation was, let alone how to solve it, the whole matter devolved on his earlier work on molecular transforms. There was an enthusiastic letter of support from R. W. James. The fellowship, worth the (not over-generous) sum of £400 *per annum,* was awarded. At last Aaron could get back to crystallography and also do something with biological significance.

Thus, in September 1953, Aaron and Liebe moved to London, and Aaron started work at Birkbeck. They found a furnished room in St John's Wood; and in the start of another new chapter to their lives, Liebe was pregnant with their first child.

6

Birkbeck-1

In 1823, George Birkbeck, a pioneer of adult education, founded the 'London Mechanics Institute' that was later to become Birkbeck College. The Institute was housed in the Southampton Buildings on Chancery Lane. From 1885, it found a more permanent home in the Breams Building on Fetter Lane. Always socially progressive, the Institute admitted the first female students in 1830. From 1858, it was allowed to bestow degrees from the University of London. Much of the teaching was done in the evenings to allow working people the chance of graduating. J. D. Bernal was appointed Professor of Physics at Birkbeck College in 1937. Throughout the Second World War, Birkbeck College stayed in the Breams Building. It suffered incendiary bombing but, somehow, teaching carried on. After the war, Birkbeck had to find a new home. Since 1952, the main building of the college has been in Malet Street in Bloomsbury.

The name Bloomsbury appears to be a corruption of the name of a manor, Blemondsbury, extending roughly from New Oxford Street to Euston Road (the New Road). Blemond's manor was given to the monks of the Charterhouse by Edward III. On the dissolution of the monasteries, Henry VIII repossessed the land and gave it to the 1st Earl of Southampton. In the seventeenth century, the 4th Earl constructed Bloomsbury Square, the first of its kind in London. In the following century, the Russell family, the Dukes of Bedford, became the main landowners, and in 1780 they built Bedford Square, one of the best-preserved examples of Georgian architecture in London. However, the major development of the squares in Blooms-bury started in 1800 when the 5th Duke of Bedford, Francis Russell, removed Bedford House and developed the land to the north as a series of urban squares grouped around Russell Square. Whereas the first developments had been patrician, the early nineteenth-century buildings were primarily intended for the middle classes. Nevertheless, although the houses were smaller, the terraces retained some architectural links with their Palladian models. Robert Smirke's British Museum, based on classical Greek architec-ture, was built on the site of Montague House to the west of this development.

During the 1930s, the University of London built its new adminis-trative centre and library in Bloomsbury to the north of the British Museum. The Senate House, large and shining-white in Portland Stone, is difficult to ignore. It is a grand art deco design by Charles Holden based on a central tower, then second only in height to St Paul's Cathedral. During the planning, there were serious objections to its position arising from the fact that the University of London was swallowing more and more of the housing that the Duke of Bedford had thoughtfully laid out for good domestic living. Evelyn Waugh described the Senate House as 'the vast bulk of London University insulting the autumnal sky'. For George Orwell it inspired the 'Ministry of Truth'. Nevertheless, the architect Erich Mendelsohn praised it highly.

Birkbeck's new building was directly to the north of Senate House in what had been the gardens of Torrington Square. Torrington Square was built by James Sim between 1821 and 1825 on an insalubrious marshy site known as the Field of Forty Footsteps after a legendary duel that had taken place there. The terraced houses on the east side of Torrington Square were retained, and Birkbeck College acquired most of these houses. Later, Birkbeck invaded much of the neighbourhood. For example, its School of History of Art and Film is housed in a former residence of Virginia Woolf in the adjoining Gordon Square. Neverthe-less, two adjacent houses in Torrington Square allowed the realisation of one of Bernal's aspirations. With the help of the Nuffield Foundation, he set up the Biomolecular Research Laboratory, where Aaron Klug was later to work.

The Biomolecular Research Laboratory was opened by Sir Lawrence Bragg on 1 July 1948. The three objectives of the laboratory were:

1. To work on the crystalline structure of proteins.
2. To develop the necessary electronic and computing skills needed for the faster and better analysis of these protein crystals.
3. To understand the fundamental nature of the constituent active materials in cements and the nature of their reaction with water.

Bernal was always aware of the need to raise money, and this certainly had some influence on his choice of themes. He had a fine office facing west on the *piano nobile* in 21 Torrington Square. He had the use of an apartment on the top floor of 22 Torrington Square. Many eminent people were entertained in his flat, including Pablo Picasso, who drew a Satyr on the wall, then said he looked lonely and added a few nymphs. Bernal's idea of management was essentially hands off: you appoint people and let them get on with it. This works splendidly (as it did for Max Perutz in Cambridge) if you have the good fortune to start with a group of smart self-motivated individuals. Unfortunately, Bernal had made some uncritical appointments and was far too good-natured to sack anybody, with the result that performance was uneven. Moreover, there was not much cohesion between the groups. For example, although Andrew Donald Booth's computer lab at Birkbeck produced the third functioning computer in the world, and pioneered magnetic drum storage and machine translation, his lab never produced a program to add up Fourier series such as had already been done by Kendrew and Bennett on the EDSAC computer in Cambridge. Such a program would have been essential for protein structure determination.

Protein structure determination, carried out by measuring the X-ray diffraction from protein crystals, was at the centre of Bernal's vision of understanding biology. Bernal was the founder of the subject and his Tilden lecture of 1939 to the Chemical Society had proposed that if you could bind heavy atoms (mercury, for instance) to proteins in crystals in a regular way, you should be able to calculate the missing information about positive/negative signs and phases from the differences in scattering with and without the heavy atom. Then you would be able to calculate the electron density of a protein and determine its structure. This was the method subsequently used in Cambridge by Perutz and Kendrew to solve the first protein structures, myoglobin and haemoglobin. It is generally referred to as the 'method of isomorphous replacement'. The protein group at Birkbeck was run by Harry Carlisle, who was attempting to determine the structure of the small crystalline protein ribonuclease. At this time no protein structure had ever been solved. During the war, Carlisle had been with Dorothy Hodgkin in Oxford and had helped solve the crystal structure of cholesteryl iodide, which was quite an achievement at the time. Bernal apparently did not offer much practical advice to Carlisle and in particular never pointed out the importance of the possible use of heavy atoms for phase determination. Unfortunately, after four years, nothing much seemed to have happened. Harry needed help.

The College moved into its new building in 1952. Bernal now had an office in the main building. Werner Ehrenberg also moved over, freeing up space at 21–22 Torrington Square. As befitted the importance of protein structure determination, Carlisle took over Bernal's splendid office on the *piano nobile*. Now he had space for a crystallographic co-worker. Bernal set up a Nuffield Fellowship, and as we have seen, as a result of this initiative, in the autumn of 1953 Aaron turned up and was duly allocated an adjoining office to Carlisle. The fellowship also included a salary for an assistant, but the assistant was allocated to Carlisle rather than Aaron.

Proteins are polymers of amino acids. A small protein consists of about 100 amino acids joined together: a large protein may contain 10,000 amino acids. Twenty different amino acids are found in nature, differing from one another in the nature of their 'side chains'. In a protein, the amino acids are joined together to yield a polymer (the units are known as peptide groups). The polymer of peptide groups, often referred to as the 'backbone', is completely uniform, but each peptide group carries one of 20 different side chains. The 20 side chains have a variety of sizes and chemical properties ranging from oily to water-friendly. They can also carry a positive or negative charge. It is the side chains that make a protein interesting. Without side chains, proteins would be something close to nylon, with long, fairly uniform fibres; instead, the protein chains fold up spontaneously to form a host of different structures and shapes with a large range of properties. This happens essentially by packing the side chains together so that there are no gaps between them, and by having the charges on the outside; a third condition is that as many hydrogen bonds as possible should be made.

Hydrogen bonds are the stuff of life. Not only does hydrogen like to join strongly with oxygen to form water but each hydrogen atom also likes to make a weaker connection – the hydrogen bond – with a neighbouring oxygen. Hydrogen bonds make water a liquid. Without the extra stickiness of hydrogen bonds, water would be a gas like methane. Proteins also like to make hydrogen bonds. Each peptide group has a hydrogen sticking out on one side and an oxygen sticking out on the other, and thus plenty of raw material for making such bonds.

Protein sequences – that is, the sequence of side chains as you go along the protein – are defined by the genetic code. Clearly, the number of different protein sequences that could be generated by stringing together 20 amino acids, even for a small protein, is astronomical. But in fact, around 20,000 different protein structures have been determined, and it is not thought that this number will grow very much bigger. Thus, of the universe of possible sequences, apparently only a handful actually fold up to make a stable protein.

In all known protein structures, two standard motifs for the folded backbone have become apparent, the α-helix and the β-pleated sheet (see Plate 1). Ten years before X-ray crystallographic structures of proteins became available, Linus Pauling wondered what regular structures would be available to proteins given that hydrogen bonds should be made between the amide hydrogen of one peptide group and the carbonyl oxygen of another. Pauling and his colleague Robert Corey showed that the peptide groups must be planar. This important property limits the number of ways in which peptide groups can come together. Pauling discovered that only a limited number of possible regular structures of the backbone can occur. Two of these, the α-helix and β-sheet just mentioned, were subsequently found to be present in crystal structures of proteins. In fact, α-helices and β-sheets form perhaps 75–80% of all the backbone structures of proteins.

The α-helix is a tight spiral of peptide groups in which hydrogen bonds are made between one peptide group and a peptide group in the next turn above. The side chains stick out sideways. This is the motif of hair, which is pure α-helix. In a β-sheet, the polypeptide chain is fully extended in a puckered format and makes hydrogen bonds with neighbouring chains to form a sheet. The side chains point up and down out of the sheet. Silk is pure β-sheet. The β-sheet is also a structure that can be assumed by some proteins in pathological assemblies. Amyloid plaques – aggregates of β-folded proteins – are thought to be a causative agent of Alzheimer's and Parkinson's disease.

In the 1950s, Armour and Co. of Chicago (then renowned for their hot dogs) purified a whole kilogram of pure bovine pancreatic ribonuclease, and gave 10-mg samples away to research workers. Because it was easy to obtain, is very stable and crystallises easily, bovine pancreatic ribonuclease became a favourite object of study for crystallographers and physical chemists. It is also relatively small (134 amino acids). As we saw earlier, if you put a protein crystal in an X-ray beam and collect the scattered radiation on a sheet of film, you get a two-dimensional lattice of spots each of which is called a 'Bragg reflection' after Lawrence Bragg. As you rotate the crystal you get different sets of spots. To measure the intensities of all the spots (that is, all the Bragg reflections) you need to turn the crystal round and put in a new film every couple of degrees. To do this was beyond the technology of the 1950s (although it later became the method of choice), so one employed devices that moved the film along as the crystal rotated so that the Bragg reflections did not overlap on the film. A widely used camera of this type was the Weissenberg camera. For proteins, a Weissenberg camera is far from ideal since the Bragg reflections are close together and overlap easily. Nevertheless, Carlisle put Aaron to work collecting the intensities of all the diffraction spots from crystals of ribonuclease with a Weissenberg camera. The camera had been modified by Carlisle so that each reflection was measured only once. This bothered Aaron, since, because of the symmetry of the crystal, most reflections are available many times. Averaging helps to get rid of errors in the data collection. Thus Carlisle was probably working with data too inaccurate for the application of the method of isomorphous replacement (testing with and without adding in heavy atoms), even if he had tried.

The amplitudes of the Bragg reflections (square roots of the measured intensities) give the size of the appropriate term to be put into a Fourier summation. In the absence of help from Booth's computer lab, Fourier summations had to be calculated by hand, which was a truly Sisyphean task. Nevertheless, if you can calculate the Fourier summation in three dimensions, then you have a 3D map of the density in the crystal, and you can fit atoms into the density and get the structure – but you cannot do this without knowing the phases. Carlisle was calculating density maps from Aaron's data. It took Aaron some time to work out where

Carlisle was getting the phases. It transpired that he started with estimates of the phases, the origins of which are lost in the sands of time,
and calculated a 3D map. Then he sliced off all the high peaks, reasoning
that proteins are made of light atoms all of about the same size, and all
peaks should therefore be the same height. He then calculated the phases
from this truncated model and applied these new phases with the
measured amplitudes to produce a new density map. By repetition of
this procedure he hoped to arrive at the structure. Such an approach is
sometimes referred to as boot-strapping, since one is attempting to pick
oneself up by pulling on one's own bootlaces. Addressing some peak in
his avowed density map, Carlisle was renowned for analytical gems such
as: 'Look at this fellow, he's trying to tell us something.' Using his fluency
with Fourier transforms, Aaron analysed Carlisle's method. He was
quickly able to show that this approach would never work. Aaron,
although a kindly fellow, was never able to compromise about truth as
it was revealed to him. Thus Aaron duly explained to Carlisle, who was
also a kindly fellow, that he was wasting his time.

Faced with an insoluble phase problem, as an act of frustration one
can compute a Fourier summation using the measured intensities
obtained directly from Bragg reflections as input, rather than the
appropriate amplitudes and phases. This action perforce yields a
density map, but it is not that of the crystal. Arthur Lindo Patterson
was able to show that this kind of Fourier summation (known as a
Patterson function – see Appendix B) gives the auto-correlation function of the object. This is not very helpful since an auto-correlation
function consists of a map of all possible vectors between all
possible atoms in a structure. For small molecules such as a benzene
ring, you may be able to work out the underlying structure, and indeed
this was the basis of Dorothy Hodgkin's Nobel Prize winning work.
Dorothy was a genius at deciphering Patterson functions. Nevertheless, for a protein with thousands of atoms the Patterson function
is simply a mess, with one notable exception: the auto-correlation
function of a helix has recognisable features. Carlisle was convinced
that ribonuclease was made up of α-helices all coiled up (this was soon
shown to be true for myoglobin), and he asked Aaron to prove that
this was true for ribonuclease. Aaron searched Carlisle's Patterson
maps for signs of the auto-correlation function of an α-helix, but he
could find none. He duly reported this to Carlisle. However, Carlisle

was not the kind of person to let facts stand in the way of a well-established prejudice, so he decided Aaron was no longer proving useful and banished him from the *piano nobile* to a servant's attic room up under the roof. More than a decade later, in 1967, two American groups solved the structure of ribonuclease to show that it is pure β-sheet, with not a turn of α-helix to be seen.

Aaron had his Nuffield fellowship but now, after barely six months, he was again without a real project. Luckily, destiny, in the form of Rosalind Franklin, intervened.

Rosalind Franklin was an authority on non-crystalline X-ray diffraction, an area of research ignored by most people because it's messy. She had been categorising coal samples as to whether or not they would form graphite when heated. This depends on estimating the kinds of graphite nuclei that are present in the coal, which can be determined by examining X-ray scattering from the sample. The samples are not crystalline, so there are no clear spots representing Bragg reflections, just a smear. France had a much stronger tradition than England in these kinds of studies, and Rosalind had spent four happy years in Paris working on coal. The secret of success in such diffraction studies of non-crystalline or amorphous materials is to have bright, clean, focused X-ray beams. X-ray optics was a strong discipline in Paris, based on small high-intensity X-ray sources and bent quartz crystals used to focus the X-ray beam, a technology developed by André Guinier. When she came back to England, Rosalind brought all this know-how with her. Later, it was of considerable importance for carrying out her virus structure studies at Birkbeck.

In 1951, John Randall at King's College London asked Rosalind to join his group to strengthen its X-ray diffraction base. Rosalind was reluctant to leave Paris, which for her had many advantages, but the King's offer looked good. At King's she was offered a three-year fellowship from Turner & Newall (a firm that mined and manufactured products from asbestos and cement). She expected to work on the structures of proteins, but when she arrived, Randall assigned her to work on the structure of DNA. The ensuing story is well known[1,2]. She and her

---

[1] See, for example, Brenda Maddox's book *Rosalind Franklin: The Dark Lady of DNA*. Harper Collins (2003).

[2] Klug, A. (1974) Rosalind Franklin and the double helix. *Nature* **248**, 787–788.

PhD student Raymond Gosling quickly established that there were two forms of DNA fibres, at low humidity (A) and high humidity (B). This finding was absolutely essential for the ensuing structural work. She found this out, not by X-ray diffraction, but by weighing the A and B forms to establish how much water each contained. Previous X-ray diffraction data of DNA fibres from Bill Astbury at Leeds were in fact a mixture of both forms and very difficult to interpret. Unfortunately, Randall omitted to inform Maurice Wilkins, his second in command, that he had given the DNA project to Rosalind, which led to grave misunderstandings that quickly destroyed the working atmosphere. Moreover, Rosalind's need for a cosmopolitan ambience found no echo in the prevailing King's culture, which was narrowly English. After two years of distressing discord, Rosalind left, leaving behind enough wonderful data to enable Jim Watson and Francis Crick to establish the structure of DNA. And thus, in March 1953, she moved to Birkbeck to work on the structure of viruses in Bernal's Biomolecular Research Laboratory. Rosalind also had her own laboratory in which to take up her coal work, and had a mature student, James Watt, working on a problem of graphitisation of coal, measuring pore sizes by the diffusion of gases.

Birkbeck was less stuffy than King's, and Bernal presented a stark contrast with Randall. Known to his close friends as 'Sage', Bernal seemed incredibly well informed about everything and, moreover, spoke fluent French. He was also an ardent Communist. Rosalind found this somewhat naïve but kept her views to herself. Bernal enjoyed the company of women and was pleased to have Rosalind around, although his secretary, Anita Rimel, did not quite share the professor's enthusiasm. Bernal had a knack for putting women at ease. Rosalind had great respect for 'Sage' and discoursed happily with him. Furthermore, she herself was respected – indeed, viewed with awe – at Birkbeck.

John Randall, in a burst of uncharacteristic clarity, had sent Rosalind a letter forbidding her even to think about the structure of DNA. Despite Randall's edict, for some months after moving to Birkbeck, she continued to work on DNA. This was necessary for Gosling to complete his thesis. At the end of 1953, her Turner & Newall Fellowship ran out. With Bernal's support, Rosalind acquired a grant from the Agricultural Research Council (ARC) to support a small group to study the structures of plant viruses. In the summer of 1954, Rosalind undertook an extended lecture tour round the United States, mostly financed by lectures on her work on coal, for which she had earned an international reputation. During this visit she met Jim Watson again, this time in an unstressed friendly environment, and discussed his work on tobacco mosaic virus (TMV). While at the Cavendish in Cambridge, Watson had shown that TMV is in effect a microcrystal built of protein molecules arranged in a helix (he had now moved on to other things). Rosalind's research started where Watson had left off.

TMV is a flat spiral of protein subunits into which a strand of nucleic acid is embedded. Because of their function, the protein subunits are called the coat protein. In the case of TMV (and many other viruses), the genetic carrier is not DNA but rather its more soluble close relative, ribonucleic acid (RNA). It is the RNA that carries the information to make more virus particles. Viruses are minimalistic, only carrying just enough information to make a virus: they rely on the host cell for all the machinery needed for replicating the RNA and for making the coat protein. Nevertheless, the coat protein is made according to the information transmitted in the virus RNA and has nothing to do with the host cell. This turns out to be medically important because antibodies generated against the coat protein precipitate the virus but do not bind to host cell proteins. Virus vaccines are usually based on making antibodies to virus coat protein.

In the autumn of 1953, Rosalind started work on the structure of TMV: this, in collaboration with Isidor Fankuchen, had been the subject of Bernal's own research until the outbreak of war in 1939. TMV was the first virus discovered. Soon after inventing the electron microscope, which had much higher resolution than a light microscope, Ernst Ruska and his brother in Berlin tried to image TMV particles in a sample they obtained from Fred Bawden and Norman W. Pirie (always known as Bill) at the ARC's Rothampsted Experimental Station in Hertfordshire. In 1938, Ruska showed that TMV particles are rods about 150 ångstroms (Å) diameter and of length 3000–5000 Å. Although a part of the Nobel Prize for Chemistry was awarded to Wendell Meredith Stanley in 1946 for crystallising TMV, the virus actually has never been crystallised (excepting perhaps for observations made by Maurice Wilkins of micro-crystals in infected tobacco leaves – somehow Rosalind never followed up on this observation). Bernal and Fankuchen obtained solutions of TMV from Bawden. They found out how to orientate the long rod-like virus particles by drying gels of the virus between inclined microscope cover slips. The dried gel formed in the crack between the cover slips showed up brightly in a polarising microscope (because of an attribute known as birefringence, where the way in which a material refracts light depends on the direction and polarisation of the light), a sign that the rod-like particles were lying parallel to each other. They discovered another remarkable phenomenon: concentrated solutions of the virus particles would orientate spontaneously, often by breaking into two layers: a lower layer that showed birefringence, where the TMV particles were spontaneously forming micelles of orientated particles; and an upper layer with no order and no birefringence. They took X-ray diffraction patterns from the dried gels in the hope of being able to discover the organisation of the subunits within the TMV particles, each of which were thought to be some sort of micro-crystal. Rosalind improved on their methods and quickly obtained very well orientated gels of TMV in glass capillary tubes. The samples were not crystals, but each virus particle was essentially a micro-crystal with helical symmetry. The virus particles are aligned along the axis of the capillary tube parallel to each other. X-ray diffraction pictures obtained from such orientated gels, where the X-ray beam is at right angles to the axis of orientation, are known as X-ray fibre diagrams. The observed X-ray fibre diagram is like the diffraction from a single micro-crystal spun around its long axis and thereby averaged over all possible orientations. Happily, the cylindrically averaged diffraction still has plenty of interesting features, particularly if the particle has helical symmetry. Rosalind had to work with these non-crystalline cylindrically averaged data. In fact, the diffraction from fibres of DNA is of the same kind, so that one could maintain that Rosalind's work at King's on DNA was a preparation for her groundbreaking work on the structure of TMV.

Rosalind was allocated an attic room in 21 Torrington Square, which served as an office and a lab. Since the roof tended to leak, she sometimes needed an umbrella on her desk to protect her papers. She managed to get the use of a micro-focus X-ray generator located on the ground floor (a speciality of the Birkbeck lab, developed by

Werner Ehrenberg and W. Spear). Four flights of stairs are good for the figure but are not necessarily good for productivity. Despite such problems, Rosalind got measurable diffraction data from well-orientated gels of TMV. Since Rosalind had intensities (but no phases) from an X-ray fibre diagram, in the summer of 1954, with Franklin-like single-mindedness, she set about calculating the Patterson function from this data (or at least a summer student from Newnham College Cambridge, Jean Taylor, did). As mentioned above, the Patterson function (autocorrelation function) of a large molecule is a mess. Moreover, as we have seen, in a fibre diagram the data are cylindrically averaged, which leads to a cylindrically averaged Patterson function. A cylindrically averaged Patterson is just that much more of a mess than a normal Patterson. However, as with the α-helix, one thing that can show up in this muddle is the signature of a helix. Rosalind's Patterson function definitely showed that TMV was a helix, confirming Jim's result in a characteristically objective way – free from any *a priori* assumptions or models.

In March 1954, Aaron also found himself ensconced in the attic of 21 Torrington Square. He became aware of someone in the next room, and inevitably they met on the narrow staircase. Somehow Rosalind felt relaxed enough to exchange greetings with Aaron (which was not her norm). They fell to talking about Rosalind's research – at this point, Aaron did not have much research to talk about – and Rosalind presented Aaron with a problem she was trying to solve. An X-ray fibre diagram presents a cut (a section) through the middle of the Fourier transform of the fibre. Most frequently, a fibre is a helix. Helical structures repeat somewhere along their long axis, usually referred to as the z-axis. An important property of Fourier transforms is that objects that repeat at intervals of c have a Fourier transform limited to planes spaced at $1/c$. In an X-ray fibre diagram, these show up as lines that, in the trade, are referred to as 'layer lines': the fibre diagram consists of a set of layer lines. Rosalind's problem was that in the fibre diagram of a strain (U2) of TMV she was studying, the layer lines were not straight. Naturally, Rosalind checked the X-ray camera for peculiarities, but there was no doubt about it, the U2 layer lines were curved. Simple diffraction theory says that this just cannot happen.

As part of his doctoral thesis, Francis Crick had worked out the Fourier transform (essentially the fibre diagram) to be expected from a helix, in particular of Pauling's famous α-helix. Because a helix is periodic along its axis, the scattering is limited to layer lines. He had shown that the way the scattering is distributed along each of the layer lines follows the form of the sums of Bessel functions. Bessel functions (see Appendix B) come in a variety of flavours that are characterised by their 'order' n and are written J_n. Zero-order Bessel functions (J_0) look rather like a cosine wave, but one that has its largest peak in the middle and then is damped (dies away) with distance. Higher-order Bessel functions J_n first rise to a big bump and then oscillate like damped cosine waves. The position of the first bump depends upon the order – the higher the order, the further out you have to go before the first peak happens. Thus, on each layer line, you get a peak (the first bump of the Bessel function) on a distance out along the layer line about the same as the distance of the layer line from the middle of the diagram. The peaks in the fibre diagram from a simple helix therefore form a very characteristic cross, often known as a 'helix-cross'. When Francis saw the sketches made by Jim Watson from Rosalind's fibre diagram of the B-form DNA, he knew immediately that DNA was a helix. Rosalind was acquainted with this theory since it had also been worked out by Alec Stokes at King's, one afternoon on the train between London and Cambridge. For Rosalind, Stokes was a respected colleague, but he was not good at writing things up. Thus he never got much credit for this important theoretical work.

TMV is a helix with a quite complicated symmetry so that the helix-cross is not immediately apparent: in particular, one gets a number of Bessel functions on any one layer line. Bernal and Fankuchen could already see that the TMV structure repeated every 69 Å. Jim Watson had the benefit of Francis Crick's helical diffraction theory and could see that TMV was a helix. Moreover, from the distribution of density along the layer lines, he could see that TMV was a helix of subunits that repeated after three turns (that is, after three turns you end up with a subunit lying exactly over the subunit three turns below). The number of subunits (later shown to be 49) proved difficult to determine and remained enigmatic for another two years. Watson's estimate was 31 subunits repeating every three turns. The three turns repeat in 69 Å, so the 'pitch' of the helix is 23 Å (69/3).

With this information at hand, Aaron looked at the problem of the bent layer lines. Careful measurement showed that some

Bessel functions (J_n) were slightly above the expected layer line position and some were slightly below. This accounted for the bent appearance. The slight displacement occurs because the number of subunits in three turns is not quite a whole number. In the case of the U2 strain, there are not 49 but 49.05 subunits in three turns. By solving this problem, Aaron was introduced to helical diffraction theory. Somewhat incidentally he produced a result concerning the effect of cylindrical averaging on the Fourier transform of a helix. This result would later lead to an important paper co-authored with Francis Crick.

Rosalind respected Aaron, and their collaboration deepened. Moreover, Aaron found a subject to work on, namely virus structure, that needed his expertise and was also close to a realisation of his boyhood dream of doing research in microbiology. Together, he and Rosalind quickly published a few papers on TMV. In this early collaboration, Aaron took on the role of the theoretical adviser. Although five years younger, even with Rosalind he could sometimes become exasperated when he felt that the obvious was not being appreciated quickly enough, which once led Rosalind to complain: 'Don't bully, Aaron.' Nevertheless, they became firm friends. Like Bernal, Aaron knew about nearly everything. Unlike Bernal, he was a pragmatic realist. Perhaps it was noteworthy that Aaron was South African, self-confident and open to the world. Being from Durban High School, Aaron was steeped in English culture but without being inhibited by the nuances of the English class system. Rosalind's rather clipped upper-class accent evoked a defensive response in an average Londoner but produced no particular emotional response in Aaron. Rosalind herself was aware of the problems of being 'English': she was much more relaxed in French. She also found release in the United States where English class accents have no meaning and might even be construed as 'cute'. Another factor that made Aaron an easy friend was that he was happily married to Liebe. Rosalind found it much easier to relate to married couples. In fact, this cut both ways, since Aaron could be heavily involved in domestic duties, which were difficult to abrogate. Aaron complained somewhat bitterly that for the good of science, his Nuffield Fellowship research assistant would be better employed washing nappies (diapers) at home for him rather than calculating useless Fourier maps for Carlisle.

Rosalind Franklin enjoyed plays and exhibitions, and had a passion for travel and mountains[3]. She was far from the bluestocking caricature later portrayed by Jim Watson[4]. Nevertheless, she was single-minded about science: if you wanted to achieve something, you gave it your undivided attention. In contrast, Aaron's wide-ranging curiosity could result in his addressing many problems at once. For example, at the time of their first collaboration, Aaron was also involved in writing a review for the BBC magazine *The Listener* of Bertolt Brecht's play 'Life of Galileo', performed at the Mermaid Theatre. Rosalind urged Aaron to more single-mindedness. Apparently her homily did not affect Aaron too deeply: he continued to pursue his interest in contemporary poetry and literature. Seductively, the British Museum with its world-famous reading room was but two blocks away. It was not unknown for Aaron to slip off for a couple of hours in pursuit of information that was not necessarily scientific. Moreover, London theatre had lots to offer. Aaron recalls seeing 'Richard II' with John Neville and 'Anthony and Cleopatra' with Michael Redgrave and Peggy Ashcroft – not the youngest Cleopatra. Then there were Christopher Fry, John Osborne and Kingsley Amis, although Aaron gave up reading the *New Statesman* after Amis came out against fighting North Korea.

About the time that Aaron and Rosalind met, Liebe gave birth to the Klugs' first son, Adam. Liebe had returned from Cape Town in May 1953. After five years of marriage, they felt that the time had come to have a child. At the start of Aaron's Nuffield Fellowship, they moved from Clare Road in Cambridge to a furnished room in Elsworthy Road near Swiss Cottage in London, where Liebe discovered that she was pregnant. There was a pressing need for more suitable accommodation. At this stage, Aaron's brother Bennie was living in London with his wife Lucy, who was also pregnant. One of Lucy's friends divorced her husband and fled back to Israel. Her flat was therefore free, and the Klugs took over the lease. Rents on unfurnished property in Britain during and after the Second World War were pegged at their pre-war levels, which made unfurnished rented property cheap but also hard to find. Leases could sometimes be handed on, often for key money. Thus, after negotiating parental help for £75 key money, the Klugs became the

---

[3] See Jenifer Glynn's book *My Sister Rosalind Franklin*. Oxford University Press (2012).
[4] Watson, J.D. *The Double Helix*. Simon and Schuster (1967).

proud (and legal) tenants of the top-floor flat of a large Victorian house north of Primrose Hill, 184 Adelaide Road. Being on the top floor had its advantages – there were two attic rooms attached that tended to fill with itinerant South Africans – but heating was with open-grate coal fires, and the coal had to be carried up three flights of stairs. On the positive side, running hot water was available from a gas-fired geyser. When the Klugs moved in, they found the interior decoration in a dreadful state. Characteristically, their circle of friends rallied round. Aaron laid new lino. Vivian Rakoff, still in London studying psychiatry, rediscovered his youthful fluency in wielding a paint brush. Liebe was painting the front door of the flat when her labour pains started. She hurried off to hospital, and the front door never did get painted. Adam was born in University College Hospital at 7.55 a.m. on 16th February 1954. The midwife remarked that he was truly a beautiful baby[5]. Notwithstanding its comeliness, this baby generated a whole new raft of domestic duties, especially the boiling and drying of nappies, which together with the carrying of coal seemed to become Aaron's speciality. Another task that Aaron took on was buying food on the way home at Camden Market. From the Market to the flat was a good mile, but getting off the bus at Camden Town rather than Chalk Farm saved 2d. To eke out their meagre income, Aaron sometimes worked as a 'ghost writer' for a professional translator, turning French and German scientific books into English. Holidays were not a realistic prospect, but the 31 bus went directly to Parliament Hill Fields, whence one had uphill access to the green open spaces of Hampstead Heath.

The Klugs continued their open-house way of life. Often the number of people in the flat considerably exceeded the number of chairs, so that Japanese-style squatting quickly became the norm. On her occasional visits to the Klugs, Rosalind sat rather straight on her chair and appeared somewhat ill at ease. Most of the Klugs' friends came from the worlds of literature, theatre, dance and medicine, but certainly not from the world of scientific research. Like the Klugs, none of them had much money. One very good friend Aaron had made at Cambridge was

---

[5] It turns out that another Klug was having a baby at the same time, sharing the same obstetrician. Brian was born more or less contemporaneously with Adam. These other Klugs could also trace their roots back to Zelva. Klug is a very uncommon name, and it is likely that anyone of that name can trace his or her family back to Zelva. (Note from David Klug.)

Dan Jacobson, who was then a budding novelist and later held a professorship in the English Department at University College London. Thus the top flat in 184 Adelaide Road acquired a remarkably Bohemian ambience, but with a South African flavouring. One of the non-South African visitors was Norman Podhoretz from Columbia University, who had taken Liebe to a May Ball in 1951. Norman was at this time working in Army Intelligence for the US Army in Germany.

A few years later, Liebe taught dance once a week in a nearby Kindergarten in St Johns Wood. In return, Adam attended the Kindergarten. Here the children of the moneyed clientele would flaunt their parents' achievements with phrases such as: 'My father's got a Bentley', to which four-year old Adam's rejoinder was: 'My father's got a *laboratory!*'

In 1956, Aaron's father and aunt/step-mother paid a visit. Lazar and Rose were on their way to the USA (by steamer) to visit Rose's other sister Sophie, whom she had not seen since 1929. Sophie lived in Cleveland, Ohio. Rose and Lazar were mildly critical of Adelaide Road, mostly on the ground that Aaron might have done better. Why wasn't he already a professor? However, they were basically hands-off parents and assumed that Aaron knew what he was doing. They stayed in a nearby flat, and Aaron showed them round London. Later, Mrs Kahan (Cohen), a long-time friend of Rose's, also visited. Her comments were more critical (she always addressed Aaron in Yiddish):

'Far dos habben wir verlossen die Slobodka?'[6]
Aaron responded: 'Did you have indoor toilets and hot running water?'
Mrs Kahan: *'Takeh takeh.'*[7]

On 4th November 1956, Dan Jacobson, Norman Podhoretz, the Klugs and a few other friends sat agitatedly listening to the radio as it reported Soviet tanks rumbling into Budapest. Within a week, the Hungarian bid for freedom had been extinguished. The reaction of young socially conscious intellectuals such as those gathered in 184 Adelaide Road that

---

[6] 'We left Slobodka for this?'
  Slobodka was a Jewish suburb of Kaunas, Lithuania. It was a centre of Jewish learning and hosted the influential Slobodka Yeshiva. Slobodka was also the main site of the Kaunas Pogrom, the brutal massacre of the Jewish population of Kaunas, in October 1941.
[7] The meaning depends on how it is intoned – in this case, somewhat grudgingly: 'I guess you're right.'

evening was a mixture of horror and despair. Brave people were being killed by the thousand while defending their right to determine their own future. Now nothing was left.

On 25 February 1956, Nikita Khrushchev, prime minister of the USSR and first secretary of the Communist party, had denounced Stalinism in a speech to a closed session of the Twentieth Party Congress. The contents of the speech were secret but were soon leaked to outsiders. While protecting Lenin, Khrushchev vigorously attacked the crimes committed by Stalin and his close associates. This speech led to the rebirth of hope in a communism not poisoned by one man's paranoia. In Western Europe these ideas were mostly restricted to a few French intellectuals, but in Poland they brought about a workers' rebellion leading to Gomulka's reformist government. The Hungarian uprising went further. The aims became a multi-party government leading to free elections. This was more than the USSR Politburo could stomach, and the tanks rolled in. They crushed not only Hungarian aspirations but decimated what was left of the west European communist parties. A few days later, the Fabian societies of Birkbeck College and the nearby University College organised a joint meeting to discuss the ramifications of the events in Hungary. Aaron went along. One of the invited discussants was Bernal, who reported that until he read a transcript of Khrushchev's speech he had not known of the atrocities that had been perpetrated by Stalin against millions of people, nor that Stalin had executed S. M. Kirov and a third of the Communist Party's Central Committee. Aaron thought to himself: 'How is it that I've known about Stalin's crimes since I was 12 years old and Bernal only for the last 6 months?' Bernal remained a lifelong communist.

7

Birkbeck-2

A place of honour in the pantheon of structural molecular biology is reserved for the plant viruses, which consist of a single strand of ribonucleic acid (RNA) enclosed in a protein coat. Such objects are highly symmetrical on account of economy of design. Plant viruses can mostly be placed into one of two categories: rod-shaped (like TMV, the tobacco mosaic virus) or sphere-like. The 'spherical' viruses can often be crystallised. Before the Second World War, Bernal and Carlisle had collaborated with Fred Bawden from Rothamsted to look at virus crystals. The war had stopped this research, but as soon as the war was over, Carlisle and Katie Dornberger undertook preliminary X-ray diffraction investigations of crystals of tomato bushy-stunt virus (TBSV – the naming of plant viruses is based on their host and the symptom) obtained from Rothamsted. Carlisle still had crystals of TBSV in his refrigerator. He also had crystals of another spherical plant virus, turnip yellow mosaic virus (TYMV), which he had obtained from Roy Markham at the John Innes Horticultural Institution in Bayfordbury, Hertfordshire.

Rosalind's Agricultural Research Council grant covered the salaries for two assistants. In the spring of 1955 she advertised the positions. She quickly recruited John Finch, who had a degree in physics from King's College London, and in the summer of 1955 she was joined by Ken Holmes, a Cambridge-trained physicist. Holmes had also taken chemistry and crystallography, and knew something about X-ray diffraction. Both Finch and Holmes planned to write doctoral theses under the official supervision of Bernal (Rosalind was not an accredited university teacher). Rosalind soon directed Finch to work with Aaron on the crystalline plant viruses. This started a collaboration that lasted 40 years. She retained Holmes's services for the TMV work, which remained her primary research interest. In the autumn of 1955, Don Caspar entered the scene. This quickly led to a successful collaboration with Rosalind on the location of the RNA in TMV. Moreover, it triggered Aaron and John Finch to start crystallographic studies on TYMV. Don became a very fruitful collaborator of Aaron's, and together they developed the rules that govern the structures of spherical viruses.

Aaron became John Finch's and Ken Holmes's teacher. A degree curriculum always leaves holes; moreover, students are not always attentive. With these two PhD students, Aaron found plenty of opportunities to exercise his didactic skills. He remarked of Holmes that his Cambridge tutors, one of whom was internationally famous and was awarded a Nobel Prize, must have had their minds on other things.

Donald Louis Dvorak Caspar was born in Ithaca, New York, in 1927. He attended the elite Bronx High School for Science from 1942 to 1944 without managing to develop much interest in science. After a year and a half at Cornell University, he was drafted into the Army, and he spent a year in the Philippines and Japan at the end of the Second World War. When he returned in the summer of 1947, his father Caspar Shapiro put him in touch with his former Cornell colleague, Isidor Fankuchen, who had been a graduate student in physics at Cornell at the same time that Shapiro was a research associate in chemical physics. Fankuchen invited Don to attend his two-week summer 'X-ray Clinic' at Brooklyn Polytechnic, designed to introduce academic and industrial scientists to the wonders of X-ray crystallography. Fankuchen always arranged to include a few undergraduates in this course, but Don found it a bit intimidating to have senior scientists as classmates, including the distinguished Harvard Professor, John Edsall.

Don Caspar decided to major in physics when he returned to Cornell that autumn where, a generation before, his father on returning from the First World War had completed a BA in chemistry[1]. After the

---

[1] Don Caspar's father Caspar Shapiro met his wife Blanche in South Hampton, Long Island, where they both had summer jobs. Caspar Shapiro completed his doctorate at Cornell in 1928 on 'Hydroquinolsulfonphthalein and some of its derivatives'. The family spent a year in

Second World War, Cornell was awash with ex-servicemen. Their upbeat lifestyle, which included drug-deluded students trying to fly off the bridge over Fall Creek Gorge, became notorious and was even the subject of a novel[2]. Don survived Cornell and went on to Yale, to do his PhD on the structure of tobacco mosaic virus in Ernie Pollard's group at the Sloane Physics Laboratory. His interest in TMV was triggered by Fankuchen. As recorded in Chapter 6, Fankuchen had worked with J. D. Bernal at Birkbeck College on the structure of TMV. This work was terminated by the outbreak of the Second World War. He returned to Brooklyn Polytechnic and published their findings, which were somewhat inconclusive because the X-ray diffraction patterns did not fit the diffraction they expected from a micro-crystal (that of Bragg reflections arising from a crystal lattice) and Fankuchen really did not know what to make of them: the observed diffraction pattern is, in fact, rather typical of a helix, but helical diffraction theory had yet to be worked out. Jim Watson first realised that the TMV diffraction pattern came from a helix and duly published his findings in 1954.

Watson's result provided Don Caspar with a new basis for analysing the structure of TMV. In the middle of the X-ray fibre diagram is the so-called zero layer line, an area arising purely from J_0 Bessel functions (see Appendix B). It consists of about 10 peaks, each of which could be positive or negative. If you know the signs, it is possible to calculate how the density of the virus varies with its radius (the radial density distribution). Don had set up an apparatus to measure the X-ray diffraction from orientated TMV gels. He felt that X-ray film was too inaccurate and preferred to measure the intensities with a Geiger counter. In 1952,

Zurich at the ETH. Blanche had been looking forward to her return to Switzerland – she was born in Geneva in 1896 – but she was somewhat preoccupied with having twins. After their return to the USA, low pay and anti-Semitism forced Caspar Shapiro to leave his university appointment at Cornell. He got an appointment as an industrial chemist with the perfume manufactory 'Charles of the Ritz'. Here he was persuaded to change the family name from Shapiro to Caspar in order to reduce the number of Jewish names on the firm's register. On retirement, Caspar and Blanche moved to Colorado Springs. Rather soon after the retirement move, Caspar succumbed to a heart attack. His widow Blanche, who joined the Peace Corps at the age of 70, lived to be 103. After Caspar's death, Blanche moved to Four Mile Canyon in Boulder, Colorado, opposite the physicist George Gamow. Four Mile Canyon was her home for nearly 40 years. A polyglot, nevertheless towards the end of her life she tended to answer the telephone with '*Allo, j'écoute.*'

[2] *Halfway Down the Stairs*, by Charles Thomson.

he found a suitable-looking X-ray spectrometer in the attic of the Sloane Physics Lab at Yale, which had probably been there since 1927. He modified this instrument for the TMV work and was able to measure the first five peaks on the zero layer line – the rest were too weak to measure with this instrument. In 1953, Max Perutz's group had published the amazing result that the signs of the Bragg reflections from even a macromolecule such as haemoglobin could be determined experimentally by comparing the X-ray diffraction from a native crystal and the same crystal with a heavy atom (such as mercury) attached specifically to each molecule in the protein crystal. Although this result had been foreshadowed by Bernal in 1939, no serious crystallographer expected it to work – but it did. Could it be done with TMV? Don tried adding lead acetate solutions to the TMV gels and could indeed see differences. He did not succeed in analysing the data until early in 1955, after he had left Yale and gone to CalTech (the California Institute of Technology). He was then able to work out the signs of the first five diffraction peaks on the zero layer line. This yielded the first radial density map of the virus. It showed a big peak at about 40 Å radius and a hole in the middle of the virus – a most unexpected result.

In the summer of 1954 at Cold Spring Harbor, Jim Watson and Francis Crick told Don of the ideas they had developed concerning the structures of spherical viruses. Francis had predicted that these viruses should have cubic symmetry (that is, that they must have at least four three-fold rotational symmetry axes), making them either tetrahedral, octahedral or icosahedral. The closest approximation to a sphere that can be obtained with a regular solid is actually a truncated icosahedron. Icosahedral symmetry has been used in the manufacture of footballs (soccer balls).

As can be seen from the figure, which shows a truncated icosahedron and a football, such objects have 12 five-fold axes (black) and 20 six-fold axes (white). You thus need 20 white and 12 black pieces. Measurements of the number of protein molecules on the surface of a spherical virus were rather uncertain but much in excess of 20+12, so there was clearly something else going on. This started Don Caspar on an Odyssey: to determine the symmetry (and later the structure) of spherical viruses. The first step was to find out whether spherical viruses really did have cubic symmetry. X-ray diffraction is a good way of finding out about molecular symmetry. Thus in the autumn of 1954, Don went to

Figure 7.1 A truncated icosahedron and a football: the five-fold axes are shown in black, and the six-fold axes are white. This design of football was used in the World Cup in 1974. (From Wikipedia; image by Aaron Rotenberg. Licence CC BY-SA 4.0.)

Pauling's lab at CalTech in Pasadena. Art Knight from the Virus Lab at Berkeley gave Don some purified TBSV, and he managed to grow some crystals – but he could not collect useful data with the X-ray equipment in Pauling's lab: all the equipment had been designed to be earthquake-proof, and no adjustments could be made to resolve the closely spaced reflections from the virus crystals. He decided to go to the Cavendish Laboratory in England to use Tony Broad's high-powered rotating-anode X-ray sets. He arrived in Cambridge on 13th August 1955 (which he discovered was the start of the grouse shooting season and the beginning of the Long Vacation, when the labs would be shut down – so he left for Athens to see the Parthenon). On his return to London, Don finally met Rosalind Franklin who had recently returned from a holiday in Yugoslavia.

Working with his TBSV crystals, Don first wished to determine whether the virus had cubic symmetry – tetrahedral, octahedral or icosahedral, as predicted by Crick and Watson. Icosahedral symmetry, being five-fold, is not found in crystal symmetry, which is limited to 2, 3, 4 or 6. In fact there was no precedent for how to determine whether a crystalline particle might have five-fold symmetry. The crystallographic data that Don obtained with the help of Tony Broad's mammoth rotating-anode X-ray tubes in October 1955 established that the virus

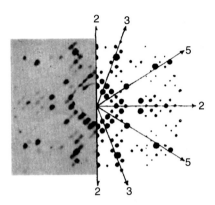

Figure 7.2 Left: A X-ray diffraction photo of a crystal from tomato bushy-stunt virus (TBSV), looking along the crystal axis[8]. Note the 'spikes' of spots (Bragg reflections) with high intensity. Right: a diagram showing idealised spots and the relationship between 'spikes' of spots and the directions of symmetry axes in an icosahedral particle. (Reprinted by permission from Macmillan Publishers Ltd. Caspar, D.L.D. *Nature* 177, 475–476 © 1956).

particle had at least tetrahedral but not octahedral symmetry. Then he was lucky. A crystal partially dried out because the capillary it was mounted in had cracked, and this destroyed the crystalline order. Instead of the expected crystalline pattern, Don saw a ring of ten smudges in the centre of the pattern corresponding to the continuous Fourier transform of the virus particles fortuitously aligned with their five-fold axes along the X-ray beam. Then he was able to show that if the hydrated crystals were appropriately aligned with respect to the X-ray beam, the Bragg reflections from the crystalline virus crystals were strong along ten 'spikes of high intensity' (Max Perutz's term), demonstrating rather graphically the underlying non-crystallographic five-fold icosahedral symmetry. Max Perutz helped Don to write up these results for a report in *Nature*[3], which accompanied Francis and Jim's paper[4] speculating that spherical viruses should have cubic symmetry.

In September, Don contacted Rosalind to discuss his radial density map of TMV and the unexpected hollow centre of the virus. After this, Don became a fairly frequent visitor to Birkbeck and kept Rosalind and

[3] Caspar, D.L.D. (1956) *Nature* **177**, 475–476.
[4] Crick, F.H.C. and Watson, J.D. (1956) *Nature* **177**, 473–475.

Aaron informed about progress in the analysis of TBSV symmetry. On one occasion, the virus group was standing outside 21 Torrington Square as Don approached. In an enthusiastic manifestation of the American disregard for the nuances of English etiquette, Don dared to shout, 'Hi Ros!' To the observers' confoundment, Zeus did not strike him down.

Aaron was particularly intrigued with Don's evidence for cubic symmetry, and immediately undertook the mathematical analysis of the Fourier transforms of such structures using spherical Bessel functions and spherical harmonics. These mathematical tools were well known to Aaron since they form the basis of solving the Schrödinger equation and can be used to represent atomic orbitals.

Don then took up contact with Bawden and Pirie at Rothamsted to try to obtain crystals of TBSV and TYMV. Bawden and Pirie were both supportive of Don's crystallographic ambitions, and they gave Don the crystals of TBSV they still had in their fridge. Furthermore, they informed him that they had given their large crystals of TBSV to Carlisle at Birkbeck about five years earlier. Thus, in November 1955, Don came to London in search of Bawden and Pirie's crystals. Aaron guided him to Carlisle's fridge. Finding the sparkling TBSV and TYMV crystals there was like finding a hoard of diamonds in a secret cavern. Don was keen to take them all back to Cambridge, but Rosalind declared that although Don could take the TBSV crystals, the TYMV crystals would stay at Birkbeck. Rosalind had been considering possible projects for Aaron, and on account of his awakened interest in diffraction from particles with cubic symmetry, a project on TYMV crystallography would seem ideal. Moreover, these crystals had actually come from Roy Markham at the John Innes institution, rather than from Bawden.

Don was disappointed at being denied the TYMV crystals, but he was more distressed that he had burned a hole in the sleeve of his brand new suit jacket while leaning on the lab bench where Rosalind kept a pilot light burning. Back at the Cavendish, Jim Watson was amused to find Don aggrieved about his visit to Rosalind, which he attributed to Don's resentment of Rosalind's commandeering the TYMV crystals. Jim later used Don's umbrage as another episode to justify the very distorted view of Rosalind Franklin that later emerged in his controversial book[5].

---

[5] Watson, J.D. *The Double Helix*. Simon and Schuster (1967).

In fact, the problem was the burn, which was soon repaired and forgotten. Furthermore, on account of the way the virus particles were packed together in the crystal, it turned out that analysing the TYMV crystal structure was much more difficult than analysing TBSV. Unravelling TYMV called for all of Aaron and John Finch's ingenuity. As these problems unfolded, Don realised that he owed a debt of gratitude to Rosalind for not burdening him with this problem.

Aaron and John Finch quickly started obtaining X-ray diffraction data from TYMV crystals. They followed Don's example and took data at the Royal Institution where there was also a powerful rotating-anode X-ray tube. Lawrence Bragg, who had recently become the director of the Royal Institution, was keen to set up a protein structure group and was interested in the virus work. One problem that John and Aaron faced was that the cell size (meaning, in this context, the distance along the crystal axes that you have to go before the pattern repeats) of the TYMV crystals was enormous, at 770 Å. This comes about because the virus particles take up two different orientations in the crystal. As a consequence of the large cell size, the Bragg spots on the X-ray film are very close together, which presented John and Aaron with the significant problem of producing a fine enough X-ray beam that they could see the Bragg reflections as separate resolved spots on the film. Diffraction data from crystals can be acquired just by putting the crystal in the beam and recording data on a film. This method is known as a 'still' because the crystal does not move. However, 'stills' do not give accurate measurements of the strength of Bragg reflections. To do this, you need to rotate or rock the crystal through a couple of degrees during an exposure. Because of the closely spaced spots, the Weissenberg camera was not adequate for this purpose. However, in 1944 the American crystallographer Martin Buerger had published the clever design of an X-ray camera that coupled movement of the crystal with the movement of the film in a way that projected an undistorted section of the 3D Fourier transform (the reciprocal-space lattice; see Chapter 9 and Appendix B) on the film. Such cameras were used extensively by Max Perutz and John Kendrew in their work on the structures of haemoglobin and myoglobin, and indeed made this work possible. Because of the nature of the movement, these cameras were known as 'precession cameras'. An improved version of the precession camera designed by Max Perutz was commercially available from the American firm Charles

Supper. John and Aaron acquired a Supper precession camera with a grant from the Wellcome Trust and used it extensively for collecting data from the TYMV crystals.

For the second half of 1955, before taking up his professorial appointment at Harvard, Jim Watson returned to the Cavendish to work with Francis Crick on virus structure. In the autumn, he and Francis visited Rosalind and Aaron to discuss the best strategy for attacking the problem of virus structure. Rosalind was now in a good position: she had Aaron as a collaborator and Ken Holmes and John Finch as productive students. There were lots of viruses one could try. The meeting was friendly, and it was agreed to inform each other of progress, to avoid overlap and to stay in touch. Rosalind's friendship with Francis Crick and his wife Odile deepened over the next couple of years. Jim Watson was very helpful and promised to ask Lord Victor Rothschild, who lived in Cambridge and was chairman of the ARC, to try to influence the Council to be more supportive of Rosalind's research. Jim wrote to Rosalind informing her of his talk with Victor Rothschild. It is not apparent that Rothschild achieved very much, because her next meeting at the end of September 1955 with the secretary of the ARC, Sir William Slater, was a catastrophe. Slater had little understanding of what Rosalind was doing and apparently treated Rosalind like a schoolgirl[6]. Rosalind particularly wanted to secure support for Aaron, whose fellowship would run out in 1957. Slater would not even consider the possibility of Aaron being paid from ARC funds. The combination of anger and frustration reduced Rosalind to tears. There was a real danger that Aaron and Liebe would be forced to give up their flat and have to move to somewhere else. For Rosalind, this was a dreadful prospect.

In March 1956, there was a meeting at the CIBA Foundation in London on 'The Biophysics and Biochemistry of Viruses'. These meetings were by invitation only. Watson, Crick, Bawden and Pirie were all there, as was Robley Williams from the Virus Laboratory at Berkeley California. The meeting with Williams was to have important ramifications for the future funding of Rosalind's research group. Don, Aaron and Rosalind were invited to give papers. Since they had collected much

---

[6] See Brenda Maddox's book *Rosalind Franklin: The Dark Lady of DNA*. Harper Collins (2003), p. 265.

of the data, John and Ken were allowed in for one morning as observers. Don and Rosalind reported on the hollow TMV and the radial location of the nucleic acid. Aaron indicated that he thought TYMV had icosahedral symmetry, but the existence of two different orientations of the virus made it difficult to be sure. It was to take about another year to establish that TYMV, like TBSV, has icosahedral symmetry.

Four weeks later, in April 1956, there was a symposium in Madrid organised by the International Union of Pure and Applied Crystallography on the subject of 'Structures on a scale between atomic and microscopic dimensions'. This was exactly Aaron, Don and Rosalind's area. Aaron talked about his work on expressing the Fourier transform of objects with cubic symmetry in terms of spherical harmonics involving Legendre polynomials and the associated spherical Bessel functions. This represented a lot of work because apparently no one had ever evaluated the selection rules for the Legendre polynomials for cubic symmetry. Unfortunately, in spite of the elegance of the theory, it never found much application. Rosalind talked about TMV. Don talked about the discovery of icosahedral symmetry in TBSV. His lecture was greeted with scorn – how could a crystallographic object have five-fold symmetry? It took some time for crystallographers to realise that particles with five-fold symmetry could still crystallise into a lattice that necessarily did not have five-fold symmetry. After the meeting, Rosalind, Don, Francis Crick and his wife Odile joined an organised tour of southern Spain. His curiosity whetted by Hemingway's novel *For Whom the Bell Tolls*, Aaron stayed in Madrid to see a bull fight.

About this time, Aaron and Rosalind speculated that the newly discovered ribosomes (then called microsomes), which like the spherical viruses were full of RNA, might also be built like virus particles. However, solution-scattering data that Don obtained in late 1956 at Yale from specimens prepared by Jim Watson and Alfred Tissiere at Harvard showed no indication of periodic structure.

Funding Aaron was a real headache for Rosalind. After a prodding from Bernal, Rothschild did successfully intervene to secure an extension of Rosalind's ARC grant until March 1958. In addition, the MRC promised to fund the two students if all else failed. Alas, Rosalind still had no funding for Aaron. Don promised to try for a year's funding at Yale, but the last thing Rosalind really wanted was for Aaron to vanish off to Yale. Robley Williams had urged Rosalind to apply for a grant

Figure 7.3 Photo taken at the meeting in Madrid (from left): Anne Cullis, Francis Crick, Don Caspar, Aaron Klug, Rosalind Franklin, Odile Crick, John Kendrew. (Photo courtesy of D. L. D. Caspar.)

from the US National Institutes of Health (NIH). These grants were occasionally awarded outside the United States for topics of medical importance that were not being adequately researched there. In the autumn of 1956, Rosalind and Aaron set about applying to the NIH for a grant for virus structure. They asked for support for the group for three years, together with some apparatus. Rosalind would be principal investigator and Aaron co-principal investigator. The grant should start in October 1957. Williams gave the application all his support and even carried out a site visit in Birkbeck to make sure that the money could be fruitfully used. On 9th July 1957, just weeks before Aaron's grant ran out, Rosalind was informed that the NIH grant had been funded – £10,000 per annum for three years. A few months later, Francis Crick came to Birkbeck to invite Rosalind and Aaron to move their work and group to the new building that was being built in Cambridge by the MRC for the enlarged Laboratory of Molecular Biology. Alleluia! The group had a future. Tragically, it was a future that Rosalind was not destined to share.

In the summer of 1956, Rosalind undertook her second lecture tour in the United States. This time she talked about virus structure. Supported by the Rockefeller Foundation, she went to a Gordon Conference on 'Proteins and Nucleic Acids' followed by an extended tour of virus labs. Rosalind was enthusiastically received wherever she went. While she was on this journey, she received an invitation from Lawrence Bragg at the Royal Institution to prepare two virus models for exhibition in the British Pavilion at the World Fair in Brussels in 1958. This was a welcome approbation. On the West Coast, she met Fred Schaffer at the Virus Laboratory in Berkeley who was working with the Mahoney strain of poliovirus. Rather unexpectedly, poliovirus had been shown to be an RNA virus like the plant viruses. Poliovirus had been crystallised by Schaffer and Carlton Schwerdt in 1955[7]. Schaffer and Rosalind discussed the feasibility of doing X-ray crystallography on poliovirus; at that time the remaining material from the earlier crystallisation was not good. A year later, in 1957, Schaffer offered Rosalind fresh poliovirus crystals. Schwerdt suggested he could take the samples with him to England. The crystals were housed in a thermos flask containing ice for cooling. Schwerdt's wife Patsy accompanied him on that trip, and she confidently carried the flask through British customs. The customs officer asked what was in the flask and she truthfully replied, 'Poliovirus.' The customs officer replied: 'Madam, you can't bring poliovirus into England.' Patsy responded: 'But it's crystalline.' Apparently this was enough reaasurance, as the customs officer let her pass. Some preliminary work was done on those crystals, but problems arose with handling them. A few months later, a second shipment was sent out – this time, commercially via Pan Am.

Members of the lab did not take kindly to poliovirus in their midst. Poliomyelitis in the 1950s was still a scourge. Indeed, Werner Ehrenberg had been crippled by it. Moreover, it would be fair to say that 21–22 Torrington Square, being 150 years old, was not the ideal venue for a

---

[7] Schaffer, F.L. and Schwerdt, C.E. (1955) *Proc. Natl Acad. Sci.* **41**, 1020–1023.

sterile lab. After some strenuous debate between Rosalind, Aaron and the rest of the world, it was agreed to keep the virus crystals in the nearby London School of Hygiene and Tropical Medicine. Only crystals mounted in capillary tubes could be brought into the lab, with adequate supplies of formaldehyde close by in case of accidents. Transport of the crystals was carefully monitored. Most of the work was actually done at the Royal Institution in Albemarle Street. The members of the group were vaccinated against polio with the newly available Salk vaccine.

Rosalind tried mounting the poliovirus crystals in standard capillary tubes as used for the TYMV crystal, but they dissolved. After some time,

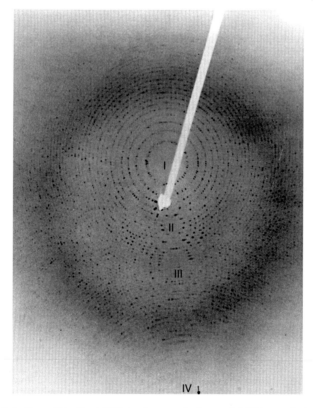

Figure 7.4 An X-ray diffraction 'still' taken from a poliovirus crystal (Mahoney strain, obtained from Carlton Schwerdt) showing closely spaced Bragg reflections out to the edge of the photograph. Wide angles of X-ray scattering mean high resolution. This virus crystal diffracts perfectly to about 2 Å resolution, a most unexpected result. (Courtesy of John Finch.)

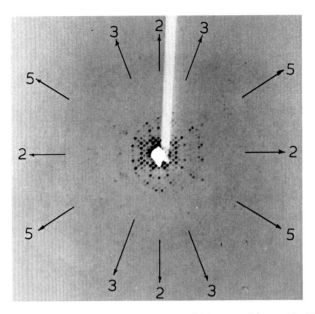

Figure 7.5 An X-ray diffraction photo of a poliovirus (Mahoney strain) crystal in dilute salt solution with the X-ray beam aligned along the crystal axis[8]. The spikes along the symmetry axes are shown. Note the spikes in the directions of the five-fold axes (the directions labelled 5 here), demonstrating the icosahedral symmetry of the virus particle. (Reprinted by permission from Macmillan Publishers Ltd. Finch, J.T. and Klug, A. *Nature* **183**, 1709–1714 © 1959.)

Aaron and Rosalind realised that the problem was the capillary tubes themselves: alkali was leaching out of the glass. Aaron and John tried again with quartz capillaries, which are neutral, and the poliovirus crystals survived. They gave the most detailed diffraction patterns yet obtained from any virus crystals. Aaron showed the photo to Bernal who was most enthusiastic. Being often obliged to think about raising funds for research, he commented: 'That photo's worth £10,000!'

Aaron and John were soon able to show that the virus particle had icosahedral symmetry. It was like its plant virus cousins; indeed, from a structural point of view you could hardly tell them apart[8]. This was a very unexpected result. How could a human virus be like a simple plant virus? This paper made an important contribution to Aaron's standing

---

[8] Finch, J.T. and Klug, A. (1959) *Nature* **183**, 1709–1714.

Figure 7.6 The structure of poliovirus capsid represented as a dodecahedron. Each edge of the dodecahedron has been replaced by a sphere made of polystyrene. There are 60 spheres in this model, which was shown at the World Fair in Brussels in 1958. The photograph was taken at the Royal Institution, London. (Courtesy of John Finch.)

in the scientific world. Suddenly he was at the forefront of topical research. With this paper, Aaron established an international reputation.

Lawrence Bragg wanted models of poliomyelitis virus and TMV for the International Science Pavilion at the Brussels World Fair in 1958 (Expo 58). The poliovirus was represented as a model by 60 polystyrene spheres arranged on a truncated dodecahedron. This was a rather crude representation, but at this time not much else was known. It was known that the protein coat acted like a bag for containing the nucleic acid (in this case RNA) but no attempt was made to represent the nucleic acid, which is inside.

In a first attempt to build a model of TMV, Rosalind sent Arthur Page, her technician, to purchase some hundreds of bicycle handlebar grips. These were duly assembled into a helix by sticking them into appropriately spaced holes in a cardboard former, but somehow the finished structure looked exactly like hundreds of handlebar grips stuck into a cylinder of cardboard. Then, in a shop window, Rosalind noticed display forms made from expanded polystyrene. This was a new medium. It was brilliant white, fresh-looking and could be moulded to any form. If you could mould a single component subunit to the appropriate shape, then they could be built to form the final structure just by stacking them on each other. The shape determined the assembly process. A child could do it[9]! This idea, which was to prove rather important for understanding how viruses come about, became known as self-assembly. Aaron was entrusted with designing a suitable TMV subunit, which was fabricated in balsa wood. Fortunately, Aaron's circle of friends and acquaintances included John Ernest, an experimental sculptor who had a studio off the Finchley Road. Ernest made a mould from the balsa wood form and produced hundreds of identical polystyrene 'TMV-subunits'. Rosalind's group was entrusted with helping Ernest to assemble the model. This turned out to be non-trivial. Although, according to Francis Crick, the problem of virus assembly should be simple enough for a child to solve, it proved frustratingly difficult for two physics graduates. It was first necessary to build a helically formed base to get the structure started. Then it transpired that the subunits were a few millimetres too wide and did not assemble properly: each had to be thinned down by hand. But at last it all worked. Spaces were left to show the nucleic acid (RNA, represented by a red plastic tube) that wound round between the TMV subunits.

After being shown at the Royal Institution, these models were duly displayed at the 1958 World Fair in Brussels, where they generated a great deal of interest. Later, the TMV model found a home in the stair well of the MRC Laboratory of Molecular Biology in Cambridge, where it remained for half a century. Sadly, Rosalind never saw the finished models.

---

[9] Crick, F.H. and Watson, J.D. (1956) *Nature* **177**, 473–475.

Figure 7.7 In 1958, John Finch took a series of photographs of the construction of the tobacco mosaic virus (TMV) and poliovirus models. Shown on the photo (left to right) are Arthur Page, John Finch and Ken Holmes. (Courtesy of John Finch.)

In the summer of 1957, Aaron and Liebe borrowed a house on the Arbury housing estate in north Cambridge belonging to their old friends Asher and Shirley Korner. Asher and Shirley were visiting the United States for the summer. At this time, Cambridge City Council had just started to develop the Arbury housing estate. In the ensuing years, the area has acquired an indifferent reputation because it was never finished. This cuts both ways: the estate still lacks infrastructure, but at least the empty lots host fenland flora and wildlife. Near the Korners' house, Vernon and Margaret Ingram had built an attractive residence. Vernon Ingram was a biochemist who worked with Max Perutz on haemoglobin. Liebe became a firm friend of Margaret and was pleased enough to be able to live in a real house for a few weeks.

Aaron needed access to colleagues and libraries in order to write two papers. As has been mentioned, the crystallographic phase problem was dominant in the 1950s, and this attracted Aaron's curiosity anew. As you might recall, Aaron had hit the phase problem already in his MSc work in Cape Town. David Sayre[10] had proposed an ingenious method of linking the probabilities of signs of reflections being the same or opposite based upon the observation that for structures made of similar-size atoms, the electron density map of the structure looks the same if you square the density. Herbert Hauptman in Buffalo and Jerome Karle at the Naval Research Laboratory in Washington DC worked out that certain combinations of three or more reflections had to be positive because the density of the crystal was positive[11]. After listening to two lectures given by E. F. Bertaut at the Cavendish Laboratory in the summer of 1956, Aaron had the idea of using the power of probability theory to work out the chance of any combination of signs (or phases) of all the reflections actually being correct[12]. This was based on simple assumptions as to how the atoms were distributed in the unit cell of the crystal. For his derivation he employed the Edgeworth series (a well-known tool in probability theory) to combine the probabilities of each of the reflections having a certain value to arrive at the joint probability of this being the right answer. Unfortunately, one of the problems with the Edgeworth series is that it gives a wrong answer if you go too far from the set of initial values that were used to set it up. Since one initially knows nothing about the positions of atoms, one perforce has bad starting values for the Edgeworth series, and its predictions are unreliable. Sadly, Aaron's elegant analysis did not really go anywhere. Later, Gerard Bricogne successfully incorporated the method in a general approach that makes as much use as possible of prior information about the distribution of density in the unit cell, as this becomes available[13]. In the meantime, Karle and Hauptman soldiered on with calculating the linkages between the sign of a reflection and a reflection twice as far from the middle of the diffraction pattern. Here, as had already been shown by Sayre, the linkages are strong. This approach to the phase problem became known as 'direct methods'. Initially, direct methods were viewed with scepticism but, after some development and a lot of programming, direct methods became a powerful tool for

[10] Sayre, D. (1952) *Acta Cryst.* **5**, 60–65.
[11] Hauptman, H. and Karle, J. *The Solution of the Phase Problem: I The Centrosymmetric Crystal.* ACA Monograph No. 3. Polycrystal Book Service (1953).
[12] Klug, A. (1958) *Acta Cryst.* **11**, 515–543.   [13] Bricogne, G. (1984) *Acta Cryst.* **A40**, 410–445.

solving the phase problem. In recognition of their work, Karle and Hauptman were awarded the Nobel Prize in Chemistry in 1985. The method is much used in organic chemistry for crystalline samples containing up to about 100 atoms. Unfortunately, direct methods are of little help with proteins, which are too big.

The second problem that Aaron wanted to tackle was how to calculate the density map in a structure from a fibre diffraction pattern. TMV gives wonderful fibre diffraction patterns extending to high angles of scattering, which means to high resolution. Thus, in principle, it should somehow be possible to deduce atomic resolution maps. However, the data are cylindrically averaged: in a solution of orientated virus, the particles all lie parallel to each other but the orientation of each particle around their common axis is random. The observed data are what one would get if one were able to take a single virus particle and collect the diffraction data as it spun around its long axis. The measured diffraction intensity is the square of the values of the Fourier transform at a certain point in the diffraction pattern (which leads to the experimental loss of the phase information), with the added problem that the value is averaged with all the other values that are the same radial distance from the axis of the diffraction pattern. Could there really be a way of sorting out this muddle? It turns out that it is possible provided the particle has high symmetry.

Because of the cylindrical averaging, it is essential to use cylindrical coordinates (in terms of angle, radial distance and height) rather than Cartesian coordinates. However, there is a catch to using cylindrical coordinates: in the Fourier transform, in place of the more familiar functions of sine and cosine, one ends up with Bessel functions (see Appendix B). Moreover, because the particle repeats periodically along the z-axis, the Fourier transform is limited to layer lines.

TMV has very high symmetry – there are 49 subunits for every three turns. This leads to the Bessel functions on any one layer line being of very different order. For example, on the zero layer line (the middle of the fibre diagram), the Bessel functions that can occur are the orders 0, 49, 98, etc. Now the measured diffraction intensity is the square of the values of the Fourier transform, which means that not only do terms such as J_0^2 and J_{49}^2 turn up but also mixed terms like J_0J_{49}. Such cross-terms are difficult to deal with, but occasionally Nature is kind. In the paper on bent layer lines that Aaron had published with Rosalind two years earlier, he had worked out that the cylindrical averaging got rid of all cross-terms like J_0J_{49}, leaving just the squared terms. On account of this result, high symmetry leads to a physical separation of Bessel functions of different orders: they turn up in different regions of the diffraction pattern. Then, in collaboration with Francis Crick, Aaron was able to show that if you could separate and measure the strengths of the different Bessel functions, you could calculate the density map from a fibre diagram by means of a Fourier transform, just as with a crystal[14]. However, instead of an ordinary Fourier transform, one had to use a Fourier–Bessel transform, which is really the same thing but in cylindrical coordinates. One still needed heavy atoms to find the phases. Indeed, heavy atoms were also useful for separating overlapping Bessel function terms.

To apply this theory, one needs perfectly orientated fibre diffraction patterns, and it has actually only been applied twice – to TMV and to orientated gels of bacterial flagella (the filaments that propel bacteria along). Nevertheless, the method found plenty of application in the analysis of electron micrographs, which in the next decade became Aaron's major intellectual preoccupation. Rosalind and Aaron quickly applied the theory to calculate a projection at limited resolution of TMV along its basic helix. This could be done by using just the inner parts of the zero and third layer lines. The resulting density map, which was never published, had one easily interpretable feature, namely a single strong peak at 40 Å radius. This peak arises from the nucleic acid and demonstrates that the nucleic acid winds around as a continuous chain embedded in the protein subunits. This result was incorporated in the model.

Rosalind had returned from her 1956 lecture tour with ominous abdominal swellings. Despite the removal of an ovarian cyst and a growth, a few months later the same problems recurred. By the summer of 1957, she knew that she was in serious trouble with cancer. She worked on heroically, subject to bouts of radiation therapy and experimental chemotherapy, all to no avail. Rosalind died on 16th April 1958. Her death stunned the group[15]. Bernal asked Aaron to take over the leadership. After some hesitation and concern, Aaron consented.

[14] Klug, A., Crick, F.H.C. and Wyckoff, H.W. (1958) *Acta Cryst.* **11**, 199–213.
[15] Maddox, B. *Rosalind Franklin: The Dark Lady of DNA.* Harper Collins (2003), p. 307.

In her will, Rosalind left Aaron and Liebe enough money to put down the deposit for the mortgage on a house in Hampstead Garden Suburb. She also left them her small car, an Austin A35, even though at this time Aaron had no driving licence. In December 1958, Aaron, Liebe and Adam moved into 8 Sutcliffe Close, giving up Adelaide Road and a way of life. Adam's comment on moving into the house was: 'Is this all just for us?' Unfortunately, the move to Sutcliffe Close was accompanied by Liebe's having a miscarriage. Aaron and Liebe were still very upset by Rosalind's death, which may have affected Liebe's pregnancy. Later Liebe suffered a life-threatening ectopic pregnancy. Her memories of Sutcliffe Close are coloured by these events.

Adam started at the Brookland Junior School. To Aaron's dismay, modern educational methods were in vogue. On a couple of occasions, he felt compelled to explain to Adam's class teacher the virtues of learning multiplication tables by singing them, as he had done at Penzance Road Primary School in Durban.

In early 1959, Aaron was honoured by an extended visit from Aharon Katchalsky, director of a division of the Weizmann Institute. Katchalsky was a very well known physical chemist who worked on many subjects, including the origin of life. A man of great erudition, he came to Aaron to learn about X-ray diffraction. This entailed climbing all those stairs to Aaron's attic room, which made Mrs Katchalsky quite nervous because Aharon Katchalsky had a heart condition. Aharon Katchalsky survived the stairs, and Aaron Klug managed to explain X-ray diffraction to him in a couple of hours, which prompted the comment from Katchalsky: 'Is that all there is to it?' He spent the rest of his mini sabbatical explaining electrochemistry to Ken Holmes, and how desalination works to anyone who would listen. He became renowned for covering the Formica tables in the Birkbeck canteen with diagrams and formulae. He was indeed an amazing teacher.

Aharon Katchalsky was interested in the structure of polyadenylic acid because, when adenylic acid is linked to an amino acid, it activates the amino acid so that it will form polymers with other activated amino acids. This was possibly a reaction involved in pre-life synthesis of proteins. Polyadenylic acid is a by-product of this reaction. Katchalsky's query about the structure of polyadenylic acid was answered in a publication from Alex Rich's group (see Chapter 12): it forms a two-stranded helix. In distinction to DNA, where the strands are anti-parallel, the two strands here run in the same direction. Later, by fibre diffraction Aaron and John Finch were able to show that poly-adenylic acid can take up two forms: Rich had been working with a mixture. This story echoes the DNA story of A and B forms.

By 1958, funding from the NIH grant was in place. Aaron now had some freedom of action. He hired a biochemist who had just finished his PhD at University College London, Reuben Leberman, to sort out virus purification and crystallisation. Reuben had green fingers for the crystallisation of viruses. In the summer of 1959, John Finch and Ken Holmes submitted their doctoral theses and duly received their titles. They both stayed on as staff members. There was now room for another doctoral student, and Aaron recruited Bill Longley, a physicist, to work with him and John Finch on crystalline viruses. Longley brought with him some early expertise in computer programming, which he trans-ferred to Holmes with notable results. Programs written by Bob Langridge, a PhD student at King's College working on DNA structure with Maurice Wilkins, were available for calculating Bessel functions on an IBM650 computer located in an office in the Strand. This early drum computer rather bizarrely used an operating system called SOAP. Nevertheless, for Holmes it was a gift from heaven. He and Finch were doing a lot of data processing, turning measurements of intensities measured on X-ray films into numbers, and they needed help. An advertisement for a computing assistant in the *Evening Standard* dredged up a strange collection of social misfits with one notable exception, Susan Fenn. Susan had graduated in biology and was very competent at everything she did. She joined the team and quickly fitted in. Within minutes of Susan's arrival, J. D. Bernal, sensing the presence of a new, attractive, woman in his department, came down the stairs of 22 Torrington Square (which for him was most unusual) to request help with computing. Susan was married to Nick Fenn who worked in the Foreign Office. Later, Sir Nicholas Fenn became Ambassador to Ireland, and British High Commissioner to Burma and then India.

The poliovirus results endowed Aaron with international visibility. He was awarded a Rockefeller travelling fellowship and with some help from Norman Podhoretz he managed at last to obtain a US visa. In the autumn of 1959, he embarked on a lecture tour of the USA. He flew to New York and was met at Idlewild (now JFK) airport by Steve Marcus and Norman Podhoretz. After a few days in New York he started out on an extensive lecture tour: Johns Hopkins, Madison, Berkeley, CalTech and Harvard. Pauling's office at CalTech was memorable: he sat ringed with molecular models and emanated a haughtiness reminiscent of a nobleman surrounded by hunting trophies. Aaron stayed in Boston for several weeks, lodging with Don Caspar. He taught a course (Chemistry III – the hydrogen bond) at Harvard in place of Paul Doty, who was away on sabbatical. Aaron stayed with Don at 52A Elliot Street, Jamaica Plain, and used the green and red subway lines to get across to Harvard. Aaron remembers that one morning after giving a lecture he was walking back to Harvard Square. While he was turning from Oxford Street into Kirkland, Jim Watson hurried past in the direction of Radcliffe College, muttering: 'I need to find a partner for tennis.' Jim's mind was apparently no longer on virus structure.

Nevertheless, following on from Jim's work, Aaron and Don Caspar were puzzled by the symmetry of spherical viruses. Symmetry plays an important role in small virus structure because a virus coat (capsid) is built from just one protein. All the proteins are equivalent. Thus, when the coat proteins assemble to form a capsid, each molecule will have the same set of contacts with each of its neighbours. As was pointed out by Jim and Francis Crick in their 1956 paper, the rules of symmetry show that the largest number of objects that can be packed together on a spherical surface so that all are in equivalent positions is 60 (Figure 7.8). But by measuring the molecular weights of the virus and of the coat protein subunits, Roy Markham had established that the number of subunits in the capsid of TYMV was around 160. How could these different numbers be reconciled? Don and Aaron surmised that each object (each left hand in the figure) would need to be replaced by a triangle of objects (shown on the right of Figure 7.8 as commas). This gives you 180 objects, which – given the possible error – would fit Markham's measurements. There is a catch: the commas round the five-fold axes do not have exactly the same environment as the commas round the three-fold axes. The angles are somewhat different. Aaron and Don argued that if the subunits were slightly flexible they could accommodate this slight mismatch. Since the protein subunits all have very similar but not quite identical environments, they referred to this kind of relationship between subunits as *quasi-equivalence*.

Aaron and Don felt vindicated in their analysis when they learned of Buckminster Fuller's geodesic radar domes, which are built of

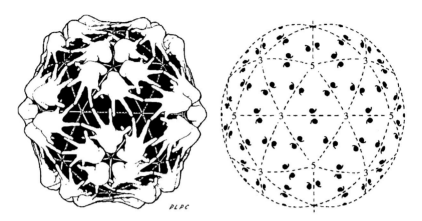

Figure 7.8 Left: Objects (left hands) are shown grouped round the five-fold axes of an icosahedron. There are 12 five-fold axes so that 60 objects can be packed on a spherical surface to have the equivalent contacts with each other. This is the maximum number that can be packed on the surface of a sphere in equivalent positions.
Right: If we replace each hand by three commas, we can accommodate 3 × 60 = 180 objects on a spherical surface. The commas are no longer strictly in equivalent positions but are actually very close to equivalence. Caspar and Klug called this *quasi-equivalence*. It predicts 180 molecules in a virus shell (capsid), which is close to the reported number of 160.
(Diagram courtesy of D. L. D. Caspar.)

interconnected struts according to icosahedral symmetry. In order to allow one to generate a spherical appearance, the triangles between the five-fold axes are subdivided into smaller triangles. The vertices of these triangles are projected out onto a sphere, and struts are used to join these points together. For simplicity of assembly all the struts should be as near as possible the same length. Thus the struts end up being quasi-equivalently related, much as in Don and Aaron's virus capsids. Aaron had come across Buckminster Fuller's work and showed it to Don, who quickly became a Buckminster Fuller enthusiast.

Although they were separated by the Atlantic Ocean during the next three years, Aaron's and Don's ideas were developed collaboratively. They met in Boston in 1959 and 1960. Otherwise their interchange was by letters[16]. The work was published in the *Proceedings of the Cold*

---

[16] A detailed account of the development of the concept of quasi-equivalence referring to the Caspar–Klug correspondence is given in: Morgan, G.J. Early theories of virus structure, in:

Spring Harbor Symposium in 1962[17]. Aaron and Don discovered that the subdivision of the triangle between the five-fold axes cannot just be any number. The icosahedron surface contains $60T$ quasi-equivalent subunits, where $T = h^2 + hk + k^2$ (h and k can be zero or any positive integer). The allowed values for T are thus 1, 3, 4, 7, 9, 13... T became known as the triangulation number. The structure of most isometric viruses conforms to this scheme, and the small viruses such as poliovirus and TYMV are $T = 3$ viruses. Later, $T = 1$ (simple icosahedral viruses) were also discovered.

During his 1959 visit, Aaron was also concerned with another problem. Don Caspar writes:

> *Regarding Aaron's 1959 visit, what I most clearly remember was his sense of triumph in setting William Moffitt right about his theory of optical rotatory dispersion (ORD) of helically arranged chromophores. Moffitt's theory included Planck's constant implying that the ORD was a quantum mechanical phenomenon. Aaron showed that the helical ORD could be calculated using the approach Lawrence Bragg had used in the 1920s to calculate the birefringence of calcite from atomic polarizabilities, which did not require Planck's constant (i.e. does not need quantum mechanics). This introduced me to Bragg's theory[18], which still amazes me.*

William Moffitt, who had studied at Oxford, was a talented young lecturer in the Chemistry Department at Harvard. Aaron felt impelled to explain to him the error of his ways; but in the end, Aaron was spared a possible confrontation because at the tender age of 33 while playing squash, Moffitt tragically succumbed to a heart attack.

Aaron also visited the Massachusetts Institute of Technology (MIT) where he met David Blow, then a postdoc with Alex Rich. David had done his doctoral thesis with Max Perutz and had contributed very substantially to solving the structure of haemoglobin. Aaron and David fell to discussing the US experience. Both admitted to being unsettled by the hustle. Nevertheless, the informality appealed to Aaron who, despite

---

Conformational Proteomics of Macromolecular Architectures, ed. H. Cheng and L. Hammar, World Scientific (2004), pp. 1–40. The correspondence has been deposited in the Jeremy Norman Archive that is now part of the collection at the J. Craig Venter Institute.

[17] Caspar, D.L.D. and Klug, A (1962) *Cold Spring Harbor Symp. Quant. Biol.* **27**, 1–24.

[18] Bragg, W.L. (1924) *Proc. Roy. Soc. Lond.* **A105**, 370–386.

Durban High School, sometimes found the English way of doing things somewhat mysterious. On balance, though, Aaron decided that he would rather live 'under the Union Jack'. He was bothered by the lack of style in your average Bostonian. New York was better, but there he was struck by the lack of any purposeful idealism amongst the intelligentsia. In a letter to Liebe, he says:

> I had to go into town to book my ticket for the West Coast and I did see a number of Fifth Avenue girls – beautiful assorted slim in Bonwit Teller clothes, and I even met one at a library party I went to with S & G [Marcus] and N & M [Podhoretz] in an expensive large apartment on Fifth Ave on Friday evening (she was a model like 'heart-talk rampant' being touted around by a Brownsville film boy, John Garfield-like, who obviously lives off accounts of Marlon, Ava etc.). The party was rather informal. Present were Robbie McCauley, William Phillips (Partisan Review), Lorca's brother – who is handsome like his brother but as pathetic in his role, the Epsteins, Richard Chase, other people from the Columbia Faculty but not the Trillings, the Kristals, and the operators on the committee for Cultural freedom. 'Too many gays' said Norman [Podhoretz], who is as assured and confident as ever in public, but even then I was depressed at the powerlessness of these people. Intellectuals with no direct influence going around excitedly and intensely with no contact with government (such as it is) or business (too much as it is) but there's nothing there. Never before was there such a great country without a centre. It's anarchy, chaos, and the wonder is that it works at all. Not only works but creates wonders. They're free men and nothing is so staggering as their efforts. But the waste, the lack of direction, the complete absence of any sense of purpose (other than abstract clap trap) singularly oppresses a European like me. No wonder Steve is violently Anglophile. Only one man stood out in that gathering, Norman Mailer.[19]

A little later, back in Boston, Aaron bemoaned the inefficiency of the postal system and secretaries who can't do shorthand:

> But I mustn't gripe – everybody is nice, people in the shops are helpful but as Steve (Marcus) says 'It's all on the surface'. There's nothing underneath, but then nobody minds that.

[19] Churchill Archive Centre: Klug collection

In 1960, Aaron was offered the chair of the newly created department of Biophysics at the Johns Hopkins University Medical School in Baltimore. In spite of his negative feelings at the end of the 1959 visit, he decided to take this offer seriously. The Dean made a big effort to get Aaron: he was invited to come with his family. Since Liebe was against flying, the family came out on the Cunard *Queen Elizabeth* and went home on the *Queen Mary*. On board ship, there was time for Aaron and Adam to indulge in their shared vice of watching movies (particularly Westerns). They landed in New York, and were met by Norman Podhoretz and his wife Midge Decter who had a gracious apartment on the Upper West Side. Aaron, Liebe and Adam stayed with them for about a week. Norman and Midge were friendly with the Kennedys and at this time were agonising as to how Jackie might take to their dining room decor. Norman Podhoretz had just become editor of the *Commentary* magazine, which then still had liberal leanings. Steven Marcus was on the English faculty at Columbia, which gave Aaron useful contacts. Aaron was keen to discuss naturalist authors such as Theodore Dreiser or John Steinbeck with people who knew something about them. Norman Podhoretz took the Klugs to Radio City. Aaron, Liebe, and Adam tried out Coney Island, where Aaron recalls being somewhat put off by the obesity of the local population.

While Aaron and Liebe were in New York, Francis Crick paid them a visit. He wanted to pre-empt the Baltimore offer. Francis was determined to persuade Aaron to come to the newly founded (but not yet built) Laboratory of Molecular Biology in Cambridge, England. He described the plans for setting up what would turn out to be one of the world's most successful laboratories. The Laboratory of Molecular Biology, or LMB, was a successor to Max Perutz's MRC Unit for Research on the Molecular Structure of Biological Systems. It would be funded by the MRC on a long-term basis. The MRC would also take the opportunity of incorporating a number of like-minded researchers in need of a home, such as Fred Sanger (already a Nobel Laureate), Aaron Klug from Birkbeck and Hugh Huxley from University College London. Aaron would be able to run a research group and would be offered a renewable five-year contract with a good salary. Aaron discussed his quandary with Lionel Trilling, professor of English at Columbia and an authority on Henry James. Trilling recommended that Aaron, given his apparently mild personality, should stay in England.

In the USA, Trilling maintained, you need to show zeal. Trilling was not convinced that Aaron was capable of showing zeal.

The Klugs went by train to Cleveland to visit Aaron's long-lost aunt Sophie and his cousins, who were most welcoming. Adam could stay with the other children while Aaron and Liebe checked out Johns Hopkins Medical School and Baltimore – but how do you get from Cleveland to Baltimore without the indulgence of air travel? Liebe discovered that the Baltimore and Ohio Railroad (B&O), one of the world's older railroad companies, still existed. Furthermore, it managed to run Pullman cars from Cleveland to Baltimore, so the Klugs went to Baltimore by train, crossing the spectacular scenery of the Appalachians.

In Baltimore, the Klugs were appropriately wined and dined. Aaron was shown the building site that would contain his fiefdom. Somewhat to his surprise – given that Aaron still did not have a driving licence – his parking lot was already laid out. The parking lot was indeed important: Johns Hopkins Medical School was in a run-down part of Baltimore, and it would be necessary to commute. Aaron was unenthusiastic about a two-car existence: his driving in from the leafy suburbs and Liebe's becoming a 'soccer mom'. In contrast, Liebe liked Baltimore a lot: the weather was warm, and Americans somehow have recognisable similarities with other ex-colonials. There was a diverse population. The Johns Hopkins Medical School was excellent. She felt safe. But then it was time to take the train back to Cleveland to pick up Adam. The reunion was muted: Adam had apparently infected the family with the childhood illness known as 'slapped cheek'[20].

On the way back to New York, Aaron made a detour to Boston so that he and Don could compare notes on quasi-equivalence. At this time, Ken Holmes was a postdoc with Don; Aaron managed to tell Ken that he would be going to Baltimore or Cambridge, and that Ken would be welcome to join him in either venue, but just for now he could not say which. Aaron took his time, but in the spring of 1961 decided in favour of Cambridge.

---

[20] Slapped cheek syndrome: a childhood virus infection characterised by a bright red rash on both cheeks.

Figure 7.9 A meeting in Caspar's laboratory at the Children's Cancer Research Foundation (now the Dana-Farber Cancer Institute), Children's Hospital, Boston, in the autumn of 1960. Left: Don Caspar, centre: Ken Holmes, right: Aaron Klug. (Courtesy of D. L. D. Caspar.)

In the autumn of 1961, Aaron, Liebe and Adam spent three months visiting Cape Town and Durban. This was a family visit that enabled Adam to get to know his grandparents. Fortunately for the family finances, the University of Cape Town was able to offer Aaron a visiting professorship. As well as displaying his skills as a teacher, Aaron was able to do some electron microscopy. He also did a structure determination from fibre diffraction on a polysaccharide gum. R. W. James had retired but still had an office in the Physics department. It pleased Aaron that for nearly three months he was able to talk things over with James. For a couple of years, as Vice Chancellor of UCT, James had been involved in daily tussles with the Government concerning the application of its laws of Apartheid, a gruelling and depressing occupation. At about the time the family returned to England, South Africa left the Commonwealth. Aaron and Liebe took British nationality.

The move to Cambridge was now imminent. In order to raise enough capital to buy a house in Cambridge, the Klugs sold their house in Sutcliffe Close and would soon be obliged to move out. *Alea iacta erat,* but finding a house in Cambridge was not simple. Aaron wanted to live in Newnham – lots of friends and a good school – but property was too expensive. Things were getting stressful when one spring morning Liebe took a call from a Cambridge estate agent offering quite a large house, 70 Cavendish Avenue, for a rapid sale, a same-day decision. Cavendish Avenue is in south Cambridge and an easy cycle ride from the new laboratory. The price was affordable. Liebe said yes. The house was one of five built at the end of Cavendish Avenue by Rodney 'Gipsy' Smith[21]. The house needed considerable repairs, but at least the Klugs would have a roof over their heads.

In the summer of 1962, Aaron went off to the United States with his splendid new UK passport to attend the Cold Spring Harbor Symposium, where he and Don presented their quasi-equivalence theory of spherical virus structure. When he returned, they moved to Cambridge.

---

[21] Gipsy Smith was an internationally famous evangelist and early member of the Salvation Army. He made his home in one of these houses. In recognition of his birth in a Gipsy tent in Epping Forest, he called the house 'Romany Tan'.

Part II

Science at the Laboratory of Molecular Biology

8

The Laboratory of Molecular Biology

On 13th February 1960, two very remarkable papers appeared in *Nature*: 'Structure of haemoglobin: A three-dimensional Fourier synthesis at 5.5-Å resolution, obtained by X-ray analysis' by Max Perutz and collaborators, and 'Structure of myoglobin: A three-dimensional Fourier synthesis at 2-Å resolution' by John Kendrew and collaborators. These papers, reporting the first atomic structure of a protein, were the harbingers of structural molecular biology. Nonetheless, these protein structures were not the first fruits of the unit that Perutz founded after the war in the Cavendish Laboratory, Cambridge. Priority goes to the structure of DNA, already published in 1953. A meeting between Kendrew and Salvador Luria had deflected Jim Watson from CalTech to Perutz's unit in Cambridge. Resisting Perutz's coaxing to work on the structure of haemoglobin, Watson ended up persuading Crick to work on DNA. Aided by data unwittingly provided by Rosalind Franklin, Watson and Crick were able to guess the structure of DNA: two anti-parallel chains arranged in a helix with stacked bases linked together by base pairs (adenine pairing with thymine, and guanine with cytosine). The structure allowed them to deduce how DNA replicates.

Perutz's unit attracted a coterie of brilliant young scientists. As a result, his unit became the headquarters of a new science, Molecular Biology. In point of fact, for the first ten years of the unit's existence, he had little competition. Few were trying to do the new interdisciplinary kind of science that was characteristic of Perutz's group; the university departments shunned it. Therefore, those screwball enough to think that you could do something about explaining biology in terms of chemistry and physical chemistry ended up going to Perutz's lab. Many of the American postdoctoral fellows that Perutz and his colleagues trained in the 1950s and 1960s became professors at leading US universities.

The Laboratory of Molecular Biology, opened in 1962 in Cambridge, grew out of Perutz's unit at the Cavendish Laboratory. It was formed by supplementing the protein structure group (Max Perutz and John Kendrew) and molecular genetics group (Francis Crick and Sydney Brenner) with researchers working on protein sequencing (Fred Sanger, already a Nobel Laureate), muscle (Hugh Huxley) and virus structure (Aaron Klug). At the Laboratory of Molecular Biology, Perutz developed a successful management style that, broadly speaking, was: 'Be nice to people, get them what they need, and let them get on with their work.' Perutz did not micro-manage, nor was there very much formal decision making. His prima-donna-like colleagues generally appreciated Perutz's low-key management style, although he could sometimes appear a little pedestrian. Competing with the repartée of Francis Crick and Sydney Brenner in one of their daily performances was not a game that appealed either to Perutz or indeed to any normal mortal. Aaron flourished in this stimulating atmosphere, and his notable achievements would be honoured in 1982 with the Nobel Prize. Later he became Director of the Laboratory.

Perutz was born in Vienna in 1914 into a moderately wealthy secularised Jewish family that owned a textile factory in Bohemia[1]. The summers were spent in Reichenau at the foot of the Rax Mountains, where the Perutzes owned a villa. Perutz loved Reichenau and the mountains. Although thought to be a sickly child, during his teens Perutz developed into a formidable downhill skier and was also a keen climber. During his third university year studying chemistry, he became intrigued by reports of the research of Frederick Gowland Hopkins at Cambridge: the uses of chemistry to study physiology. This approach was singularly lacking in Vienna, and Perutz resolved to go to Cambridge. The famous Viennese physical chemist Hermann Mark

---

[1] Blow, D. (2004) *Biogr. Mem. FRS* **50**, 227–256;
Georgina Ferry, *Max Perutz and the Secret of Life*. Cold Spring Harbor Laboratory Press (2008).

happened to be visiting Cambridge and was entrusted with presenting Perutz's petition to Hopkins. However, during his visit to Cambridge, Mark was overwhelmed by a meeting with John Desmond Bernal, who had taken the first good X-ray diffraction images from a protein crystal. As a result, Mark forgot to mention Perutz to Hopkins and instead ended up arranging for Perutz to go to work with Bernal at the Cavendish Laboratory.

In 1934, Bernal and Dorothy Crowfoot (Hodgkin) had managed to obtain high-resolution X-ray diffraction data from crystals of pepsin through the simple expedient of keeping them wet. The implications of this finding were profound: protein molecules of the same type were identical in structure. The prevailing view at the time was that proteins were random copolymers and colloids. This view came about because biochemists at the time could neither understand how genetic information was passed from parent to offspring nor envisage the idea that genetics might control protein sequences. Bernal had the vision that by solving the atomic structure of proteins by X-ray crystallography, one could understand how life worked. Perutz, in turn, was fired by Bernal's vision. After a shaky start, in 1938 he obtained crystals of haemoglobin from Gilbert Adair. At this time Bernal moved to Birkbeck College London, and Lawrence Bragg came to Cambridge as Cavendish Professor. Perutz stayed in Cambridge with haemoglobin for the rest of his life. Despite wartime hardships (he was interned as an enemy alien and transported for some months to a camp in Canada) and distractions (he was recruited to work with Magnus Pyke on the Habakkuk Floating Ice Airfield Project), Perutz worked to get better haemoglobin crystals and to collect X-ray diffraction data from them. It was entirely unclear how one would interpret these data. Nevertheless, he had the unswerving support of the Rockefeller Foundation: Warren Weaver, a director of the Rockefeller Foundation, who coined the phrase 'molecular biology', was a believer.

In the autumn of 1941, under the threat of Rommel's relentless advance on Cairo, John Kendrew had been sent to set up an 'Operational Research' unit in Cairo whose job was to evaluate and improve the success of air attack on shipping, Rommel's supply lines. In 1943, after Rommel had been driven back, Kendrew was sent to India to help plan the attack on Japan. While carrying out tests on the efficiency of the use of bombs for jungle clearance he met Bernal. Between explosions, Bernal told Kendrew about proteins and Perutz's small group at the Cavendish, which intrigued Kendrew. Thus, in the autumn of 1945, Kendrew journeyed to Cambridge to consult with Perutz.

Perutz was housed in a room in the first floor of the Austin Wing of the Cavendish Laboratory. In spite of Kendrew's turning up unannounced, Perutz was impressed by this efficient young man in a Wing Commander's uniform. On his side, Kendrew decided to take the risk of joining the uncertain-looking venture. It was agreed that Kendrew should come and work with Perutz on an X-ray diffraction study of the differences between foetal and adult sheep haemoglobin for a PhD thesis, officially under the direction of Sir Lawrence Bragg. Kendrew had the residue of a college research fellowship at Trinity from before the war, so he did not cost anything. Nevertheless, by 1947 the continued existence of the group was in jeopardy. David Keilin, Head of the Molteno Institute and an authority on haem-containing proteins (also known as haemoproteins), suggested to Bragg that he might approach Sir Edward Mellanby, Secretary of the Medical Research Council. In October 1947, in one of the world's more far-sighted acts of funding, the MRC took over Perutz and Kendrew's nascent group as a unit 'For the Study of the Molecular Structure of Biological Systems'. Perutz was Director and John Kendrew an independent part-time member. The main aim was to elucidate the chemical structure of proteins – the molecules of life.

Kendrew and Perutz were at the same time members of the Molteno Institute. There, Kendrew worked with Joan Keilin, who introduced him to horse heart myoglobin. Myoglobin, like haemoglobin, is an oxygen-carrying haem protein. It has the advantage for analysis that it is only a quarter the size of haemoglobin. This was to become Kendrew's research project.

Protein crystals yield tens of thousands of X-ray reflections, each of which needs to be measured for the calculation of the electron density map of the crystal by means of a Fourier summation. There were three problems: measuring 20,000 or more reflections; determining the phases of the reflections; and calculating a Fourier summation for each of perhaps 100,000 lattice points. When the MRC group was founded, none of these problems had been solved. The phase problem was particularly difficult. Such was the faith of the MRC that they funded the group despite such manifest shortcomings.

Kendrew engaged Hugh Huxley as a research student. Huxley had interrupted his studies to work in radar development during the war. His notable success in this field was honoured with an OBE (Order of the British Empire). Kendrew asked Huxley to carry out by hand a tedious calculation (a Patterson summation – see Chapter 6 and Appendix B) from the sheep haemoglobin X-ray diffraction data. Huxley was affronted by the drudgery of the process and quickly consulted with John Bennett, one of his friends from Christ's College. Could Bennett devise a program to sum the Fourier series on EDSAC I, the Cambridge University electronic computer? He could, and an active collaboration ensued between Bennett and Kendrew. Together they forged the first ever program for computing a Fourier summation for X-ray structure analysis.

In 1953, Perutz and his co-workers showed how to solve the phase problem for protein crystals by using heavy atom markers. This idea had been floated already by Bernal in 1939, but the prevailing prejudice was that the scattering from one (say) mercury atom containing 80 electrons would hardly show up against a protein containing more than 10,000 electrons. Perutz and his student David Green established experimentally that the effects of the mercury on a haemoglobin molecule are quite easy to see. After much hard work from Max's student, Anne Cullis, with a library of crystals that Howard Dintzis, a visiting American postdoc, had soaked in various heavy-atom solutions, suitable diffraction data emerged. It took the ingenuity of Michael Rossmann, who joined the group in 1958, to devise a computational scheme to find the coordinates of the heavy atoms in the unit cell (the fundamental repeating unit) of the crystal — the first step towards solving the phase problem. Hilary Muirhead, Perutz's next student, was then able to apply some theory developed by David Blow and Francis Crick to carry out the step needed to obtain phases from those coordinates, and Rossmann used his fluency with computer programming to calculate the density of the haemoglobin molecule by Fourier summation. Haemoglobin is a tetramer (formed from four subunits), and the map showed that all four of the component protein subunits, two of one kind and two of another, looked nearly the same. Indeed, all four look very like myoglobin — a striking similarity that Rossmann noticed when he first examined the density. Each was a coiled-up sausage made of α-helices wrapped around the haem group that held the iron. The strong family resemblance of haemoglobin and myoglobin was the first indication that the evolution of proteins had been extraordinarily conservative, redeploying an effective functional module whenever possible.

Kendrew was carrying out the same research programme with myoglobin. He took over Perutz's methods. Dintzis, who had originally been employed by Kendrew to find isomorphous derivatives of myoglobin, soaked myoglobin crystals in all manner of heavy-atom salts in the hopes that some of the heavy ions would stick in some 'cosy corner', and indeed some did. Kendrew had the advantage over Perutz that myoglobin was smaller than haemoglobin. There were fewer data to collect, not so many phases to be determined and the Fourier summations necessary to get the density were smaller in size. John Kendrew was a skilled manager who organised and inspired a team of visitors and postdocs to obtain the high-resolution structure of myoglobin by 1960 – the first glimpse of the atomic structure of a protein. Perutz was disappointed that haemoglobin did not get to atomic resolution first. On the other hand, he was delighted to see the close relationship between the two proteins.

When he retired in 1954, Lawrence Bragg left little free space for the incoming Cavendish Professor, Neville Mott. Mott wanted to re-establish the Cavendish Laboratory as a pre-eminent physics laboratory and needed to make new appointments. Thus there was considerable pressure on Perutz's unit to move. In 1957, Mott moved the unit out of the Austin Wing into the infamous huts in the courtyard, but even the huts were needed. Given the tangible success of the unit, the MRC fortunately made the far-sighted decision to found a new Laboratory of Molecular Biology to be built on the new Addenbrooke's Hospital site to the south of Cambridge[2].

Queen Elizabeth opened the new building on 28th May 1962. Jim Watson came over from Harvard for the event, but Aaron was in London that day and could not make it. Given the honour of explaining to the Queen the atomic structure of haemoglobin and myoglobin, Kendrew felt it appropriate to wear his crimson silk DSc gown. Since the myoglobin models were composed of brightly coloured plastic balls, one lady-in-waiting dared to enquire whether we all had these coloured

[2] For a history see Soraya de Chadarevian's *Designs for Life*. Cambridge University Press (2002).

balls inside us! After the educational part of the visit, the Queen departed to other parts of the Hospital site for more education. As she descended the stairwell, the bulk of the lab dutifully lined up to do homage to their sovereign. One cannot help wondering whether the Queen felt that her subjects might have looked more endearing without cameras stuck to their noses.

The architecture of the new laboratory could be generously described as functional: a four-storey rectangular building 60 m long and 14 m deep, lying roughly east–west on the eastern end of the new Addenbrooke's site. The stairwell, about 20 m in from the western end, came out at the unprepossessing main entrance. The area to the west of the stairwell housed the University Department of Radiotherapeutics. The west end also contained the common seminar room. The ground floor housed services such as stores and Tony Broad's fearsome X-ray generators. The copious workshop was built out at the back to the north on the ground floor level. Floor one was offices, floors two and three were labs. Each floor had a central corridor. There was a pent-house on the roof that housed the canteen.

Figure 8.1 The Laboratory of Molecular Biology (LMB) in 1962 (architects: Messrs Easton & Robertson, Cusdin, Preston and Smith – London). The Department of Radiotherapeutics is to the left of the main entrance. (© MRC Laboratory of Molecular Biology.)

The Birkbeck group moved to Cambridge in the first half of 1962. Excluded from Cambridge by the prevailing house prices, Ann and John Finch[3] and Mary and Ken Holmes purchased adjoining houses in the unprepossessing village of Hauxton, to the south of Cambridge. Hauxton, once the head of navigation of the Granta, was a remnant of Danish settlement founded some 1200 years earlier, with a substantially larger number of inhabitants recorded in the Domesday book of 1086 than existed when Holmes and Finch moved there. Hauxton was within cycling distance of the LMB. Pleasingly, the route encompassed the neighbouring village of Little Shelford which, unlike Hauxton, was blessed with some old-world charm.

By November 1961, the Finches had already moved to Hauxton. John Finch worked for a while in the huts in the Cavendish using the Tony Broad X-ray tubes housed in the basement of the Austin Wing until the whole lab moved into the new Hills Road building in Feb 1962. The Holmes family waited until the move was complete and came in April 1962. Reuben Leberman came with them as a lodger, which awakened the interest of the village gossips. Bill Longley, who was finishing his doctorate in London, came to Cambridge later. Aaron, Liebe and Adam moved to 70 Cavendish Avenue on 26th July 1962. Aaron's office had been moved from Birkbeck somewhat earlier. This move called for some care because Aaron's filing system was essentially sedimentary. Now the whole group was settled in Cambridge and the new life could start. Naturally, this was not without its complications.

In the autumn, Adam started at the Cambridge Grammar School for Boys, which in Queen Edith's Way was conveniently close to 70 Cavendish Avenue. There was much work to be done in the house, and for some months it seemed to be full of people. Remembering Cambridge winters, Liebe insisted on installing central heating. The house soon became enmeshed in a tangle of copper pipes. To add to the confusion, Liebe was soon pregnant and took to her bed to avoid another miscarriage. Thinking it might lighten the household burden, Aaron took on a Dutch *au pair*, Margot, but romantic complications ensued: the heating engineer fell for the *au pair*, which resulted in the desolate wife and

---

[3] John Finch started as a doctoral student with Aaron at Birkbeck College London in 1955. He came with Aaron to the MRC Laboratory of Molecular Biology in Cambridge in 1962 and worked there until his retirement in 1995.

children coming round and weeping on Aaron's shoulder. The ensuing fracas called for all Aaron's negotiating skills. He ended up returning Margot in an unsullied state to her most respectable Dutch parents, and the families remained friends. In 1964, Margot joined the Klugs for a summer holiday in Brittany, a month in a run-down mansion near St Malo.

Aaron also managed to retain the heating engineer. Whether his disappointment in love or other factors caused the delay is not clear, but the heating installation was not finished before winter struck. The winter of 1962–63 was one of the coldest of the century. In Cambridge, the temperature stayed well below freezing for about six weeks. Temperatures below -10 °C were common. The river Cam froze, even under the bridges, which gave Perutz the chance to demonstrate his consummate skills as a skater. When water was finally injected into the Klug's heating system, it froze instantly. Thus it was another three or four weeks before Liebe got her heating. Nor were Finch and Holmes spared. In the New Year, the Holmes threw a party for Don Caspar and his wife Gladys who were their visitors and houseguests. During that night, the water input froze. Luckily, the next morning, Holmes and Finch were able to locate the stopcock and to unfreeze it by applications of boiling water. The Finches themselves were less fortunate: in the meantime, their stopcock had frozen solid. After six weeks of carrying water from next-door, Finch decided to try using a welding transformer connected to two available points on the inlet pipe (luckily copper) straddling the stopcock. After a couple of minutes, the Finches had water again.

In October 1962, Jim Watson and Francis Crick (for Physiology and Medicine) and Max Perutz and John Kendrew (for Chemistry) were all awarded Nobel Prizes. Enthusiastic jollification ensued. Since the notifications of the two prizes were spaced by a week, the celebration spread out over about 10 days and included Hugh Huxley letting off fireworks from the roof of Crick's house, the 'Golden Helix'. Kendrew and Crick both ordered such quantities of champagne that they got onto the mailing lists of notable champagne wholesalers.

Aaron had an office on the first floor on the north side of the laboratory building. Finch, Holmes and Leberman shared the office next door. To the west of their room was Huxley's office and then a large room dedicated to model building. Next to this was a larger office shared by Brenner and Crick. On the south side was Perutz's office and the office of the administrator, Audrey Martin. Opposite Finch and Holmes was Kendrew's office. Leberman spent most of his time in his lab for virus chemistry on the second floor, which he had designed. Other members of the structural studies division found themselves in a difficult situation because no lab space had been allocated for their biochemistry. It took some time to sort out this oversight, which perforce led to Leberman sharing his facility with other structural biologists.

Hugh Huxley had moved to Cambridge early in 1962, and by the time Aaron arrived, Huxley had already set up his new electron microscope – a Siemens Elmiskop 1. Leberman thought that this would be a great way of checking his virus preparations and suggested that Finch might do some electron microscopy (often shortened to EM). The idea was already in the air because Huxley, together with Geoff Zubay, while still in London had produced rather good electron microscope images of turnip yellow mosaic virus (TYMV, as mentioned in Chapter 7) from virus material supplied by Aaron. Huxley organised a two-day instruction course for Finch, who quickly became a talented electron microscopist. Subsequently, John Finch and Aaron worked together in a most effective partnership for nearly four decades. At this stage they were primarily interested to see how many different examples of the Caspar and Klug quasi-equivalent structures they could find. For this purpose, the resolution of the electron microscope was adequate. It was clear it would take a very long time to work out the 3D structure of a virus by X-ray crystallography, whereas electron microscopy seemed to offer a quick and easy route to the overall structure, even if not to atomic detail. Through trying to understand the images of spherical viruses, Aaron was led to devise the methods for 3D reconstructions of structures from EM images. This finding made a large contribution to his receipt of the Nobel Prize in 1982. It is noteworthy that this was not his avowed research aim; rather, his goal remained the understanding of spherical virus structure.

Right from its inception, the LMB established an international reputation and attracted a cluster of very talented postdoctoral fellows. A number of these chose to work with Aaron. With their help and John Finch's expert electron microscopy, Aaron followed his chosen aim. Slowly, a game plan evolved, but it took a decade.

9

Image Analysis in Electron Microscopy

The game plan involved the electron microscope. The Hungarian-American physicist Leo Szilard held a patent on an electron microscope but never actually constructed one. Ernst Ruska, as the subject of his doctoral thesis, built the first electron microscope in Berlin in 1931. Ruska demonstrated that the focal length of the magnetic solenoid lenses used to focus electrons could be made very short (necessary for a powerful microscope) if one used a toroidal iron case (a *Polschuh*) round the solenoid with a narrow gap on the inside of the toroid – a form that is still used today. During the 1930s, Ernst Ruska and his brother Helmut tried the new microscope out on biological samples, including TMV (the tobacco mosaic virus mentioned in previous chapters), and managed to see the virus. Thus they were getting substantially better resolution than could be obtained with the light microscope. In 1939, Siemens in Berlin marketed the first commercial electron microscope. After the war, Ruska worked with Siemens to produce the Elmiskop 1, which for many years was the workhorse of electron microscopy.

In the early electron microscopes, electrons from a heated filament were accelerated up to 80 kV and focused by condenser lenses onto the specimen. After passing through the specimen, the electrons were focused by the objective lens and subsequent projector lenses onto a fluorescent screen or film to give a very highly magnified image of the object. Magnifications of 100,000 could be achieved. Naturally, there were also problems. Electrons are readily absorbed by air, so the whole microscope, including the specimen, has to be in vacuum. Biological specimens do not like a vacuum. In addition, electrons interact with thin biological specimens rather weakly, so that it is difficult to see much in the magnified image. The electrons' interaction with the specimen has two aspects: as particles they can be absorbed or scattered, or as waves they can suffer a retardation of phase. In 1960, there had not been much success in registering these phase changes, and even today, 'phase-contrast' electron microscopes are still far from routine. Nevertheless, a phase change is all that light biological materials can manage. A way round this is to use the microscope out of focus, which to a limited extent turns the phase changes into intensity changes and thus, by enhancing contrast, makes a phase-contrast object visible. Unfortunately, in an out-of-focus image, edges are surrounded by fringes and other artefacts. Between 1965 and 1975, Aaron and his co-workers transformed the analysis of electron micrograph images from an art into a science.

There was also the problem of protecting the specimen from the high vacuum: the drying-out induced by high vacuum usually destroys a biological specimen. Here TMV is an exception; the particles are so stable that they can even survive high vacuum without too much distortion. Hence TMV was a lucky choice for the Ruska brothers. The way round the vacuum problem is to embed the sample in a thin sheet of amorphous ice at a low temperature, but this demanding technique was not developed for another two decades.

In the 1960s, the way forward was to stain biological materials with a heavy metal, which gives lots of contrast. For viruses, the method of choice is to immerse the virus in a 2% aqueous solution of uranyl acetate or phosphotungstic acid on the microscope grid (a fine copper grid is used to support the specimen in the microscope). The grid is then blotted nearly dry so that the specimen finally becomes immersed in a uranyl acetate film that acts as a cast, much like plaster of Paris. Subsequently, in the microscope image, one can see virus-shaped holes in the uranyl acetate film. The embedded virus itself hardly shows up. The resolution is limited – you cannot see atoms – but structures about ten times coarser than atomic structures can be seen rather well. Since one is really looking at a cast of the object, this approach is limited to displaying surface features. The method is referred to as negative staining and was developed at about the same time by Sidney Brenner and Bob Horne at the Cavendish, and by Hugh Huxley and Geoffrey Zubay in London. It turns out to be rather good for viruses. Moreover, Huxley and Zubay found a nice trick for improving visibility: holey grids. The microscope supports used for electron microscope (EM) work are thin carbon films produced by sputtering[1] carbon onto a thin film of collodion or formvar (a thermoplastic resin) adhering to the copper grid. The collodion is then dissolved away to leave the carbon film as a specimen carrier. The carbon film itself adds substantially to the random speckling (referred to as 'noise') in the image. However, if you breathe on the collodion films, some of the tiny droplets in your breath react with the film to produce holes. These in turn result in holey carbon films on the grid. The negative stain happily spreads across these nano-holes to produce thin films of unsupported stain. Thus, negative stain on holey grids gives the best visibility.

John Finch and Aaron started by looking at poliovirus but could see very little because there is not much contrast on the surface, whereas with TYMV or TBSV (turnip yellow mosaic virus and tomato bushy-stunt virus, respectively) they could see strong surface features. Huxley and Zubay had already seen 32 knobs or protrusions on the surface of TYMV. These 32 knobs could readily be reconciled with the Caspar–Klug classification of icosahedral virus shells by T-number if rings of six subunits and rings of five subunits each clustered to form a knob. For a $T = 3$ virus lattice (as mentioned in Chapter 7), there are 12 five-fold and 20 six-fold clusters giving 32 clusters in all, with a total of 180 subunits in the virus capsid.

Roy Markham and Bob Horne were also taking EM images of TYMV. Markham was sceptical of the Caspar–Klug theory. He developed a photographic method of analysing images for rotational symmetry (and later for translational symmetry)[2]. If an object had, for example, five-fold symmetry and you kept rotating it through 72° (360/5) and projecting it onto a sheet of photographic paper in a dark-room projector, then the resulting superimposed image would show detail if the object really had five-fold symmetry but not if it had six-fold or seven-fold symmetry. According to Markham, all the knobs on

---

[1] Sputtering is a technique used to deposit thin films of a material onto a surface.
[2] Markham, R., Frey, S. and Hills, G.J. (1963) *Virology* **20**, 88–102.

TYMV were actually five-fold (even the should-be-six-fold knobs), and the virus contained 160 subunits.

Finch's images using holey grids showed enough resolution (detail) to check out the symmetry of the knobs, so it should have been easy to sort out this ambiguity. In fact, there was a serious problem. It soon became clear that for nearly all of the observed images, negative stain was engulfing the entire virus particle. Thus the electron micrographs of the virus were showing a superposition of the front and back surface views. If the negative stain was thin, you occasionally obtained one-sided views. If a particle lies so that the view is along a two-fold axis, then the back and front images are in the same orientation so on superposition you get a clear (less-complicated) image. Otherwise, because of the superposition of front and back, the images are messy and cannot be interpreted simply by looking at them. Aaron and Finch then approached the problem of front–back superposition experimentally: they obtained a specimen holder for the electron microscope that allowed the specimen to be tilted, and they used this holder to gather pictures of single TYMV particles at various degrees of tilt. They fabricated a model from wire and rubber tubes that was based on their interpretation of hundreds of single-sided images and made shadowgraphs of the model at various angles to show that one could get moderate agreement between the projected views of the model and the observed images as one varied the tilt[3].

Sometimes the virus coat protein assembles into a particle without the RNA. Then the virus fills with negative stain demonstrating the internal location of the RNA in the intact virus. More detailed information on the configuration of the RNA could be gleaned from the X-ray diffraction studies of crystals of TYMV that were continued by Bill Longley[4]. Don Caspar summarised the work on TYMV in an elegant drawing (Figure 9.2).

Markham's result bugged Aaron. He was certain his analysis of TYMV was right, but Markham was irritatingly self-confident in his wrong-headedness. Moreover, his superposition method had a certain claim to being an impartial method of analysis, whereas Aaron and John Finch spent their time picking a few images from hundreds in a manner

---

[3] Finch, J.T. and Klug, A. (1966) *J. Mol. Biol.* **15**, 344–364.
[4] Klug, A., Longley, W. and Leberman, R. (1966) *J. Mol. Biol.* **15**, 315–343.

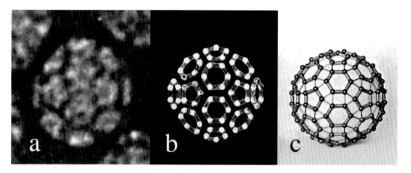

Figure 9.1 a, A negatively stained electron micrograph of a single particle of turnip yellow mosaic virus viewed over a holey grid (the protein is white). The particle is aligned looking along a two-fold axis. In this view, the front and back surface features are the same, which makes the image easier to interpret.

b, An idealised drawing derived from the inspection of hundreds of images as shown in **a**. The five- and six-fold groupings of the virus coat protein can be seen. There are 12 five-fold and 20 six-fold rings, accounting for the 32 knobs seen in low-resolution electron micrographs.

c, A ball and spoke model based on the drawing in **b** that was used for generating shadowgraphs at various tilts to be compared with images from virus particles tilted in the electron microscope. (Reprinted from Finch, J.T. and Klug, A. *J. Mol. Biol.* **15**, 344–364 © 1966, with permission from Elsevier.)

Figure 9.2 A drawing of turnip yellow mosaic virus by D. L. D. Caspar. (Reprinted from Finch, J.T. and Klug, A. *J. Mol. Biol.* **15**, 344–364 © 1966, with permission from Elsevier.)

that could easily be construed as subjective. Markham's superposition method produced startling improvements in visibility when applied to two-dimensional periodic objects called polyheads, which are aberrant structures derived from TYMV coat protein or other spherical virus coat proteins. However, to use Markham's method one first had to guess what the periodic repeat of the object was, which was a subjective procedure. Was there a general way of dealing with such problems in a way that the data could speak for themselves? Aaron realised that there was. To turn Markham's superposition method into something much more powerful, one should work with the Fourier transform of the object. Aaron wrote:

> Markham's method, as well as our own problems in the electron microscopy of viruses and other biological particles, have stimulated us to consider other methods which might reveal periodicities present in the image of the object under investigation. Clearly what is required is a decomposition of the image into its harmonic components, i.e. a complete Fourier analysis. The Fraunhofer diffraction pattern of the image would automatically provide just such an analysis.[5]

Charles Taylor and Henry Lipson at Manchester University were authorities on Fraunhofer diffraction patterns, which are named in honour of the remarkable optical inventor Joseph von Fraunhofer. These are the diffraction patterns produced by shining a parallel beam of monochromatic light onto a specimen. They are often referred to as 'optical transforms', since the operation is essentially a way of producing the Fourier transform of the object. Aaron was in contact with Taylor because he and his collaborators had corrected Aaron's triphenylene structure. They had shown that Aaron had got the molecular orientation in the crystal correct but the molecule was in the wrong place. Aaron asked Taylor if he could help to make some optical transforms of electron micrographs. Taylor responded by recommending to Aaron an American postdoc, Jack Berger, who had been visiting Manchester for about a year from his home laboratory in Buffalo, New York. Berger knew the Manchester set up rather well. To further the work, Berger and his wife Rae later moved to Cambridge for a year and rented a house in

---

[5] Klug, A. and Berger, J.E. (1964) *J. Mol. Biol.* **10**, 565–569.

Cherry Hinton. Aaron was always somewhat absent-minded about how he had arrived at the lab in the morning: he could have used his bike, or come up in the bus from town, or even have come by car, in which case there could be problems locating the vehicle in the large car park. Since Berger lived close to Aaron it became traditional for Aaron to ask, 'Oh, do you have your car here? I'd be quite grateful for a ride. . .' with such a surprised tone in his voice that one could never decide if he was totally absent-minded on such matters or simply a skilful actor. About five years later Berger was visiting the United Kingdom from Buffalo and called in at the lab to visit his old playmates. As evening approached, Aaron looked round and, noting that Jack Berger was there, asked if he just happened to have a car. Berger drove Aaron home.

While the scale of the electron micrographs was wrong for the Manchester apparatus, with photographically reduced images it all worked well. To try out the method, Berger and Aaron first looked at the images of negatively stained crystals of the protein catalase made by Bill Longley. Catalase is an important and ubiquitous enzyme that gets rid of dangerous hydrogen peroxide in the body by turning it rapidly into oxygen and water. Catalase can be induced to form two-dimensional crystals on the microscope grid that are just one molecule thick. At limited resolution, the object appears to be two types of interleaving column. Because the object is periodic in both directions, the diffraction pattern is a two-dimensional lattice. The horizontal lattice spacing is about half the vertical value, showing that the unit cell of the catalase crystal is twice as wide as it appears on first inspection. Apparently the zigzag columns (Figure 9.3) are of two different kinds, leading to the wider unit cell. This shows up easily in the optical transform, although it is not immediately apparent to the eye. The optical transform takes an average over the whole of the object, showing up such features more clearly than local inspection does.

Berger and Aaron then tried the method out on TMV. The electron micrographs of TMV showed very little detail because of the superposition of the front and back surfaces. The optical diffraction patterns showed that there was indeed detailed information in the micrograph images and that one could begin to sort out the contributions from the front and back surfaces of the virus, but the problem was too complex for further analysis. Then Jack and Rae Berger went back to the USA, and the work of analysing the contributions from the front and back

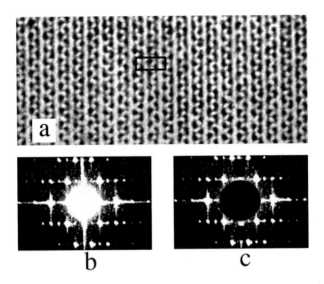

Figure 9.3 The diffraction pattern or optical transform of a lattice is a lattice (the reciprocal lattice).

a, A negatively stained two-dimensional catalase crystal. The repeating pattern or unit cell of the lattice is shown (black-edged rectangle). Note that in the horizontal direction this includes two zigzag motifs because only alternate motifs are alike (the unit cell is 2.4 times as wider as it is high).

b, The optical transform of a. This displays two artefacts: 1. Scattering from the strong incident beam – the beam is scattered from each of the glass surfaces it passes. 2. Spikes in the vertical and horizontal direction that are diffraction effects from the edges of the illuminated area.

c, Same as b, with the artefacts masked out. Note the lattice-like structure of the diffraction pattern. This is called the reciprocal lattice because the lattice spacing stands in a reciprocal (that is, inverse) relationship to the lattice spacing of the scattering object. In the object, the repeating lattice is 2.4 times as wide in the horizontal direction as in the vertical direction. Therefore, in the reciprocal lattice, the horizontal spacing is smaller (narrower) than the vertical spacing, in the ratio 1/2.4. Note: the resolution is given by distance from the centre of the outermost reflections. (Reprinted from Klug, A. and Berger, J.E. *J. Mol. Biol.* **10**, 565–569 © 1964, with permission from Elsevier.)

surfaces of periodic objects by filtering diffraction patterns was taken up by David DeRosier.

David DeRosier completed his doctorate thesis in biophysics at Chicago with Bob Haselkorn on a light-scattering method of measuring the sizes of viruses and virus coat proteins. His data gave 180 TYMV coat protein molecules per virus. DeRosier was intrigued that 180 was exactly the number predicted by the Caspar–Klug theory, and, after a discussion with Haselkorn, decided that he would like to work with Aaron. Thus in the summer of 1965, full of expectations, DeRosier accompanied by his wife Anne came to the Nobel-Laureate-laden Laboratory of Molecular Biology. He arrived full of the naïve expectation that seasoned electron microscopists could interpret their images. The ensuing acrid controversy concerning the structure of the polyoma–papilloma viruses soon dispelled this illusion.

Aaron and John Finch's work on TYMV had shown that one could interpret two-sided images of viruses by tilting the specimen and comparing the tilted images with shadowgraphs of a model tilted in the same way. Could this become a general method? Aaron and Finch decided to try this out on negatively stained electron micrographs of rubisco, a large octahedral enzyme complex. Rubisco is involved in carbon fixation, the process that plants use for turning atmospheric carbon dioxide into sugars. The method of interpreting the images came from Don Caspar. A balsa wood model was manufactured by the workshop incorporating whatever features could be seen by eye. Then DeRosier was given the grunt work of immersing this model in plaster of Paris to simulate negative staining and taking X-ray radiographs of the plaster of Paris at various degrees of tilt to compare with the EM images. The way forward was trial and error improvements to the model – and more plaster of Paris. This strategy did not lead to improvement, and the project was abandoned.

After this abortive apprenticeship, Aaron urged DeRosier to turn his attention to the problem Aaron had started with Berger: how to get better-averaged images out of their optical transforms. Consider a rectangular repeating structure (as in Figure 9.3). This gives diffraction maxima that form a lattice. If in the optical diffractometer, instead of a film, one had a sheet of cardboard with holes cut at the places where the lattice points occur, one could just select out the diffraction maxima from the lattice points and let them go on through another lens to form an image. Then one would obtain a noise-free averaged structure at least as good as Roy Markham's. Also, if the object were built on a lattice but with superimposed back and front surfaces, then the method would allow one to separate the back and front surfaces of an image. The basic idea was that the two surfaces would give rise to two diffraction patterns consisting of grids of spots, but these two grids would, in general, be at an angle to each other. By preparing a suitable mask you could let through diffraction from just the front or just the rear. DeRosier decided to try out this process of 'optical filtering' with the cylindrical tails from bacteriophages.

Bacteriophages are viruses that attack bacteria. In contrast to the simple plant viruses and small animal viruses, which are basically a spherical bag with some nucleic acid inside, the bacteriophages are complex machines consisting of a large DNA-containing bag with an attached syringe-like structure for injecting the nucleic acid into the bacterium. The phage tail is the syringe. It starts out as a long cylinder that shortens on contact with the bacterium. The cylinder has an inner and outer component. The interplay of these two structures finally drives the contraction. The cylindrical phage tail alone occurs as a substantial fraction of T4 bacteriophage preparations and is easy to purify. On the microscope grid, the phage tail is generally completely immersed in stain: both top and bottom surfaces show up in the image.

Optical filtering could not be done in Manchester. DeRosier needed to build his own optical diffractometer, which necessarily would be some metres long. Luckily, the LMB building contained a small unused lift shaft that was ideal for housing such an optical diffractometer. DeRosier and Aaron consulted with Edward Linfoot, the head of the Cambridge Observatory, on what would be the best optics and lenses to use, and DeRosier then built the diffractometer. This was longer than the Manchester apparatus since you needed two lenses: one to make the diffraction pattern and the second to form an image from the filtered diffraction pattern. The apparatus worked, the filtering worked, and the first fruits of the method were published in *Nature*[6]: Aaron and DeRosier were able to form separate images from the front and back surface of a phage tail.

Naturally, there were problems. One of the worst was dust. Aaron and DeRosier had been able to harness a new technology, namely lasers, as a light source. Up until this time the best available sources for optical diffraction experiments were mercury arc lamps, which were small bright sources. Such sources, however, were not uniform nor were the beams truly monochromatic. These properties limited the resolution of

---

[6] Klug, A. and DeRosier, D.J. (1966) *Nature* **212**, 29–32.

the diffraction pattern produced – but they did make such experiments relatively insensitive to dust. With laser-light things are quite different: the incident beam is very bright, and exposure times are much shorter. Moreover, the beam is monochromatic, parallel and completely phase-coherent. One effect is that the beam registers every particle of dust it passes, and these turn up as smudges or fringes in the final image. Cleanliness in the converted lift shaft was next to Godliness.

Bob Horne at the Cavendish and his collaborators had used negative staining to look at the surface structure of polyoma virus and came up with 42 capsomeres (surface bumps) arranged with icosahedral symmetry. Other workers rather uncritically took over the result of Horne *et al.* in allocating the number 42 to the number of capsomeres of polyoma and the closely related papilloma viruses, which thus became the established norm in the literature for this class of viruses. Aaron was certain the number 42 was wrong. It had already been questioned by Caspar and Klug in their Cold Spring Harbor paper. Moreover, Carl Mattern at the National Institutes of Health thought that the number was not 42, settling instead on the number 92. John Finch and Aaron decided that to get a sensible answer they would have to do some experiments themselves. They decided to start with human wart (papilloma) virus.

Unlike the plant viruses, which are packages of single-stranded RNA, polyoma and papilloma viruses contain double-stranded DNA. Polyoma viruses are very common in all mammals. In humans, they generally cause mild infections, although they can lead to cancer in immuno-suppressed patients. Papilloma viruses mostly cause benign warts, but some strains do cause cancer. The recognition that papilloma virus causes cervical cancer was the subject of the Nobel award to Harald zur Hausen in 2008. Human warts were freely available from the wart clinic in the newly built hospital just across the road. To extract the virus, one first needs to grind up the warts with carborundum powder in a mortar with a pestle. Then the juice is subjected to filtration and centrifugation, which finally yields masses of virus particles. Excised warts are not objects of great natural beauty, and thus this extraction procedure is not to everyone's taste. Nikolai Kiselev, a visitor from Moscow who later worked on the wart polytubes, steadfastly refused to have anything to do with the preparation on the grounds that the excised warts were an affront to his sensibilities. Moreover, he thought the virus particles might be infectious – as indeed they were.

Negatively stained images of the wart viruses not only showed double-sided but also a few single-sided images, which were not engulfed by the stain. Working with single-sided images, where they could easily resolve the surface bumps, Aaron and Finch established without doubt that the surface lattice was a $T = 7$ structure with 72 surface bumps[7]. They also worked with rabbit papilloma virus and got the same result[8]. Polyoma was very likely to be the same, but Horne adamantly maintained that polyoma virus had 42 surface bumps. Jo Melnick from Baylor College strongly supported 42 over 92. The Finch and Klug publication of the result that papilloma viruses had 72 surface bumps provoked not only Bob Horne but also Carl Mattern. Aaron and John came under immediate crossfire from all camps. Each thought that their method was just as valid. The critique of the Klug and Finch team was mostly that their analysis was based on a small number of selected images (single-sided). Aaron pointed out that the majority of images were two-sided and therefore too complicated to be interpreted by eye. One needed to simulate the appearance by projecting a model. The adversaries argued that their models predicted the appearance of two-sided virus particles just as well as Aaron's models. While this claim was true in a rough and ready way, it did not stand up to detailed inspection. In response to the battery of criticism, Aaron felt moved to write a riposte[9]:

> *The correctness of the 42-structure was first questioned by Caspar and Klug and independently by Mattern. Mattern himself, in a later paper, went on to suggest that the most likely structure was with 92 capsomeres. I do not accept either the 42- or the 92- capsomere structure... The purpose of the present paper is to show that the 72-structure can be identified in the published electron micrographs of several of the authors listed above.*

This was Aaron's first experience of a serious scientific controversy, but certainly not his last. He was very bothered by his opponents' apparently deliberate lack of attention to detail. If you looked at the detail, it was clear that they were wrong. Their critique was not only unfair, it was also miserably unprofessional.

[7] Klug, A. and Finch, J.T. (1965) *J. Mol. Biol.* **11**, 403–423.
[8] Finch, J.T and Klug, A. (1965) *J. Mol. Biol.* **13**, 1–12.
[9] Klug, A. (1965) *J. Mol. Biol.* **11**, 424–431.

Buckminster Fuller had designed plastic rods and connectors (Geodestix) for building models of his geodesic domes. Aaron and Don Caspar soon found these invaluable for building models of the Caspar–Klug quasi-symmetrical icosahedral surface lattices, which they called icosadeltahedra. Models for $T = 1, 3, 4$ and 7 are shown in Plate 2. The five-fold axes are marked with red connectors, the six-fold with green. Each of the yellow sleeves represents a subunit. It is noteworthy that the $T = 7$ icosadeltahedra come in two kinds that Aaron called *dextro* and *laevo* (right and left), since they are related as mirror images. This property can be seen in the models. In the $T = 7$ surface lattice, to get from one five-fold axis to the next, you have to go over two green axes and then make a jump to the left or a jump to the right to reach the neighbouring red axis, rather like a knight's move in chess.

Since $T = 7$ comes in *dextro* and *laevo* forms, which is it in papilloma? To resolve this question, one needs to know which side of the virus one is looking at: in projection, both surface lattices look the same. Finch managed to find out whether he was viewing an upper or lower surface of a virus by taking stereo views. Two photos of a scene taken at about 6° to each other form a 'stereo pair'. To view them, they must be printed at a size that allows them to be spaced by 6.5 cm (the average separation of human eyes), one view for each eye. Once you have aligned the two images so that the brain realises that the two eyes are looking at the same thing, then, if you are good at stereo, the scene springs into 3D (on account of his weak right eye, Aaron was not in fact very good at seeing stereo images).

Finch obtained stereo pairs of the negatively stained virus by tilting the microscope stage through 6° between taking photos. In a few cases, stereo pairs showed the dome of the spherical virus sticking out towards the viewer (meaning that it was being viewed from above). Here one could see that the knight's move between a pair of five-fold surface bumps went to the right: thus the papilloma virus is $T = 7$ *dextro*.

It seemed likely that the controversies would diminish if one could see the structures in 3D. However, Aaron had not quite reached this mindset. He was still thinking in terms of separating the front and back surfaces of images, as he could do for periodic objects such as bacterio-phage tails, by using optical filtering. The results were somewhat frus-trating because you really want to see inside the phage tail and not just look at the surface. Could one somehow calculate a 3D map of the density in the phage tail? Aaron saw a way forward. The bacteriophage tail was a helix; TMV was a helix. Thus it should be possible to use the theory Aaron had developed in 1958 for TMV to reconstruct a 3D density map from a helical diffraction pattern.

Aaron's idea was that you could use Holmes's program, which was really a Fourier transform program for data set up in cylindrical coordinates, to compute the 3D density of a phage tail with input data from optical transforms of the electron micrographs. Furthermore, there was no need to go through all that stuff with heavy atoms that Holmes had to do with TMV to find the phases: with some ingenuity, you could measure them directly from the optical transform. This ingenuity involved forming an interference pattern and counting fringes at each point to obtain the phase. Just in time to save David DeRosier from going crazy with the tedium of counting fringes, the project was finished, and the first 3D map of a phage tail was computed.

In the meantime, technology was making strides. DeRosier realised that you could do it all much more easily (at least with a lower chance of going insane) by using a scanning device attached to a computer to turn the electron micrograph into a grid of numbers and simply compute the Fourier transform of this grid of numbers with a fast Fourier transform program (a so-called FFT). This had the supreme advantage that the program printed out the phases – no fringe counting! Moreover, the computed data were more accurate than the optical transform data. Also, in the computer you could implement various correction factors that should be applied to the density observed in the electron micrograph. Thus the whole calculation was repeated (the details of this nerve-wracking process are described in some detail in a historical memoir by DeRosier[10]). The calculated diffraction pattern (after some appropriate manipulations) was stuffed into Holmes's program, and out came a 3D density map of a phage tail.

Displaying the output density was not trivial, because computer graphics had not yet been invented. Using the then-current technology, a balsa wood construct was made by gluing together sections through the map. Each section was cut out with a fretsaw, making a cut at a certain contour level. The resulting model was the first 3D reconstruction from an

---

[10] DeRosier, D.J. (2010) *Methods Enzymol.* **481**, 1–24.

electron micrograph[11]. It represented an important change of mindset. No more talk of separating front and back surfaces: now the world was in 3D.

The DeRosier–Klug paper is both profoundly original and rather unusual in that it is really two papers sewn together. One wonders how it survived the editorial process. Both halves are concerned with 3D reconstruction from EM images. The first half is a completely novel demonstration that you can obtain 3D reconstructions from EM images of helical objects, in this case a phage tail. The second half, which is a general recipe for calculating 3D reconstructions from sets of projections, is independently very important. Although no applications were presented, this part contained real enlightenment and was the harbinger of a new branch of science, tomography. The second part of the paper arose from wondering why the scheme that DeRosier and Aaron had just carried out on a phage tail had worked at all. Helical particles are special because they offer lots of views of the same thing from various angles. It is as if one had tilted an object many times and taken a new micrograph at each setting. But you could do this for anything; it does not have to be helical. Keep turning any old object through (say) 2° and each time take a new EM image: each photo gives you a projection of the object in a different direction. Then the question arises, how does one turn a lot of projections back into an object in three dimensions? This is the central problem of tomography, which will be discussed in Chapter 13.

The first usable solution to the tomography problem was that proposed by David DeRosier and Aaron Klug in the *Nature* paper in 1968. Apparently they both independently thought of the method at about the same time and then spent an interesting hour trying to explain it to each other. Their method was based on two important properties of Fourier transforms:

1. The Fourier transform of a projection in real space is a section in reciprocal space[12];
2. The Fourier transform of a Fourier transform is the thing you started out with.

[11] DeRosier, D.J. and Klug, A. (1968) *Nature* **217**, 130–134.
[12] This known as the projection-slice theorem, central slice theorem or Fourier slice theorem. It was actually first used in 1956 by Ronald Bracewell for a radio astronomy problem.

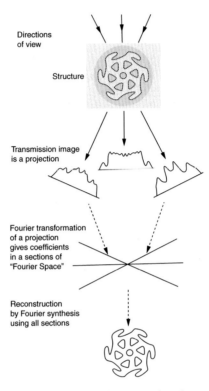

Directions
of view

Structure

Transmission image
is a projection

Fourier transformation
of a projection
gives coefficients
in a sections of
"Fourier Space"

Reconstruction
by Fourier synthesis
using all sections

Figure 9.4 Scheme for the general process of reconstruction of a structure from projected images (Reprinted by permission from Macmillan Publishers Ltd. DeRosier, D.J. and Klug, A. *Nature* **217**, 130–134 © 1968.)

Thus if we take the Fourier transform of an EM projection we end up with a section in reciprocal space. By collecting together sections calculated from many projections at different angles, we could get all the values of the Fourier transform in 3D. If we Fourier transform these data back, we have got the object in 3D – Figure 9.4 (in the example shown below, the reconstruction is limited to a two-dimensional slice: you get three dimensions by assembling lots of slices).

DeRosier and Klug wrote:

The electron microscope image represents a projection of the three dimensional density distribution in the object at all levels perpendicular to the direction of view. According to a theorem familiar to crystallographers, the Fourier coefficients calculated from a projection of a three-dimensional density distribution form a section through the three-dimensional set of Fourier coefficients corresponding to that distribution. By collecting many different projections of a structure in the form of electron microscope images it should therefore be possible to collect, section by section, the full set of Fourier coefficients required to describe that structure. The number of projections needed to fill Fourier space roughly depends on the size of the particle and the resolution.

The idea is quite simple and, in retrospect, it is not clear why it did not occur to them earlier. Both must have been too hooked on the idea of separating the front and back images. After all, Fourier transforms were bread and butter to Aaron – he spent time teaching their properties to the likes of Ken Holmes and John Finch and to students at Peterhouse. Indeed, Holmes, together with David Blow, had written a rather successful introductory textbook incorporating some of Aaron's teaching, which explained the relationships give above[13]. Moreover, all X-ray crystallographers knew that if you took a section through the 3D diffraction data and computed the Fourier series from just this section, you ended up with a map of the electron density projected along the axis at right angles to the plane of the section. It was standard practice for small-molecule crystallography: three projections derived from three succinctly chosen sections, although the crystal diffraction pattern usually told you all you wanted to know. Fourier transforms are just a generalisation of Fourier series for a non-periodic object (that is, not a crystal) and the same rules hold: if you take the transform of a section in the diffraction pattern, you compute a projection of the density. Conversely, if you take the transform of a projection of the density, you end up with a section through the 3D transform (essentially, the 3D diffraction pattern). Fill in the values of the 3D diffraction pattern with the transforms of lots of views; transform this back (equivalent to a Fourier synthesis) and you're done!

---

[13] Holmes, K.C. and Blow, D.M. (1965) *Methods Biochem. Anal.* **13**, 116–239 – also published by Wiley as a standalone book, *The Use of X-ray Diffraction in the Study of Protein and Nucleic Acid Structure.*

10

Spherical Virus Structure

Although the reasoning about Fourier transforms in the previous chapter is all theoretically sound, Aaron and DeRosier were still not out of the woods: the various sections in reciprocal space give a very uneven sampling of the values of the Fourier transform, so that in the middle (low resolution) there are far too many values and at the outside (high resolution) there are far too few. To compute the Fourier transform, you need the values on a regular grid. There are well-known standard numerical solutions to this problem, but in 1968 there were no computers big enough to solve a problem with so many points.

Tony Crowther, an applied mathematician, joined the Laboratory of Molecular Biology in 1964 as a doctoral student working with David Blow. Crowther knew about the DeRosier and Klug work since he had programmed the flying-spot densitometer that DeRosier used for scanning the micrographs. In January 1968, at the end of his doctorate, Crowther went to work in Edinburgh. When he came back to Cambridge for his PhD viva, he indicated to Aaron that things were not working out in Edinburgh. Aaron arranged for him to come back to the LMB, and Crowther rejoined the laboratory in November 1968. DeRosier went to Texas early in 1969, so they overlapped for a few months. They discussed the idea of using cylindrical coordinates rather than a Cartesian grid. This reduced the size of the computation to manageable proportions[1]. The icosahedral symmetry could be included in the data reduction by searching for 'common lines' where two sections related by symmetry intersected each other. Crowther was indeed able to set up this procedure. In cylindrical coordinates, the Fourier transform becomes a Fourier–Bessel transform. This presented no particular problem because Aaron had already worked out this theory for TMV, the tobacco mosaic virus previously mentioned. Thus cylindrical coordinate interpolation became the method of choice. With this method, Crowther and DeRosier, together with a new addition to Aaron's team, Linda Amos, were able to calculate from Finch's EM images the first 3D reconstructions of two spherical virus particles: the tomato bushy-stunt virus TBSV, and human wart virus (Figures 10.1 and 10.2)[2].

The resolution of the human wart virus reconstruction was about 60 Å, just adequate to resolve the surface knobs (morphological units). The map showed 72 knobs, exactly as predicted from Finch's and Aaron's earlier studies. Moreover, the relationship between the five- and six-fold axes was as in a knight's move in chess. This result provided a complete vindication of Aaron's allocation of human wart virus to the $T = 7$ surface lattice. Higher resolution was later obtained which showed that the knobs were in fact rings of subunits. There was no more talk of 42 or 92 knobs[3].

Tomato bushy-stunt virus looked quite different. Aaron had already shown with Finch and Reuben Leberman that TBSV was a $T = 3$ lattice and that the subunits clustered around the true two-fold and local two-fold axes. There are 30 two-fold axes and 60 local two-fold axes, giving a total of 90 surface bumps or morphological units. These were nicely displayed in the 3D reconstruction.

In 1964, after taking his degree at Harvard, Stephen Harrison came to LMB as a visitor before starting his PhD. Harrison was offered the

---

[1] Crowther, R.A., DeRosier, D.J. and Klug, A. (1970) *Proc. Roy. Soc. Lond.* **A317**, 319–340.

[2] Crowther, R.A., Amos, L.A., Finch, J.T., DeRosier, D.J. and Klug, A. (1970) *Nature* **226**, 421–425.

[3] Twelve years later, a higher-resolution structure of polyoma virus (which has a very similar structure to human wart virus) showed that both the six-fold and five-fold vertices (knobs) were occupied by five-fold rings. The polyoma subunit has a flexible tail to accommodate the symmetry mismatch. This result disturbed Aaron deeply, but it was correct. Nature has no respect for elegance of design. (Rayment, I., Baker, T., Caspar. D.L.D., Murakami, W.T. (1982) *Nature* **295**, 110–115.)

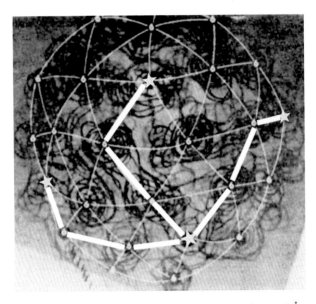

Figure 10.1 Contour map of a 3D reconstruction of human wart virus at 60 Å resolution. The contours were traced on stacks of plexiglass sheets and photographed over a light box. High contours indicate absence of stain. The $T = 7$ surface lattice is superimposed. Note that the maxima (knobs – morphological units) lie on the five-fold and local six-fold axes. The five-fold axes are highlighted with white stars. The pathway (white lines) between five-fold axes through intervening six-fold axes is like the right-handed knight's move in chess. (Reprinted by permission from Macmillan Publishers Ltd. Crowther, R.A., Amos, L.A., Finch, J.T., DeRosier, D.J. and Klug, A. *Nature* **226**, 421–425 © 1970.)

chance to work with Leberman on how turnip crinkle virus assembled. He chose rather to take X-ray diffraction pictures from crystals of turnip crinkle virus, which is structurally related to TBSV. He returned to Boston and did his PhD with Don Caspar on the X-ray crystallography of TBSV. Nearly a decade later, after Harrison had become a faculty member at Harvard, he and his collaborators solved the structure of TBSV at 2.9 Å (near-atomic) resolution[4]. Their structure showed that the coat protein monomer was composed of two separate domains joined together by a flexible hinge. The monomers formed dimers

---

[4] Harrison, S.C., Olson, A.J., Schutt, C.E. Winkler, F.K. and Bricogne, G. (1978) *Nature* **276**, 368–373.

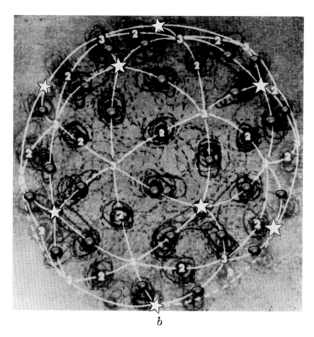

b

Figure 10.2 Contour map of a three-dimensional reconstruction of tomato bushy-stunt virus (TBSV) at about 30 Å resolution. Contours were traced on stacks of Plexiglas sheets and photographed over a light box. High contours show absence of stain. The $T = 3$ surface lattice is superimposed. The five-fold axes are marked with stars. The three-fold and two-fold axes are labelled. The local two-fold axes are marked with small grey spheres. Note that the high protein density lies on the two-fold and local two-fold axes. (Reprinted by permission from Macmillan Publishers Ltd. Crowther, R.A., Amos, L.A., Finch, J.T., DeRosier, D.J. and Klug, A. *Nature* **226**, 421–425 © 1970.)

around the two-fold and local two-fold axes using the same contacts between subunits for both kinds of two-fold symmetry axes, but the hinge accommodated the difference of tilt angle across the two-fold and local two-fold axes. This result explained a lot. Don Casper and Aaron had done some serious soul-searching about the quasi-equivalence theory, because you need highly specific interactions for a virus to self assemble; it will not do for the protein subunit to aggregate with any old protein it comes across, but specificity is not compatible with flexibility. How do you get both? Harrison's structure showed that the virus coat protein was solidly built for specificity but that flexibility was also there – as a hinge between two domains of the protein. Moreover, another structural element switched from *disordered* (on the subunit dimer in solution and on the local two-fold axis) to *ordered* on the true two-fold axis. So both flexibility and its positional regulation had evolved conjointly into the virus coat protein monomers[5].

In 1971, the annual Cold Spring Harbor Symposium on Quantitative Biology was devoted to protein structure. In 1960, all the protein crystallographers in the world could comfortably meet round a dining room table. By 1970, there were already a couple of hundred. Even the virus structure community was big enough to warrant its own session, although half the papers were devoted to TMV. Aaron presented a paper with his new co-worker, Tony Durham, on the polymorphism of TMV coat protein and its variant structures, examining under which conditions of salt and temperature you find disks or helices. Another new co-worker, Jo Butler, who had replaced Leberman after he left for Heidelberg, presented data on the mode of assembly of TMV from its RNA and disks. Holmes presented the 10-Å-resolution structure of TMV.

The next paper was given by Bror Strandberg from the prestigious Wallenberg Laboratory in Uppsala, Sweden. Earlier, Strandberg had been a postdoctoral fellow with John Kendrew in Cambridge. He was now working on the X-ray crystal structure of the smallest known spherical virus, satellite tobacco necrosis virus (STNV). This virus was so small and contained so little RNA that it could not even organise its own reproduction. It needed help from tobacco necrosis virus (TNV), a larger spherical virus rather like TYMV. STNV could only infect a plant that was co-infected with TNV. Chemical studies showed STNV to have 60 subunits in the virus coat (the capsid). Electron microscopy showed it to be roughly spherical. The virus particles often packed on the microscope grid to form an oblique lattice in which the two axes were at 72° to each other (this is the icosahedral angle). Moreover, virus clumps were often seen on EM grids with five particles grouped round one central particle. All this argued strongly for STNV being the archetype icosahedral virus: in the Caspar–Klug notation, a $T = 1$ virus.

---

[5] In ensuing decades, many 'spherical' virus structures have been solved. While nearly all show icosahedral symmetry, the ways in which the coat proteins cope with the symmetry are varied (see Stephen Harrison for a lucid account of virus structure): https://www.youtube.com/watch?v=KoJWuWzVgqQ

Michael Rossmann and David Blow at LMB pioneered the use of non-crystallographic symmetry to help solve the phase problem. Since a crystal lattice can only have two-, three-, four- or six-fold symmetry, how do you find particle symmetry of a kind that cannot show up in the crystal symmetry? This non-crystallographic symmetry had been of great interest to Don Caspar when he showed that TBSV had icosahedral (five-fold) symmetry. As we discussed in Chapter 7, Caspar's solution was to look for spikes in the diffraction pattern (Figure 7.2). The Rossmann–Blow solution had more in common with Roy Markham's ideas of turning the object through a certain angle in the darkroom and superimposing the images. If it is a five-fold axis and you turn it through 72°, the image should look stronger. In place of EM images, Rossmann and Blow had diffraction data. They were not able to compute the structure without knowing the phases, but they could compute the Patterson function (see Chapter 6 and Appendix B) from the X-ray diffraction intensities. So they searched for symmetry in the Patterson function. They produced diagrams rather like sky-maps showing the positions of non-crystallographic symmetry axes. This was the rotation function. It was notably heavy on computer usage.

Michael Rossman left the LMB in 1964 to take up a professorial appointment at Purdue University, Indiana. He had become very interested in virus structure because it would provide an excellent area for the application of the rotation function to detect particle symmetry. In 1971, he spent a sabbatical summer with Bror Strandberg in Uppsala and tried out his rotation function program on Strandberg's diffraction data from STNV. The answer appeared to be unequivocal: STNV had octahedral symmetry! However, if it were octahedral, there should only be 24 subunits in the capsid rather than 60 subunits. Strandberg quickly organised new molecular weight determinations of the subunit size. The physical chemists seemed to agree rather quickly that there were really about 24 subunits in the virus particle rather than 60. The Cold Spring Harbor meeting was imminent: the stage was set for a denouement. During the day leading up to the virus session, members of the meeting were urged in a rather conspiratorial manner to turn up for an evening session when all would be revealed. Since scientific meetings are normally about as exciting as Tupperware tea parties, this news did have its appeal.

There was a meeting at the beach that afternoon between Aaron, Crowther and Strandberg, at which Strandberg told Aaron of the result but could not show the data because the results were on a slide, so Aaron had some preparation. In the session, Strandberg gave an introduction to STNV, described the crystals and then showed the rotation function plots. As he was about to go to the next slide on which the interpretation (octahedral symmetry) would be outlined, Aaron spoke up from the front row, saying: 'Leave that slide up! You are about to make an interpretation that I'm sure is wrong, and the reason for it is in those plots, but I need another minute to figure out where the problem is.' Strandberg obligingly stood there for a minute of painfully awkward silence, after which Aaron conceded that it would take a bit longer to find out what was wrong and allowed him to continue. Aaron's interjection put off many of the audience, and had you taken a poll, you would have found a huge majority siding with Strandberg, but those who knew the virus field knew that it was likely that Aaron was correct.

Strandberg and Rossmann were indeed wrong, but it was to take more than a few minutes' effort to find out why and how. A general backlash against the rotation function ensued. Moreover, Aaron's strong reaction gave him a reputation for being truculent. His outburst had in fact been atypical: Aaron eschewed public displays of emotion. Nevertheless, *manifest wrong-headedness* could elicit considerable displays of public irritability.

Later, in didactic mode, Don Caspar pointed out to Rossmann that it was very important to look directly at the virus diffraction data: the symmetry shows itself as spikes in the diffraction pattern, as he had shown with tomato bushy-stunt in his famous *Nature* paper. Caspar added, 'That's the way to look at the symmetry. It's a primitive rotation function.'

The resolution of the problem was given by Aaron in a companion paper that was published in the final proceeding of the symposium. It turned out that STNV had crystallised with one of its icosahedral two-fold axes at 45° with respect to a crystallographic two-fold screw axis. This disposition of two-fold axes is typical of octahedral symmetry. Indeed, the apparent octahedral symmetry peaks in the rotation function were actually twice as high as the intrinsic icosahedral peaks. On reflection, Rossmann came to the same conclusion: the rotation function had worked, but Rossmann and Strandberg had interpreted the answers in the wrong way. It took some time for the noise from this unfortunate confrontation to die away.

11

Tobacco Mosaic Virus

After Rosalind Franklin's death, Holmes and Aaron continued the X-ray analysis of X-ray fibre diagrams from tobacco mosaic virus, TMV. Specimens for X-ray work were gels in which the rod-shaped virus particles were oriented parallel to each other, but randomly rotated about their own axes. These gels give excellent X-ray diffraction patterns to high resolution, but because of the rotational averaging, the three-dimensional X-ray diffraction information is projected into two dimensions. Unscrambling these data to reconstruct the 3D structure proved to be a major undertaking necessitating the production of half a dozen heavy-atom derivatives of the virus. The first 3D Fourier maps to a resolution of about 10 Å were not published until 1971[1], after Holmes and his colleagues had moved to Heidelberg in Germany. A resolution approaching 4 Å was later achieved[2]. At this resolution, while it is still not possible to identify individual amino acid residues in the virus coat protein with any certainty, the positions of the RNA bases (adenine, guanine, cytosine and uracil) show up rather clearly, as do the protein α-helices. Thus this map, taken together with the detailed map of the virus coat protein subunit that Aaron's group later obtained from the double disk structure (see below), showed how the RNA bound to

[1] Barrett, A.N., Barrington Leigh, J., Holmes, K.C. *et al.* (1972) *Cold Spring Harbor Symp. Quant. Biol.* **36**, 433–448.

[2] Stubbs, G., Warren, S. and Holmes, K.C. (1977) *Nature* **267**, 216–221.

Figure 11.1 Tobacco mosaic virus: each coat protein subunit is depicted as a clog. The protein subunits assemble into a right-handed helix with $16^{1}/_{3}$ subunits per turn (49 in three turns; Franklin, R.E. and Holmes, K.C. (1958) *Acta Cryst.* 11, 213–220). The diameter of the helix is about 180 Å (18 nm). The centre is hollow, and is 40 Å in diameter. A single strand of RNA is embedded into the protein coat at a radius of 40 Å. There are three RNA bases per protein subunit. About 1/20 of the length virus is shown. Drawing by D.L.D. Caspar. (Reprinted from Caspar D.L.D. *Adv. Prot. Chem.* 18, 37–118, © 1964, with permission from Elsevier.)

the coat protein and yielded a considerable amount of information about the assembly process by which the viral RNA gets incorporated into the protein helix.

By the late 1960s, the TMV project was in a production phase. Aaron had already produced the theoretical framework for calculating a 3D map of TMV from measurements of the X-ray fibre diffraction pattern in 1958. Measurements of X-ray diffraction data from film had continued with Susan Fenn at Birkbeck and were now being extended by Brenda Wayne, and later by Angela Mott. Holmes and Bill Longley together with Tony Woollard were producing better fine-focus rotating anode X-ray tubes and better X-ray optics for collecting data. Whilst in Boston with Caspar, Holmes had taken advantage of the Massachusetts Institute of Technology 'project MAC' to program Aaron's theory in the IBM programming language Fortran for an IBM 7090 (IBM's first transistor-based computer). Back in England, a commercial IBM

7090 was available near Tottenham Court Road in London, and the LMB set up daily access (via a data link in the form of steam trains and couriers carrying stacks of IBM punched cards). Later, the Cambridge Astronomy Laboratory installed an IBM 360 a mere cycle ride away.

Obtaining an electron density map of TMV from the X-ray fibre diffraction patterns depended on solving the X-ray phase problem, more demanding for this case than for crystals because the data were also cylindrically averaged. A solution required finding a half-dozen different ways of attaching heavy atoms (such as mercury) in well-defined positions to the virus coat protein. Reuben Leberman, Leslie King, Richard Perham and even Ken Holmes indulged their chemical fantasies in trying to persuade TMV to react with heavy-atom reagents. The coat protein has evolved to be unreactive: various chemical reactions and mutational changes were tried. Gradually, six heavy-atom derivatives of TMV were produced.

The remaining challenge was to locate the heavy atoms: that is, to work out their spatial coordinates within the helical array. The radius was easy to find, but to determine the relative heights and azimuths of the heavy atoms proved quite difficult. Finally, Holmes and Aaron used a method that had been proposed by Lawrence Bragg for solving a similar problem in the structure determination of myoglobin. Aaron and Holmes referred to this as the 'method of Bragg ellipses'. Once they knew where the heavy atoms were, they could use Perutz's method of isomorphous replacement to calculate an electron density map of the virus.

The intrinsic difficulties of working with cylindrically averaged X-ray intensities limited the resolution one was likely to achieve, but there was a complementary approach. As well as assembling into helices, the TMV subunits will also form flat disks containing 17 subunits in a turn. The disks are actually formed from a pair of rings coming together with the same relative orientation (a double disk). Leberman found out how to crystallise the double disks, which later became the subject of X-ray crystallographic investigation. The resulting high-resolution X-ray crystal structure could contribute to interpreting the helical diffraction structure.

The history of the disk crystals begins in 1963, when Aaron and Leberman started out to crystallise the TMV coat protein monomer. Crystallisation was attempted by adding a concentrated ammonium

Figure 11.2 In June 1970, Ken Holmes organised a workshop in Heidelberg, Germany, that brought together all 13 of the scientists in the world who were known to be interested in the structure of tobacco mosiac virus, TMV. Participants at the EMBO Workshop were (from left to right): John Barrington Leigh, Tony Barrett, Don Caspar, Peter von Sengbusch, Aaron Klug, Richard Perham, Ken Holmes, Jo Butler, Eckhard Mandelkow, John Champness, Tony Durham, Peter Gilbert and Reuben Leberman. In the autumn of 1968, Holmes and Leberman moved to the Max Planck Institute for Medical Research in Heidelberg to open a new Department of Biophysics. The Heidelberg and Cambridge groups stayed in close contact, with the aim of calculating an electron density map of the virus from orientated gel data to as good a resolution as possible and then interpreting this with the help of atomic models derived from the crystalline disks. (Photo: Ken Holmes.)

sulphate solution. In order to frustrate the natural tendency of the protein to aggregate into a helix, Leberman introduced various chemical modifications in the hope of blocking the normal contact sites, but none of these modified proteins crystallised. It was known from the work of Gerhard Schramm and Wolfram Zillig in Tübingen, Germany, that the protein on its own, free of RNA, can aggregate into a number of distinct forms as well as the helix. Thus, in 1965, Aaron and Leberman tried using crystallising conditions under which the protein appeared to be mainly aggregated in a form identified by Caspar as a trimer of subunits (the so called 'A protein'). The results of these experiments seemed unpromising until, a few weeks later, Leberman discovered some quite large crystals.

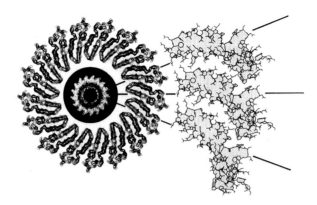

Figure 11.3 The disk viewed from above at successive stages of resolution. From the centre outward: 1 (middle), an electron microscope image at about 2.5 nm resolution; 2, a slice through the 5.0 Å electron density map of the disk obtained by X-ray analysis, showing rod-like α-helices; and 3, part of the atomic model built from the 2.8 Å map. (Figure 3 in Klug, A. (1983) *Les Prix Nobel en 1982*, 93–125, reproduced with permission. © The Nobel Foundation, 1983).

These were entrusted to Chang Yu-Shang, a postdoctoral fellow from Shanghai[3], who was very unsure that the crystals were protein because they were hard and robust, more like salt. The X-ray diffraction picture of the crystals demonstrated they were indeed protein and, moreover, showed strong 34-fold symmetry. EM analysis by Finch showed that the crystals were built from two rings, each of 17 subunits, referred to as the two-layer disk (or double disk)[4]. The geometry of the disk was clearly related to that of the virus particle. The cylindrical rings contained 17 subunits each, compared with $16^1/_3$ units per turn of the virus helix, so that the lateral bonding within disks was likely to be closely related to that in the virus. Electron micrographs showed that the two-layer disk was polar, meaning that its two rings faced in the same direction, as do successive turns of the virus helix. Aaron therefore determined to solve the structure of the 17-fold double disk by X-ray crystallography.

[3] Chang Yu-Shang (Zhang Youshang) returned to China in time to become embroiled in the Cultural Revolution. After a very difficult interim period, he was finally appointed an Academician at the Academia Sinica Shanghai Institute of Biochemistry, where he later became Director of a division in the Institute.

[4] Finch, J.T, Leberman, R., Yu-Shang, C. and Klug, A. (1966) *Nature* **212**, 349–350.

The molecular mass of the double disk was 630,000 daltons, 34 times that of one protein subunit (18,600 daltons). Thus it was a very demanding crystallographic problem. Indeed, along with TBSV (see preceding chapter), these were the first very large structures ever to be tackled in detail by X-ray crystallographic analysis. It took more than 12 years to carry through the analysis of the double disk to high resolution. Data collection for a structure of this magnitude necessitated the development of special X-ray cameras and computer-linked densitometers. John Champness was entrusted with the technical side of the data collection. Longley, Woollard and Holmes had already much improved the X-ray sources. In 1971, Aaron recruited Anne Bloomer, an experienced protein crystallographer from Oxford, to work full time on the project.

While the 17-fold rotational symmetry of the double disk made the technical side of the problem much more difficult than might have been

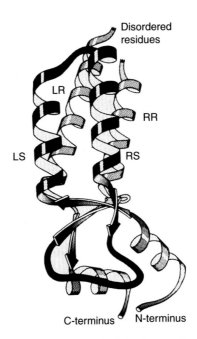

Figure 11.4 Tobacco mosaic virus protein subunit is a bundle of four α-helices (labelled here as LS, RS, LR, RR). Both the C-terminus and N-terminus are on the outside of the disk where they form a β-sheet (the N-terminus being a chain end terminated by an amino group, the C-terminus by a carboxyl group). Towards the lumen of the disk (top) there are 25 disordered residues. (© MRC Laboratory of Molecular Biology.)

expected, this high symmetry gave rise to lots of redundant information in the X-ray data, which was cleverly exploited by Gerard Bricogne to improve and extend the resolution of the electron density map[5].

The end result showed that the structure of the subunit was rather simple: a four α-helix bundle with a β-sheet on the outside. Both the N-terminus (beginning of the polypeptide chain) and C-terminus (end of the polypeptide chain) were on the outside of the disk. Towards the lumen of the disk, 25 residues of the structure were disordered and not visible. The orientation of the two layers of the disk were similar but different. A comparison with the helical structure of the virus is shown in Figure 11.5.

TMV can readily be taken apart to yield its coat protein and nucleic acid (RNA), neither of which originally appeared to be infectious. In 1955, Heinz Fraenkel-Conrat, at the University of California at Berkeley, made scientific history by recreating a virus from its component coat protein and nucleic acid. The newly created virus was infectious.

This accomplishment brought Fraenkel-Conrat considerable attention along the lines of 'life created in a test tube'. While scientists before Fraenkel-Conrat had taken viruses apart, he was the first to put one back together – in effect, making the world's first synthetic virus. Fraenkel-Conrat later showed that the RNA alone could be infectious, if handled carefully, which took some of the drama out of his discovery but reinforced the notion that it was the nucleic acid that carried the genetic information. The coat protein was just the coat.

Studies of the coat protein showed that it could form various aggregates ranging from A-protein (a trimer) to a reconstituted helix that looked just like the virus but contained no RNA. These aggregates were characterised by their speed of sedimentation in an ultra-centrifuge in units called svedbergs (S), after the inventor of the ultra-centrifuge. The sedimentation rate depends on the mass but also on shape and other factors. Studies of the effects of temperature, salt concentration and acidity (pH) by Tony Durham, together with EM studies by John Finch, were used to systematise the stability of the aggregates of the coat

---

[5] Champness, J.N., Bloomer, A.C., Bricogne, G., Butler, P.L.G. and Klug, A. (1976) *Nature* **259**, 20–24.
Bloomer, A.C., Champness, J.N., Bricogne, G., Staden, R. and Klug, A (1978) *Nature* **276**, 362–368.

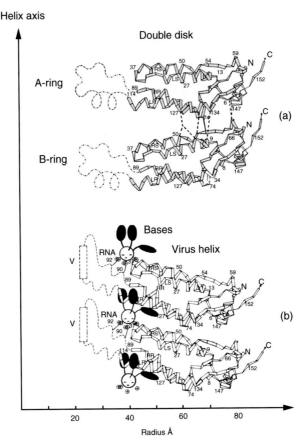

Figure 11.5 Comparison of the structures of the double disk and the viral helix. The ribbons show the path of the polypeptide chains of the protein subunits.

a, The double disk. Note that the two rings A and B are in contact toward the outside of the disk but not towards the centre. The dashed lines at low radius indicate the disordered mobile portion of the protein in the disk, extending in from the RNA binding radius.

b, The virus structure: adjoining subunits from two adjacent turns of the helix. The relative orientations of the protein subunits are quite different from a. The subunits are in contact along their entire length. The residues that are mobile in a are ordered and form extensions of the RR and LR helices (shown dashed) towards the lumen of the virus helix, providing an essential part of the RNA binding site. These are joined together by a vertically orientated structure (V) that is part of a wall that shields the RNA from possible chemical attack via the virus lumen. The lateral contacts (not shown) between subunits are very similar in the A- and B-rings and the helix. Three bases of the RNA bind round the LR α-helix, rather like a claw. Positively charged arginine residues, from a subunit one turn down in the helix, cluster round the negative charges of the three phosphate groups to form a very stable electrostatic interaction. (Diagram by K.C.H.)

protein in a phase diagram[6]. The most abundant aggregate at neutral conditions and at room temperature had the sedimentation constant 20 S. This aggregate was shown by EM to be a disk or perhaps a short helix. Aaron took it to be the crystallographic double disk.

Aaron was interested in the mechanism of virus assembly. In Fraenkel-Conrat's experiments, many hours were required to produce a virus. This seemed far too long. One of the coat-protein aggregates must provide a nucleus for viral assembly: but which one? Because under normal conditions it was the majority species, the 20 S aggregate looked like a good candidate. The first reconstitution experiments carried out by Jo Butler and Aaron to test the 20 S aggregate nucleation hypothesis proved dramatic[7]. When viral RNA was mixed with a 20 S preparation, complete virus particles were formed within 10 to 15 minutes, rather than over a period of hours. The idea that 20 S aggregates are involved in the natural biological process of initiation was strengthened by companion experiments in which assembly was carried out with RNAs from different sources. These showed also that 20 S aggregates have a preference, by several orders of magnitude, for the viral RNA over foreign RNAs.

Aaron identified the 20 S nucleating aggregate as the double disk[8]. However, a disk is not a helix, so how does viral growth along a helix take place? Aaron assumed the existence of a 'lock-washer' intermediate, a dislocated form of the double disk that has the same geometry as the viral helix, that results from binding RNA. Such a 'lock-washer' would provide an ideal basis for nucleating RNA incorporation and virus assembly. Indeed, when solutions of 20 S particles were suddenly made more acid, which mimics the effect of binding RNA, lock-washers appeared[9].

This mode of nucleation of helix assembly could readily explain the specificity of recognition of the virus RNA by the protein. The surface of the lock-washer presents a set of 51 (17×3) nucleotide-binding sites,

[6] Durham, A.C.H., Finch, J.T. and Klug, A. (1971) *Nature New Biol.* **229**, 37–42.

[7] Butler, P.J.G. and Klug, A. (1971) *Nature New Biol.* **229**, 47–50.

[8] Caspar and co-workers later re-examined the structure of the 20 S particle (see Casper, D.L.D. and Namba, K. (1990) *Adv. Biophys.* **26**, 157–185). They came to the conclusion that the 20 S aggregate is not a double disk but rather a two- or three-turn helix (rather more like the 'lock-washer' described here).

[9] Klug, A. and Durham, A.C.H. (1971) *Cold Spring Harbor Symp. Quant. Biol.* **36**, 449–460.

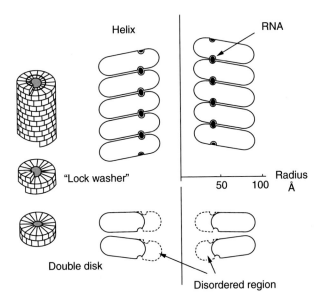

Figure 11.6 Illustration of the disk nucleation hypothesis for the formation of tobacco mosaic virus, TMV. The binding of RNA to the double disk induces the formation of a helical-like 'lock-washer'. The virus assembly proceeds by the binding of further 'lock-washers'. Note that the disordered region in the lumen of the double disk becomes ordered in the virus. (Diagram by K.C.H.)

which can interact with a long run of bases leading to much increased specificity. Specificity must arise from the presence of a unique sequence on the viral RNA optimised for interaction with the protein disk. After some effort, David Zimmern and Butler isolated the nucleation region – 60 nucleotides long, just over the length necessary to bind to a single turn of the helix. This discovery uncovered a new puzzle: the nucleation sequence occurs about one-sixth of the way along the RNA from the 3' end[10]. This means that that over 5000 nucleotides have to be coated in the major direction of elongation (from 3' to 5') and 1000 have to be coated in the opposite direction. Apparently, 'lock-washers' permit the RNA to enter through the central hole and

---

[10] The 3' end of a strand is so named because it terminates at the hydroxyl group of the third carbon in the ribose or deoxyribose sugar-ring; its opposite number, the 5' end, has the fifth carbon in the sugar-ring at its terminus.

intercalate between the two layers. The nucleation sequence binds tightly enough to permit nucleation of virus assembly, but as growth proceeds, the RNA is dragged through the lumen of the nascent virus[11] (a fuller account is given in Aaron's Nobel Lecture, 'From macromolecules to biological assemblies'[12]).

Don Caspar offers the following short historical summary of Aaron's disk nucleation hypothesis:

In 1963 Caspar[13] had speculated, on the basis of hydrodynamic and X-ray fiber diffraction data (much later recognized to have been misinterpreted), that the 20 S aggregate was a polar two-layer disk, which could dislocate to a lock washer form to assemble into the virus helix with or without RNA or alternatively to pack in the polymorphic stacked disc assembly. When the X-ray diffraction data from Leberman's 1965 TMV protein crystal grown at high ionic strength and alkaline pH revealed a four-layer disk consisting of a dihedrally related polar A-ring–B-ring pair, Aaron intuitively identified this experimentally established two layer disk with Caspar's hypothetical disc model for the 20 S aggregate formed under more physiological conditions. Electron microscopy of negatively stained TMV protein aggregates formed under different conditions and hydrodynamic studies of the assembly process in Aaron's lab appeared to prove the model of the two-layer disk as an essential intermediate in the virus protein assembly. The 1970 discovery by Butler and Klug that the 20 S aggregate – named the 'disk' – is the critical nucleating aggregate for the virus assembly eventually led to the formulation of an ingenious model of how the interaction of a specific viral RNA sequence offset from one end of the strand triggered switching of the double disk to a lock washer pair whose jaws grasped the RNA strand looped through its lumen, leading to incorporation of more disks at different rates at the two ends of the growing virus helix. Description of this intricate disk model of nucleation and assembly of the TMV helix figured prominently in Aaron's Nobel Laureate address; and accounts of this model in elementary biology texts and advanced biochemistry, molecular biology and cell biology tomes

---

[11] Butler, P.J.G., Bloomer, A.C., Bricogne, G. *et al.* In *Structure–Function Relationships of Proteins* (R. Markham and R. W. Horne, eds.), 3rd John Innes Symp. pp. 101–110. North-Holland/Elsevier (1976).

[12] Klug, A. In *Les Prix Nobel en 1982*, pp. 93–125. Imprimerie Royale, Norstedt (1983).

[13] Caspar, D.L.D. (1964) *Adv. Prot. Chem.* **18**, 37–121.

made it clear to a wide audience that assembly of highly ordered virus particles and cellular organelles does not proceed like sticking together rigid molecules in a regimented crystal lattice but does involve intermediate aggregates whose parts may have structures and interaction properties quite different from those in the completed assembly. Although the accumulated experimental data on the structure and assembly of TMV and its polymorphic protein aggregates now point clearly to a dynamic and irregularly ordered short helix instead of the crystalline two layer disk as the model for the TMV nucleating aggregate, Aaron's imaginative analysis of the pioneering studies on TMV assembly he guided in his laboratory over 40 years ago has opened the path to understanding the structural biology of the remarkable variety of macromolecular assemblies involved in health and disease.

12

From Thon Rings to Modern Dance

Harold Erickson visited the LMB as a postdoc with Aaron in the years 1968–1970. He came from Johns Hopkins University in Baltimore where he had taken a PhD in biophysics with Michael Beer. He and his wife Jacqueline were to interact quite significantly with the Klug family: Erickson worked with Aaron to understand the physical basis of image formation in the electron microscope; Jacqueline worked with Liebe to start up a modern dance group.

Erickson's contribution was concerned with what happens as you go through focus in an electron microscope. As you vary the focus, as with a light microscope, the objects you observe become blurred. Moreover, sharp objects are surrounded by rings that vary in size with the focus. In the ideal case, at the real focus (the Gaussian focus), all these rings vanish. In practice, electron microscopes also have aberrations caused by the imperfections in the electron lenses. Untrammelled by any understanding of the physics, electron microscopists varied the focus in order to get nice bright images. These images were always underfocus images, and some of the details they showed were certainly aberrations.

Erickson's thesis project with Michael Beer had been to develop methods to label specific bases of DNA with heavy atoms and then to use high-resolution electron microscopy to sequence the DNA. The conclusion of his thesis was that this was a flawed idea, and that the promising results of previous graduate students were in fact artefacts. From this work Erickson developed a strong technical interest in EM.

To get the highest visibility, he and Beer always took a through-focus series and picked the best. Erickson noted that all of the images at different degrees of focus had different features (that is, different aberrations), and he wondered how these might be combined to get an image free from aberrations, maybe with better resolution and contrast than could be obtained from any single image.

On his way to Cambridge, Erickson made a detour to attend the Fourth European Regional Conference on Electron Microscopy in Rome in August 1968. Here he came across a two-page abstract that described how to reconstruct the true image with improved resolution from a through-focus series. This was exactly the problem that Erickson had been worrying about during his thesis work. Unfortunately, the abstract was in German, and contained some heavy mathematics that Erickson could not really understand. Erickson attended the talk and met the speaker, Peter Schiske, but since Schiske spoke no English and Erickson little German they were unable to communicate effectively. When Erickson arrived in England he explained to Aaron as well as he could that one could use a through-focal series to improve signal and resolution, but at this moment Aaron was not interested. He was working with negative stain images at 20 Å resolution, and he did not need any improvement in resolution.

Earlier, Aaron had collaborated with A. V. Grimstone, in the university Zoology Department, on the structure of microtubules. Microtubules are filaments that form the scaffolding of cilia and flagella. The beating motions of flagella propel single-cell organisms. Microtubules also form during cell division. Each microtubule is a hollow cylinder consisting of 13 protofilaments arranged around a ring, and each protofilament is a linear polymer of tubulin molecules. Tubulin comes in two types: α and β. In a protofilament, α and β alternate. The spacing of the tubulin molecules along the protofilament is 40 Å. Using the optical diffraction methods that he had developed with Jack Berger, Aaron could see a reflection corresponding to 40 Å on optical diffraction patterns of EM images of microtubules, but he could also see a spacing of 80 Å, which is twice the size of one tubulin molecule and would imply that the α and β subunits were substantially different in structure. Erickson was set to work to purify microtubules from tetrahymena cilia and take EM pictures. However, the better Erickson purified the microtubules, the more the 80 Å line faded, leaving only the 40 Å line, which showed that the basic lattice repeat was 40 Å. Thus there appeared to be no evidence for α and β having distinct structures. Indeed, later work showed that the two versions of tubulin are so similar in structure that 80 Å spacing would not be visible. The project was progressing, but was not really what Erickson had hoped for in coming to Cambridge.

One afternoon, Aaron returned in an agitated mood to the LMB from a seminar he had presented at the Cavendish on the general concepts of Fourier image processing. He was disturbed because Ellis Cosslett had challenged his basic concept: Cosslett noted that the EM produces an image primarily by phase contrast, and questioned how the phase modulation in the image formation played out in the phases from the optical or computer diffraction pattern of the image. These phases in the Fourier transform of the image were the basis for noise filtering and 3D reconstruction. Aaron remembered his earlier discussion with Erickson about image formation and through-focal series, and wondered if this might resolve the question as to how focus and related phase contrast affect the phases (and amplitudes) in the computed Fourier transform (diffraction pattern) of the EM image.

As mentioned earlier, electrons interact with the specimen in two ways: they either suffer a phase retardation, which leads to a phase-contrast image, or they are scattered out of the beam so that there are fewer electrons, which leads to an amplitude-contrast image (rather like an X-ray radiograph). In fact, most electrons travel through a biological specimen without being scattered. They experience the changes of electrical potential as they pass an atom, which leads to a phase retardation of the electron wave. This is analogous to the phase retardation suffered by light as it passes though a glass lens or prism: the thicker the glass, the greater the retardation. In EM, the specimen behaves in the same way – the thicker the specimen, the more the phase retardation.

The physics of image formation of phase and amplitude objects in the EM had been worked out theoretically by O. Scherzer[1], who was the earliest to describe how defocus and aberration affected the image of an atom. The effects are most readily described in terms of how they affect the Fourier transform of the image. As one varies the defocus the Fourier transform of the phase-contrast image is modulated (multiplied

---

[1] Scherzer, O. (1949) *J. Appl. Phys.* **20**, 20–29.

by) by a function called the phase contrast transfer function (CTF) that varies with resolution – that is, it takes different values as you go out from the centre (low resolution) to the outside (high resolution) of the Fourier transform. Moreover, these values alter considerably as one varies the focus. At the middle of the Fourier transform, the CTF starts at zero, so all the low-resolution, large-scale components of the image are missing in the phase-contrast image! For underfocus images, it then declines to form a broad negative peak. After this first peak, it crosses the zero line and starts oscillating like a cosine wave. Friedrich Thon showed that optical diffraction patterns of defocused images of carbon films produced a series of rings, which correspond to peaks and zeros in the oscillating part of the CTF. You can use an observation of the positions of the rings of the CTF to measure how far you are from focus.

Thon's analysis predicted that if you were observing a crystalline specimen, for which the Fourier transform consists of well-defined diffraction spots, within the first cross-over of the CTF, as you varied the focus the amplitudes of the spots would change but the phases would not. This answered the question posed by Cosslett in the Cavendish seminar. Since DeRosier and Aaron had limited their analysis to a resolution of about 20 Å, which lies inside the first zero of the CTF, the phases they had calculated with DeRosier's Fourier transform program were fine.

Besides engendering a large sigh of relief, this result whetted Aaron's appetite for more. At larger defocus, the CTF would shrink towards the middle, so that some higher-resolution diffraction spots would find themselves on the other side of the first zero of the CTF. Here the CTF is positive. Since the CTF multiplies the diffraction pattern, Thon's analysis predicted that the observed phases of the higher-resolution diffraction spots would change sign (phase change by 180°) at large values of underfocus. Erickson and Aaron decided to test these ideas experimentally.

Aaron suggested using our old friend the tobacco mosaic virus (TMV) as the test image, but Erickson had an aversion to Bessel functions, probably shared by many readers of this book, so they decided to use catalase crystals. These two-dimensional crystals produced a simple rectangular lattice of diffraction spots (see Figure 9.3), with well-behaved phases that were multiples of 90°. Just as the theory predicted, as the focus varied, the phases remained constant so long as

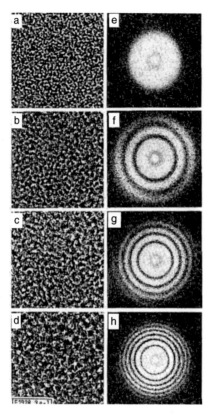

Figure 12.1 Electron micrographs of a carbon film at various degrees of underfocus
(a, b, c and d) and the optical transforms (Fourier transform) of each micrograph (e, f, g and h)[2].
The same specimen was used for all four micrographs. Underfocus produces rings (Thon rings)
in the optical transforms. At higher underfocus, the rings become closer together. Panel a
has been taken close to the Scherzer focus where the contrast transfer function, CTF,
remains more or less constant out to a certain radius – see e. (Reproduced from Thon, F. and
Siegel, B.M. *Ber. Bunsen Gesellsch.* **74**, 1116–1120 © 1970, with permission.)

all spots remained in the first peak of the CTF. Furthermore, they
changed by 180° when greater defocusing placed them in the second
peak of the CTF.

The lens aberrations produce another factor that modulates the
Fourier transform. The effective CTF is produced by multiplying these

---

[2] Thon, F. and Siegel, B.M. (1970) *Ber. Bunsen Gesellsch.* **74**, 1116–1120.

two functions together. It turns out that the lens aberrations are beneficial. Scherzer showed that by judicious choice of underfocus one can produce an effective CTF that is roughly constant up to the radius in Fourier space at which the CTF crosses the zero line and begins to oscillate (see Figure 12.1a and e). At this optimum focus, phase objects produce enhanced visible contrast with little distortion up to the value of the resolution given by the CTF's first crossing of the zero line. Aaron and Erickson reproduced the Scherzer underfocus in their practical example and realised that most of the published micrographs were in fact close to this degree of underfocus. Pragmatically, EM practitioners had done the right thing. Interestingly, you need an imperfect electron lens to get an image: without the aberrations there is no Scherzer focus.

Furthermore, their analysis demonstrated that the images contained a substantial contribution from amplitude contrast, especially at low resolution. This effect is greater for heavy atoms such as uranium than for light atoms such as carbon and is therefore particularly pertinent for negative stain imaging. The presence of a significant amount of amplitude contrast is fortunate because these low-resolution data are missing from the phase-contrast image. Indeed, the presence of low-resolution amplitude contrast contributes significantly to the visibility of objects in the image: these would otherwise appear as empty shells.

In summary, to quote from Erickson and Aaron's ensuing publication[3]:

> . . .it is comforting to be able to confirm quantitatively that the moderately underfocussed micrographs used in most biological microscopy are valid images, frequently the best possible in terms of resolution and contrast, with no artifacts in the low and medium resolution range of interest.

Erickson and Aaron also considered how one could recover data from the high-resolution part of underfocused images by extending the analysis into the region where the CTF was oscillating. Clearly, if you can compute the CTF then dividing the Fourier transform of the image

---

[3] Erickson, H.P. and Klug, A. (1970) *Ber. Bunsen Gesellsch.* **74**, 1129–1137.

by the calculated value of the CTF (taking account of possible changes of sign) corrects the Fourier transform to its proper in-focus values. Subsequently, a second Fourier transformation can be used to recover the sought-after undistorted image, but there are problems: what do you do at the cross-over point of the CTF where its values are zero or very small? Dividing by small numbers amplifies the errors in the data, and dividing by zero is catastrophic. The problem had already been considered by Peter Schiske, who proposed using various degrees of underfocus so that the zero points of the CTF moved around, and then merging all the data. In each case, by selecting only those parts of the Fourier transform of the image where the value CTF was well away from zero, one could fill in values for the whole of the Fourier transform without introducing large errors. Using this method, Erickson and Aaron were able to recover in-focus images from a series of underfocus images.

There were two conferences on the developments in EM technology in 1970: one was a Discussion Meeting at the Royal Society in London on the 12th to 13th March, and the second a Discussion Meeting of the Deutsche Bunsen Gesellschaft on 16th to 18th March at Hirschegg in the Kleinwalsertal. The Kleinwalsertal, which is actually in Austria, lies about 20 km north of Lech in the Vorarlberg but is separated from it by the bulk of the Widderstein, a solid chunk of dolomite 2300 metres high. Crossing into the Lechtal is a climb in summer and a ski tour in winter. Thus, although it is part of Austria, the only access to the Kleinwalsertal is via the Allgäu in southern Germany.

Walter Hoppe was director of a department in the Max Planck Institute for Eiweiß and Lederforschung in Munich (this was later absorbed into the Max Planck Institute for Biochemistry). Hoppe was the German pioneer of protein structure research, and Robert Huber was his most notable co-worker. Hoppe and Max Perutz had organised two rather popular workshops on protein crystallography at Hirschegg in 1966 and 1968 (a third was to follow in 1970). The programmes were organised so that adequate time was available for skiing.

Hoppe was much more interested in techniques than in results. Having decided that the problems of protein crystallography were in principle solved, he turned his attention to electron microscopy. Here he was a visionary. In 1968, independently of Aaron and DeRosier, Hoppe had proposed the same method of using EM projections to calculate sections in Fourier (reciprocal) space. By taking lots of projections to calculate sections in Fourier space, one can fill up the Fourier space with values and get a map of the object by means of a second Fourier transformation. Aaron and DeRosier had published this method in the journal *Nature* in English and were greeted with enthusiasm. Hoppe published in *Die Naturforschung* in German and, to his chagrin, was ignored. He decided that to publicise the things that were going on in his department, he should organise a meeting at Hirschegg on EM techniques. He arranged this as a Discussion Meeting of the Bunsen Gesellschaft to enable the publication of the proceedings.

Aaron, newly elected to the Royal Society and enthusiastic about the Society's work, organised a Discussion Meeting at the Royal Society in collaboration with Hugh Huxley. The Royal Society had recently moved to its present site in Carlton House Terrace (earlier the German Embassy), where it has the large and well-appointed Wellcome Trust Lecture Theatre. This was a substantial meeting with many scientists from the USA. DeRosier came over from Texas for the occasion. New technologies were represented; there were three papers on scanning electron microscopy. Hoppe presented a scholarly paper on the use of zone plates to correct lens aberration. Erickson presented his work with Aaron on Fourier processing of electron micrographs to compensate for the effects of defocusing and aberrations[4]. Nigel Unwin presented work on the use of a spider's web (they are very thin) strung across the aperture of the electron microscope to induce phase shifts of the direct beam so as to produce phase contrast. A second section was devoted to specimen preparation, and a third section to analysis and 3D reconstructions. All these papers were duly published in the *Philosophical Transactions of the Royal Society.*

Hoppe's Hirschegg Discussion Meeting started three days later. For those who attended both meetings, this entailed a hurried journey across Europe. Aaron did not make it; he sent Erickson as his emissary. Hugh Huxley, a keen skier, was certainly not going to let this opportunity go to waste. He duly motored across Europe in his splendid blue Jaguar sports car, only stopping at Heidelberg on the way to pick up Holmes. Erickson was apprehensive that somebody would already have all the answers:

---

[4] Erickson, H.P. and Klug, A. (1971) *Phil. Trans. Roy. Soc. Lond.* **B261** 105–119.

after all, the kind of work he was about to report had a strong German tradition, starting with the invention of the electron microscope by Ernst Ruska. However, Schiske had indicated that he was working on other problems. Hoppe himself was preoccupied with his zone plates, but Hoppe's name appeared on half a dozen papers from his group. Perhaps, thought Erickson, he had a graduate student who had looked at the effects of the CTF on images. Indeed, this was the case: Joachim Frank's thesis was concerned with measuring the CTF and correcting for it. He and Erickson met at the meeting. Frank's work was not just theoretical: he had examined and corrected EM images of bacteria[5]. The microscope he was using showed strong astigmatism (this means that the focal length is different for different angular directions, so that you get a different underfocus just by looking in another direction in the image). All in all, Erickson's work was easier to comprehend. The choice of specimen, a crystal of catalase, turned out to be ideal: it was easy to compare the corrected signs of Bragg reflections calculated from images at underfocus with the signs obtained from an in-focus image.

Aaron and Erickson's work together with Joachim Frank's studies put imaging in biological electron microscopy on a quantitative footing. It soon found an application. Richard Henderson and Nigel Unwin managed to get high-resolution electron microscope data from the 'purple membrane' – a two-dimensional array of ion pumps embedded in a bacterial membrane – by immersing the membrane in a dilute glucose solution. These unstained specimens yielded data with 7 Å resolution, but the contrast was very low. Going to extreme defocusing could increase the contrast, but then the higher-resolution features were in the second or third peaks of the CTF. The phases have to be corrected for underfocus in order to reconstruct a high-resolution image[6]. Schiske's method of combining images of different defocus to obtain an optimal image was used. Subsequently, a 3D image reconstruction from tilted specimen maps as proposed by DeRosier and Klug showed each molecule to be composed of seven rod-like structures roughly perpendicular to the membrane, the first time that α-helices could be visualised by electron microscopy[7]. A further two decades were to elapse

---

[5] Frank, J., Bußler, P., Langer, R. and Hoppe, W. (1970) *Ber. Bunsen. Phys. Chem.* **74**, 1105–1115.
[6] Unwin, P.N.T. and Henderson R. (1975) *J. Mol. Biol.* **4**, 425–440.
[7] Henderson, R. and Unwin, P.T.N. (1975) *Nature* **257**, 28–32.

before a quantitative correction for the CTF became commonplace for the reconstruction of images because a new technique was needed for preserving specimens. This technique was electron cryomicroscopy – known generally as cryo-EM – in which the specimen is embedded in vitreous ice at very low temperatures. Unlike negative staining, where the resolution is limited to about 20 Å, cryo-EM images can have data out to atomic resolution.

While Aaron was involved in a decade of science, Liebe was concerned with bringing up her two sons. Their second son David was born on 16th July 1963 in the Evelyn Nursing Home on Trumpington Road, Cambridge, which Liebe had chosen because they would allow her to go home after 48 hours. Liebe attended the medical practice of David and Shirley Emerson. Her pregnancy was problematic, and she spent long periods in bed in order to avoid another miscarriage. At the time of David's birth, thalidomide was being handed out to women liable to miscarriage. Fortunately, the Emersons would not use thalidomide. Instead, they offered daily encouragement and support and managed to get Liebe through her pregnancy. Shirley Emerson supervised the birth, which Aaron attended. Liebe was so impressed by the Emersons' doctoring skills that she named her son David.

David was a strong-minded chap, who soon exhibited his extroverted character. In this, he contrasted with Adam, who was more like Aaron and tended to keep his feelings to himself. To Liebe's delight, her parents Annie and Alter came in 1963 to stay for a few months. In addition to helping with the new baby, this stay gave Alter the chance of preparing Adam for his Bar Mitzvah. Nor was this Alter and Annie's only visit. They quite liked Cambridge and turned up from time to time. Alter and Aaron were both scholars with wide interests and enjoyed each other's company. Aaron's parents Lazar and Rose were more infrequent visitors, but one visit was particularly notable. In June 1967, Huw Wheldon and his wife Jacqueline Clarke were staying with the Klugs for the weekend when the Six Day War started. Huw worked with the BBC and very early got wind of this surprise development. The news engendered great concern in the Klug household because at this moment Aaron's parents were in Israel visiting Aaron's half-sister Robin. Some frenzied telephoning managed to get Rose and Lazar on a flight from Tel Aviv to Heathrow on the following Tuesday. Aaron drove up to London Airport and brought them back to Cambridge,

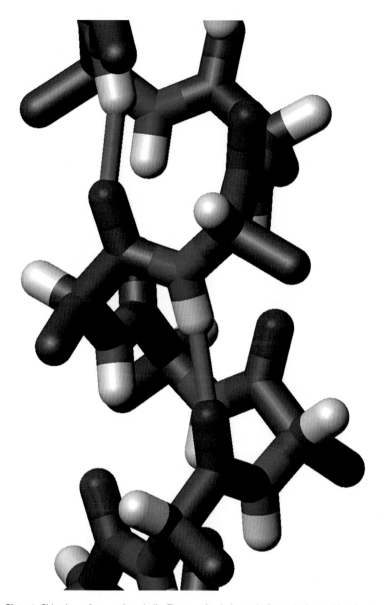

Plate 1 Side view of a protein α-helix. The protein chain made from alanine residues* in a diagrammatic representation in which the bonds between atoms are shown. The atoms lie at the intersection of the bonds. They are colour coded: dark grey – carbon; blue – nitrogen; red – oxygen; white – hydrogen. Two of the hydrogen bonds between a hydrogen attached to a nitrogen (an amide nitrogen) and oxygen attached to a carbon (a carbonyl oxygen) are shown (thin); the H to O distance is about 2 ångstroms (0.20 nanometres)†. Hair is composed of protein chains organised in α-helices. (From Wikipedia, image by WillowW.)

---

* When amino acids combine to form a peptide, the elements of water are removed: what remains is called an amino acid residue.

† 10,000,000 ångstroms (Å) = 1 mm. 1 Å is about the size of an atom.

Plate 2 Models of icosadeltahedra for $T = 1, 3, 4$ and 7, built with Geodestix (Buckminster Fuller). The five-fold axes are marked with red connectors, the six-fold with green. Each model has twelve five-fold axes and an increasing number of quasi-six-fold axes. The total number of connectors (subunits in a virus) is 60. The total number of surface bumps (five-fold + six-fold) is 12 + 10 $(T - 1)$. $T = 7$ comes in left- and right-handed (*laevo* and *dextro*) forms. (© MRC Laboratory of Molecular Biology.)

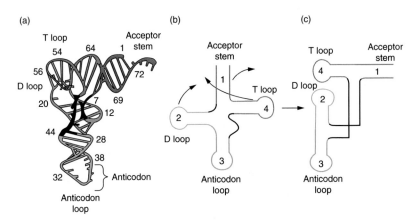

Plate 3 a, The 'L'-like structure of tRNA shown here in its final form. The sugar phosphate chain is shown as a worm, the Watson–Crick base pairs as solid lines. The colours correspond to the loops of the Holley cloverleaf structure. b, c, How a is related to the four arms of the Holley cloverleaf structure. The black worm shows the section of chain that forms a triple helix. (Source unknown.)

Plate 4 View from Aaron's room in Peterhouse, showing Little St Mary's Lane and Church. Drawn by Rosalind Franklin's niece Sarah Glynn, then aged 16.

Plate 5 Aaron Klug (right) and Brian Heap (left) in conversation with Zhang Cunhau at the National Natural Science Foundation of China, Beijing. (Courtesy of Ling Thompson.)

Plate 6 A quinquennial dinner to mark the end of Aaron Klug's presidency was held in the presence of the Queen and Duke of Edinburgh at the Royal Society on 22nd November 2000. From left to right: Lady May, the Duke of Edinburgh, the Queen, Sir Aaron Klug, Lady Klug, Sir Robert May (president-elect). (Courtesy of the Klug family.)

Plate 7 Setsuro Ebashi (left) and Aaron Klug (right) enjoying a shabu shabu meal in Okasaki. (Courtesy of the Klug family.)

Plate 8 The obverse side of the gold medal of The Royal Swedish Academy of Sciences bears an image of Alfred Nobel. On the reverse side of the Chemistry Prize, Nature is represented as a goddess emerging from the clouds holding a cornucopia; the Genius of Science holds up the veil that covers Nature's face. The inscription reads: *Inventas vitam juvat excoluisse per artes* –'They who bettered life on earth by new-found mastery' (Virgil). (© The Nobel Foundation. Photographer: Lovisa Engblom.)

Plate 9 Vice-Chancellor's Gold Medal of the University of Cape Town awarded to Aaron Klug in 1983. The two previous recipients were Chris Barnard and Allen Cormack. (© MRC Laboratory of Molecular Biology.)

Plate 10 Durban High School awards the 'Dux' to the best scholar of the year. It was withheld in 1941 because of the War. Aaron received his award in 1983. (© MRC Laboratory of Molecular Biology.)

Plate 11 Aaron Klug's Badge of the Order of Merit. The Order, which is limited to 24 members, is bestowed by the Queen. The badge is held for the lifetime of the recipient. The previous holder of this badge had been Dorothy Hodgkin, also a Nobel Laureate in Chemistry. (© MRC Laboratory of Molecular Biology.)

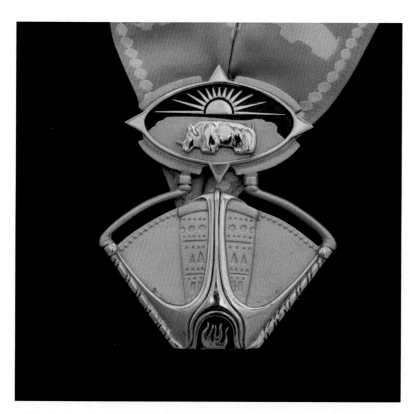

Plate 12 Aaron was awarded the Order of Mapungubwe, South Africa's highest honour, in gold. The medal consists of an oval frame above an inverted trapezium. The oval frame contains a golden rhinoceros lit by the sun rising above Mapungubwe Hill. In the centre of the trapezium is an ornate crucible with molten gold flowing down to a red furnace.
(© MRC Laboratory of Molecular Biology.)

Plate 13 John Finch, Aaron's closest collaborator. John studied Physics at King's College London and moved to Birkbeck College for his PhD on the crystallography of 'spherical' viruses. He moved to the MRC Laboratory of Molecular Biology in Cambridge with Aaron Klug where he took up electron microscopy. John was a talented experimentalist whereas Aaron was more theoretical; their complementary abilities were the basis of a fruitful collaboration that functioned most successfully for four decades. (Photo © MRC Laboratory of Molecular Biology.)

where they stayed for a couple of weeks before winging their way back to Johannesburg and Durban. Lazar was becoming increasingly frail and forgetful and never managed the journey again. He died in 1971.

At the end of the decade, Adam was preparing for university and David becoming quite independent so that Liebe had time to think about dance and theatre again. Jacqueline Erickson was a gifted classical ballet dancer who had also studied Martha Graham-style modern dance with Dale Sehnert at the Peabody Conservatory in Baltimore. On arriving in Cambridge, she quickly joined the Cambridge University Dance Group. When she discovered Liebe's expertise she requested Liebe to come and give them a one-off class. The one-off became many-off until Liebe turned into the group's teacher and choreographer. At that stage, classes were held at Homerton College and rehearsals wherever one could find space, sometimes in the Klugs' garden. After a couple of years, the name was changed to the Cambridge Contemporary Dance Group, and membership was made available to all who wanted to come. They found a new home at Parkside Community College. Teachers came from 'The Place', a well-known Contemporary Dance school in London. Liebe approached the Gulbenkian Foundation who gave them a grant to cover the travel costs for the teachers from London.

The Cambridge Contemporary Dance Group was integrated into the Parkside Community College as part of the adult education programme. Liebe's intention was to make contemporary dance available to all in Cambridge. She was fired by the idea of dance being part of society, as it had been in her youth. There were classes starting with children and going on to whatever age and ability. Soon, a small performing group emerged, called 'Helix', with Liebe as choreographer. In the next couple of years this group travelled widely, trying to introduce Contemporary Dance to small communities, even if there was only an audience of three or four. The Eastern Arts Association often funded performances.

Andrée Blakemore, a very talented dancer, joined the group in 1969 and quickly established a rapport with Liebe. Andrée's husband Colin was a lecturer in neurophysiology. They had just returned from Berkeley, California, where Colin had completed his doctorate and where Andrée had been enhancing her reputation as a dancer. Andrée was remarkably beautiful, which certainly helped spur interest in the group. The first time they performed was at the Churchill College

Theatre in 1969. This piece was danced to a poem by Marina Ivanovna Tsvetaeva, translated and spoken by Elaine Feinstein who had just published a book of translations of Tsvetaeva's poems. Aaron and Colin Blakemore were out front of house selling tickets. Later they worked the lights – under Liebe's meticulous instruction!

Adam, who, like his father, tended to know about everything, told his mother about the Underground rock group 'Soft Machine' who performed in a Proms concert in 1970. Liebe became interested in their frenetic music. Then she discovered the avant-garde rock group 'Henry Cow' at the University. Henry Cow were more intellectual and less frenetic than Soft Machine. They were influenced by modern composers, in particular by Oliver Messiaen and Kurt Weill. In 1970, the University held a drama week, and Liebe approached Henry Cow to ask whether they would collaborate with her group to put on a show. They agreed, so ten dancers, four musicians and Liebe went to work. They danced to the music of a piece called 'Dance You Monster to my Soft Song', named after a Paul Klee painting. This all came to pass in the Cambridge Union – no front stage and no back, just an all-round view. For the second half of the show, Owen Pegrum directed Murray Carlin's 'Not Now, Sweet Desdemona', which he described as a duologue for Black and White within the realm of Shakespeare's *Othello*. Murray Carlin, who during his student days had introduced Liebe to Norman Podhoretz, was South African. Owen Pegrum, who also happened to be from Cape Town, was Adam's English master at the Cambridge Grammar School.

In 1972, Henry Cow were invited to Edinburgh for a series of repertory concerts at the Traverse Theatre, which was followed almost immediately by writing and performing music for a ballet with artist Ray Smith and the Cambridge Contemporary Dance Group at the Edinburgh Festival Fringe. Liebe commissioned 'With the Yellow Half-Moon and Blue Star', which was composed by Fred Frith and again named after a painting by Paul Klee. Part of the music can be heard in their album 'Legend'. The venue was a basketball pitch in a school gym. In keeping with the surroundings, costumes were gym-like: dark green tights and pale green leotards. The group had good audiences and excellent reviews. Andrée Blakemore was especially praised. Aaron and young David worked as 'roadies'; their main job turned out to be keeping the company supplied with adequate quantities of fish and chips. The dancers also paraded through the town in weird costumes lent by the costume designer Yolanda Sonnabend, who was from Rhodesia and part of the South Africa clique.

Liebe was keen on using contemporary music. In the mid-seventies she choreographed a dance to Stockhausen's 'Stimmung'. There was a lot of interpretation and improvisation. This was duly performed in Ely, perhaps to the dismay of the local population. She worked with Peter Britton who conducted Stravinsky's 'The Soldier's Tale' at the Mumford Theatre in the Anglia Polytechnic. Liebe choreographed the meeting of the soldier and the girl. There were other performances with Britton in which he played the marimba. Liebe also choreographed the chorus for *Murder in the Cathedral*, directed by Gerald Chapman at the ADC Theatre, for *The Trojan Women*, also at the ADC, and for the biannual Greek play at the Arts Theatre. Her last piece was 'World Music', composed by Tim Souster, a student of Stockhausen. This was large-scale work in the Mumford Theatre with electronic and live music. It was acclaimed a stunning piece.

In 1981, Liebe gave up control of the Contemporary Dance Group and moved on to other things.

13

Tomography

Since its inception, X-ray radiographic imaging has been limited by the fact that any radiograph gives only a projection through a limb or body. Interpreting the resulting wraith-like images became an art requiring skilled radiologists. An early attempt to get round this limitation was to connect the X-ray source and the film holder by a mechanical link with a fulcrum in the middle so that the source and film could be moved in contrary directions during an exposure. The effect of this linkage was to blur the image except for parts near the plane of the fulcrum. The result was called a 'tomogram' from the Greek *tomo*, slice. The method was not very good, but it did give the method a name. Much better solutions were needed. In the 1960s, three groups, quite unbeknownst to each other, were trying to solve the tomography problem.

As soon as Aaron and David DeRosier had worked out a way of getting 3D data out of projections via Fourier transforms, Aaron realised that this could have implications for X-ray radiology, and he consulted with nearby radiologists. The response was generally discouraging.

For some years, Aaron's erstwhile climbing buddy Allan Cormack[1] had been working on the problem of making 3D maps of an object by taking a series of X-ray photos while spinning the object round.

---

[1] The following information concerning Allan Cormack is taken from *Imagining the Elephant*, a biography by Christopher L. Vaughan, Imperial College Press (2008).

Reconstructing the 3D map confronted Cormack with exactly the same theoretical problem that had beset Aaron. Cormack had arrived at a different solution.

In 1949, R. W. James offered Cormack a lectureship at Cape Town. Cormack accepted the offer gladly and abandoned his Cambridge doctorate. At the University of Cape Town (UCT), he lectured in the physics department together with Aaron's old friend John Juritz. In 1956, Cormack acquired an extra appointment as part-time radiation physicist at the Groote Schuur Hospital, the UCT teaching hospital. Here he encountered the problem of X-ray dosage and how to measure absorption coefficients, which seemed to him to be based on rather crude estimates: he wondered if one could not do better. However, his deliberations were cut short because in the same year, to the delight of Barbara, his wife, he was offered an appointment at Harvard for a sabbatical year. The family moved to Cambridge, Massachusetts, where they were able to stay in an apartment on Brattle Street belonging to Barbara's parents. Harvard had a working cyclotron (one of the earliest types of particle accelerator), which allowed Cormack to pick up his work on nuclear physics, but on the side he was thinking about the problem of reconstructing three-dimensional pictures from a series of two-dimensional X-ray micrographs. At the end of his sabbatical year, Cormack accepted a lectureship at the nearby Tufts University rather than return to apartheid-ridden South Africa. However, he had to return for a few months to avoid having to pay back his salary for the sabbatical year. At UCT, he conducted experiments on collecting X-ray absorption data from a crude model of a head (a 'phantom') as it was rotated. He refined these experiments when he returned to Tufts and built a better apparatus for measuring data as a phantom was rotated. This was perhaps the world's first CAT scanner. He worked out a sophisticated theory for recovering the density from the projections. This theory[2] and its application to data obtained with the Tufts scanner was published in 1964 and completely ignored.

---

[2] Cormack, A.M. (1964) *J. Appl. Phys.* **35**, 2908–2913. Cormack's theory involved the expansion of the projections of the slices in circular harmonics. The expression of the radial part of the circular coefficients in Tschebycheff polynomials yielded coefficients which, when applied to the appropriate Zernike polynomials, regenerated the two-dimensional density of the slice.

Unknown either to Cormack or Aaron, during the late 1960s Godfrey Hounsfield[3] was working on the same problem. Hounsfield was an engineer working for the recording firm EMI, who were then making good money with the Beatles. Hounsfield's allotted project had just gone bottom-up, and no one seemed particularly interested in what he was doing, giving him time to explore an idea he had had: could one get 3D pictures of the inside of an object (for example a human head) from lots of 2D X-ray radiographs taken at a series of angles as one turned the object round an axis at right angles to the X-ray beam? Hounsfield was not an academic and did not study the literature, so he had absolutely no idea that Cormack had worked on the same problem. Nor was he aware that Aaron and DeRosier had published their Fourier transform solution to this class of problems.

Hounsfield used a lathe bed to rotate a phantom (model head) in an X-ray beam in order to collect radiographic projections of the object every 1°. This gave him lots of data. The problem was solved in slices taken along the axis of rotation. For each slice, there are 180 or so projections of the trial object projected down a series of parallel lines. Hounsfield represented the density in each slice as an array of numbers. Each column of numbers in the slice needed to add up to the measured value for that column and for that particular angle of rotation. There are lots of rotations which give rise to lots of equations. It should be possible to recover the set of numbers – a bit like doing a Sudoku puzzle. Hounsfield worked out an iterative procedure in which the field of numbers is first assumed to be all the same. Alterations were then introduced in a systematic way so that the field of numbers gradually added up to give all the right projections. Such procedures can easily become unstable and can yield a sheet of numbers that fluctuate wildly. Hounsfield's method was cautious enough to be effective. It turned out that he could actually make 3D maps of people's brains from a set of radiographs taken over a range of angles. He called this computer assisted tomography (hence the term CAT scanner). Hounsfield contacted the Ministry of Health in London, where he received enthusiastic support from the radiological officer, Dr Evan Lennon. Preliminary clinical studies were successful. Brain tumours could readily be seen,

---

[3] See 'Sir Godfrey Newbold Hounsfield' by P.N.T. Wells (2005) *Biogr. Mem. FRS* **51**, 221–235.

and in 1972 Hounsfield obtained a UK patent on his results that was closely followed by a US patent[4]. The discovery quickly became a commercial success, although EMI never really understood why they were selling medical equipment. Finally, indifferent management led to their losing the whole project.

In 1979, Cormack and Hounsfield were awarded the Nobel Prize for physiology or medicine for their concurrent – although completely independent – development of CAT scan theory and technology. Their selection as recipients of the prize was unusual. Neither Cormack nor Hounsfield held a doctorate in medicine or science. Their discovery was awarded the prize only after the Medical Faculty of the Karolinska Institute vetoed the first choice of the selection committee, reportedly because of a split between factions, with one side favouring discoveries in basic medical science and the other those in medicine. Moreover, it was unusual to award the prize to two men who had never met.

In the meantime, Aaron, Crowther and DeRosier analysed the Fourier transform method rigorously[5]. Although it was computationally cumbersome, the Fourier transform solution to the problem of reconstruction became the gold standard for judging other methods. They also analysed the algebraic method and proposed a way of finding solutions by what they called the 'back projection method'. This method had been first suggested for electron micrographs by Roger Hart in 1968, but the name 'back projection' was proposed by Crowther, DeRosier and Klug. The idea is shown in Figure 13.1.

The projections are arranged on a plane at the angles corresponding to the projecting direction. Rather naively, one can attempt to reconstruct the projected object by drawing lines at right angles to each projection. The weight of each line is the weight of the place on the projection where it originates. At each point in the plane, you add up the contributions from all the lines intersecting at this point. One can see that this simple recipe does indeed reproduce the object. Unfortunately, it introduces all sorts of spurious solutions as well (Figure 13.1).

---

[4] *A method of and apparatus for measuring X or gamma radiation.* UK Patent 1283915 (1972); *Method and apparatus for measuring X or gamma radiation absorption or transmission at plural angles and analyzing the data.* US Patent 3778614 (1973).

[5] Crowther, R.A., DeRosier, D.J. and Klug, A. (1970) *Proc. Roy. Soc. Lond.* **A317**, 319–340.

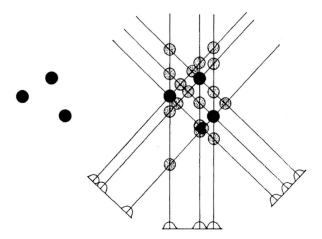

Figure 13.1 Illustration of the idea of 'back projection'. A simple object consisting of three black disks is shown on the left. On the right are shown three projections of this object. A back projection is produced by drawing lines at right angles to each projection. The value of the projection at the base of each line is used as the value of the reconstructed density along the line. The values of all the lines are added together. One sees that the strongest spots (arising from the overlap of three lines) do reproduce the object, but chance coincidences produce an extra spot. Even with many more projections, this simple method does not reproduce the object accurately. However, if the projections are suitably modified (weighted back projection) and if there are lots of projections, then the object can be accurately reconstructed. (Reproduced from Figure 6 of Crowther, R.A., DeRosier, D.J. and Klug, A. *Proc. Roy. Soc. Lond.* **A317**, 319–340 © 1970 Royal Society.)

In 1967, Aaron acquired a new PhD student, Peter Gilbert. Gilbert played at being 'cool', with a penchant for velvet smoking jackets, but in fact he was a rather serious fellow. Gilbert set out to see how you could get rid of the spurious solutions to the back projection algorithm. He showed that in favourable cases you could modify the back projection method to produce a reconstruction as good as the Fourier method but with half as much computing. The basis of Gilbert's 'weighted back projection' method lay in ideas well known in optics, namely the 'point response function' and 'deconvolution'. Optical systems, especially the human eye, tend to be imperfect. If you look at a star, which is actually a very small point of light, you tend to see something with rays – the classical graphical representation of a star – that just shows how bad your sight is getting. Thus a point in the object ends up as some complicated smudge in the image called the 'point response function'.

If a camera is slightly out of focus, then each point in the object ends up as something close to a Gaussian error function in the image (in this case, the point response function is approximately an error function). Can you do anything about this? The problem is called 'convolution': each point in the object gives rise to a point response function in the image, and then all these are added together. If you knew what the point response function looked like, then perhaps you could calculate the perfect image. This process is called 'deconvolution'. It works, up to a point. For example, if you play with Photoshop and you ask it to sharpen up an out-of-focus image, then it will calculate a 'deconvolution' to get rid of the blurring. Too much deconvolution, and the calculated image acquires spurious fringes on all the edges.

The deconvolution or sharpening is carried out with the help of Fourier transforms. One of the remarkable properties of Fourier transforms is that if you calculate the Fourier transform of two functions that are convoluted with each other (for example the camera image and an out-of focus smudge – the point spread function) you end up with the transforms of each of the functions multiplied with each other. This is a wonderful simplification. You then simply divide the Fourier transform of the object by the Fourier transform of the smudge (point response function) and then calculate another Fourier transform to get back to the image: one of the miracles of mathematics. But there is a problem. The Fourier transform of the point response function may become very small or have zeros. Since one cannot divide by zero, one must take precautions and remove such parts of the transform, which leads to some degradation of the final image.

Aaron and Peter Gilbert recognised that the problem with back projection was that each point in the image was convoluted with all the lines of the back projection (that is, the point response function was a fan of lines passing through each point). If the fan of lines goes all the way round (lots of projections taken at some small angular interval) then the point response function is something like the classical representation of a star. The Fourier transform of this star has a simple form – it goes like $1/r$, where r is the distance from the middle. Now to get rid of the effects of convolution you divide the Fourier transform of the back projection by the Fourier transform of the point response function. Since the Fourier transform of the point response function has the simple form $1/r$, to *deconvolute* you just multiply the Fourier transform of the back projection by r – the distance from the middle. Then a Fourier transform of this weighted Fourier transform gives the reconstructed image. This works well, and the method of 'weighted back projection' became one of the most widely used methods of reconstruction both in electron microscopy and in medical tomography. Gilbert published this elegant result in his Cambridge PhD thesis in 1970, but the full publication was not made until 1972[6], by which time others had published the same result. Moreover, Aaron and Gilbert became aware of the fact that the same result had been published earlier in an astronomical context[7]. They did not get much credit for their insight.

Aaron was frequently consulted by other biological electron microscopists who were increasingly eager to make 3D images from their data. In response to the many calls on their time, Tony Crowther and Aaron decided to organise an advanced study course on 'Image processing of electron micrographs of biological structures'. This was duly financed by the European Molecular Biology Organisation. The course took place at the end of September 1973. Since it was out of term time, the participants could be housed in Peterhouse, where Aaron was a teaching fellow. Besides Aaron and Tony Crowther, the lecturers were Linda Amos, David DeRosier (from Austin, Texas), Harold Erickson (from Duke, North Carolina), John Finch and Nigel Unwin: altogether, an amazing collection of talent. For those involved, it was also a reunion, with dinner in college at the termination of the week's work.

While the 3D reconstruction theory was primarily aimed at electron micrographs, it was clear to Aaron that the method should have medical applications. Aaron consulted the radiologists on the Addenbrooke's Hospital site but encountered the same lack of interest that beset Allan Cormack. Thus the Hounsfield patent was a bombshell for Aaron. How could anyone patent the obvious, particularly when Aaron had published a much better solution to the problem a few years earlier? Moreover, Hounsfield was making unjustifiable claims about the structure of the brain, which were complete nonsense. The apparent details in the brain were noise produced by Hounsfield's iterative method of

[6] Gilbert, P.C.F. (1972) *Proc. Roy. Soc. Lond.* **B182**, 89–102.
[7] Braceweel, R.N. and Riddle, A.C. (1967) *Astrophys. J.* **150**, 427–434.

getting the answers. Ugh, these engineers! On the positive side, Allan Cormack contacted Aaron to tell him about his work, thus renewing an old friendship. While retaining confidence in the superiority of his Fourier method, Aaron was struck by the mathematical elegance of Cormack's method, which Aaron likened to peeling an onion.

Aaron was no longer alone. In 1970, Gabor Herman's group published the 'Algebraic Reconstruction Technique for three dimensional electron microscopy and X-ray photography'. They called this technique ART[8]. It was an iterative method of solving the problem rather like the one devised by Hounsfield. Gabor Herman, who held a PhD from King's College London, was director of the Medical Image Processing Group in the Department of Computer Science at the State University of New York at Buffalo. In the paper, ART was applied to electron micrograph images of the bacterial ribosome. The Buffalo group claimed that they could get a 3D reconstruction from very few views over a narrow range of angles. Moreover, they claimed it was a much better method than Aaron's Fourier reconstruction. This claim elicited a strong response from Aaron, who with alacrity penned a reply: 'ART and Science or conditions for three-dimensional reconstructions from electron microscope images'[9].

He and Tony Crowther wrote:

> *In a recent paper in this journal Gordon, Bender & Herman make what we believe to be extravagant and misleading claims for a method ("ART") for three-dimensional reconstructions from electron micrographs. Comparing their method to the Fourier method developed in this laboratory they say "It is clear that we are doing considerably better than the Fourier method". They make this claim without any reliable criterion to justify its validity. They simply ignore the criterion adopted by DeRosier and Klug. . .*
>
> *It is unfortunate that the authors themselves seem unable to assess just what they have accomplished. For it seems to us that, despite their emphasis on using a limited range of data, they have produced a valid algorithm for direct reconstruction. This they have done by using*

---

[8] Gordon, R., Bender, R. and Herman, G.T. (1970) *J. Theor. Biol.* **29**, 471–481; Bender, R. Bellman, S.H. and Gordon, R. (1970) *J. Theor. Biol.* **29**, 483–487.

[9] Crowther, R.A. and Klug, A. (1971) *J. Theor. Biol.* **32**, 199–203.

back-projection and some form of correlation to arrive at a trial structure,
which is iteratively refined until it satisfies the projection equations.
However satisfactory this procedure, the fact remains that no amount of
ART can make up for missing data.

This was fighting talk, and it elicited a robust reply: 'ART is Science'[10].
Unfortunately, this paper continued to ignore the criteria for resolution
established by Klug and DeRosier and therefore missed the point. Moreover, the authors, rather flippantly, reconstructed a picture of Aaron's
face from a small number of projections. Perhaps significantly, Gabor
Herman, the group leader, did not put his name on the 'ART is Science'
paper[11].

The introduction to 'ART is Science' quotes a comment made by Aaron
in his first version of the manuscript. This was later edited out in Aaron's
finally published paper. Initially Aaron had written: 'Progress will come
with experience and the use of rigorous and objective criteria; it will not be
assisted by the extravagant claims and misleading comparisons made by
Gordon, Bender and Herman.' Posterity has shown that Aaron was right.
The episode demonstrated once again that Aaron could be deeply
troubled by what he considered 'irremediable wrong-headedness'.

At the time, the controversy stirred some muddy waters. It was
becoming clear that tomography as a technique would be of considerable importance. With this in mind, Herman's group was being funded
by the US National Science Foundation. However, when this institution
got wind of the controversy they urged R. B. Marr of the Brookhaven
National Laboratory at Upton on Long Island to organise an international meeting in order to evaluate what was happening in the field.
The organising committee included Aaron and Allan Cormack. The
meeting at Upton duly took place in July 1974. Tony Crowther and
Aaron presented a paper, which was read by Crowther. However, by
this time the heat had gone out of the controversy. The bickering of a

---

[10] Bellman, S.H., Bender, R., Gordon, R. and Rowe, J.E. (1971) *J. Theor. Biol.* **32**, 205–216.

[11] Crowther points out that the main problem was that Gordon *et al.* were using 'pseudo-
projections' made by just adding up the density numbers in the pixels, not interpolating, so
their starting equations were completely consistent. As soon as you have real projections
with noise, the equations are inconsistent and the ART procedure rapidly blows up. This can
be fixed by using Gilbert's SIRT or Crowther's EFIRT algorithm.

couple of pioneer groups was already largely forgotten. The field was being taken over by professionals who were interested in computing speed, noise analysis, maximum entropy, maximum likelihood and such things. Tomography quickly established its own international journals. The field was now one of 'methods' and their improvement. This was no longer Aaron's scene. Aaron established 3D electron microscope reconstruction to solve a biological problem. He was not interested in developing methods for the sake of developing methods, although he was rather good at it.

Over the next few years, it became clear that a number of results discovered in electron microscopy had already been established in other fields. The scientific literature was large, and computer searches were not available. The earliest solution to the problem of reconstructing two-dimensional data from one-dimensional projections had been given by Johann Radon, an Austrian mathematician, in 1917. The act of projecting a function in a plane along a set of parallel lines for all possible orientations of the lines is known as the Radon transform. Radon gave a method for recovering the original function from the Radon transform. This method is allied to the weighted back projection method but involves a differentiation, which makes it difficult to use with noisy data.

14

Transfer RNA

The elucidation of the structure of DNA was possible because DNA occurs as long double-stranded polymers that can be orientated simply by physically pulling fibres from a gel. X-ray fibre diagrams from orientated gels yielded enough information to establish the low-resolution structure. RNA, in contrast, occurs mostly in single-stranded forms or looped forms that do not lend themselves to making orientated fibres. Thus, for some time after the structure of DNA was determined, the structure of RNA remained enigmatic. This was one factor contributing to Rosalind Franklin's decision to work on the structure of tobacco mosaic virus. TMV and many small viruses contain RNA: perhaps the structure of RNA could be determined in viruses.

RNA is found in the cytoplasm of cells, much of it in the form of ribosomes, particles too small to be seen by the light microscope. George Palade at the Rockefeller University used electron microscopy and the ultracentrifuge to characterise ribosomes. They were shown to be roughly pear-shaped particles 200–300 Å across, made up of two subunits consisting mostly of RNA. Ribosomal RNA makes up 80% of the cytoplasmic RNA. It was established that ribosomes were the sites of protein synthesis. Jim Watson and others initially thought that small spherical viruses might be models for ribosomes. This turned out not to be the case, and in fact another 50 years elapsed before the intricate structure of the ribosome was revealed by a spectacular application of X-ray crystallography.

In the cell nucleus, the DNA code is 'transcribed', or copied, into RNA (messenger RNA – mRNA), which is exported to the cytoplasm. In the cytoplasm, the RNA code is 'translated' into a protein – that is, it gives rise to a polymer of amino acids of which the sequence is determined by the RNA (for an enthralling account of how all this was discovered, see Horace Freeland Judson's book, *The Eighth Day of Creation*[1]). Translation is carried out on the ribosomes. Francis Crick canonised this flow of genetic information from DNA to RNA to protein as 'The Central Dogma', which shows why acquired characteristics cannot be passed on to the next generation.

The theoretical physicist George Gamow first proposed that the message for making a protein was carried in DNA as a three-letter code. Gamow founded the 'RNA Tie Club', a discussion group in which each member was named after one of the 20 known amino acids that occur in proteins. Each member was presented with a necktie and a tie clip carrying the symbol of the appropriate amino acid. Sydney Brenner (VAL – valine) introduced the name 'codon'[2] for the three contiguous bases that define an amino acid on the protein chain.

In a letter sent in 1955 to the RNA Tie Club, Francis Crick (TYR – tyrosine) proposed the existence of an 'adaptor' molecule that reads the genetic code and selects the appropriate amino acids to add to the growing polypeptide chain. Since the genetic code is made up of nucleic acid bases, to read the code it would seem reasonable for the adaptor molecule to be made of RNA carrying the complementary bases. The adaptor would also carry the related (cognate) amino acid. These adaptors, which are indeed made of RNA, were discovered by Paul Zamecnik and Mahlon Hoagland in 1958. They were named transfer-RNAs (tRNAs). It was found that each tRNA had the same sequence at the 3' end, CCA, to which the cognate amino acid could be attached by specific enzymes (amino-acyl tRNA synthetases). This process, which derives energy from the hydrolysis of adenosine triphosphate (ATP), is known as *charging*. Each amino-acyl synthetase is able to recognise its amino acid and the corresponding anticodon on the tRNA. The 'charged' tRNA subsequently binds to the appropriate codon on

---

[1] Judson, H.F. *The Eighth Day of Creation*. Penguin Books (1995), first published by Jonathan Cape (1979).

[2] Brenner, S. (1957) *Proc. Natl Acad. Sci. USA*. **43**, 687–694.

the messenger RNA while attached to the ribosome. The ribosome then attaches the amino acid to the nascent polypeptide chain, discharges the empty tRNA and steps along to the next codon on the messenger RNA.

There are 20 amino acids, and most have more than one codon. Moreover, there are special codons for starting and stopping a protein, each with its own tRNA so that any cell contains 40 or more different kinds of tRNA. They are all very alike, so purifying a particular tRNA calls for considerable biochemical ingenuity. Robert Holley and his co-workers in the US Plant, Soil and Nutrition Laboratory at Cornell University did the pioneering work in tRNA purification and chemistry. They first had to find a fractionation technique for tRNAs. For this, they adapted the 'countercurrent distribution' procedure developed by Lyman Creighton Craig in the 1940s, in which substances are separated on the basis of their different solubilities in immiscible liquids flowing in opposite directions.

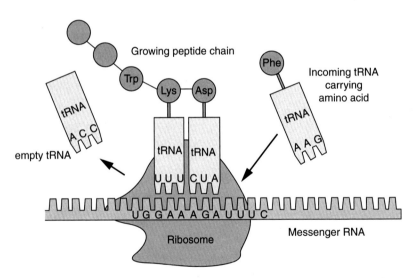

Figure 14.1 Protein synthesis on the ribosome: the sequence is determined by messenger RNA, which contains a sequence of three-base (or three-letter) codons. Each codon (for example GAU, standing for guanine, adenine, uracil; or AAA, adenine, adenine, adenine) specifies one of 20 amino acids. The codons are recognised by tRNA molecules that carry the appropriate complementary sequence (anticodon) of three bases (for example CUA, for cytosine, uracil, adenine; or UUU for three uracils). Each tRNA also carries the corresponding amino acid, which is then inserted into the growing protein chain. (Diagram modified from Wikipedia; image by Boumphreyfr. Licence: CC-BY-SA-3.0)

In this way, three tRNAs, alanine, tyrosine and valine RNAs, were obtained in a relatively homogeneous form. Next, one had to determine their sequence. Holley used a method similar to that developed by Fred Sanger for sequencing proteins: first find some enzyme to cut the tRNA into smaller pieces and then sequence the pieces. Some 20 graduate students later, Holley was able to report the sequence of the tRNA that coded for the amino-acid alanine. It contained 77 nucleotides, nine of which had non-standard bases produced by chemical modification. A short time later (in 1968) he was awarded a Nobel Prize for this work.

Holley found that the sequence contained four pairs of short sequences where the bases could make Watson–Crick type base pairs. This led to the suggestion of the cloverleaf secondary structure of tRNA. Later sequences showed that the cloverleaf was common to all tRNA molecules.

For Crick, the structure of tRNA had a special importance since it was his adaptor molecule. If one could obtain crystals, the atomic structure could be obtained by X-ray crystallography. Since Aaron had some experience with the structure of RNA, Francis urged him to get involved with tRNA. Unfortunately, in the matter of crystallisation tRNA proved particularly refractory. Thus a verbal report in 1968 from Hasko Paradies, then a postdoctoral fellow in Bror Strandberg's lab in Uppsala, that he had crystallised tRNA by using organic solutions and copper ions excited considerable interest at the LMB in Cambridge. Paradies duly turned up at LMB with a strange tale about how all his results were locked away in Strandberg's safe. Those who knew Strandberg thought it unlikely that Strandberg even owned a safe and found the story difficult to swallow. Aaron questioned Paradies closely and came to the conclusion it was all a fraud. Later, it was indeed shown that Paradies had made up the whole story and had never done an honest day's scientific work in the whole of his life. Nevertheless, he was charismatic, and his ideas were catalytic. Organic solvents as agents for crystallisation seemed rather interesting and worth a try. Brian Clark led a group at LMB working on the initiator tRNA (the one that starts a protein). His collaborators could produce initiator tRNA in pure form. The method of purification was similar to Holley's and involved the use of a countercurrent machine consisting of an awesome oscillating battery of glass mixing vessels. Clark urged his colleague, Bhupendra Doctor, to crank up the countercurrent machine, and after a week or two of cranking and clanking he managed to produce about 100 mg of rather pure initiator tRNA. Shirley Morris, a co-worker of Ken Holmes, dissolved some of this sample in a water/dioxan mixture. Then she increased the dioxan concentration by evaporation and, *eureka,* microcrystalline spherulites appeared. Holmes took an X-ray diffraction photograph of the spherulites. This showed rings. An X-ray diffraction photograph of this kind is known in the trade as a 'powder photo' because, as in a powder, the sample is made of masses of tiny crystals in all possible orientations. Rather than a single spot, each Bragg reflection thus gives rise to a ring on the film as if the single-crystal diffraction pattern has been spun round an axis. The result demonstrated without doubt that the sample was microcrystalline. Moreover, the rings were crowded round the direct beam, showing that the crystals had a large unit cell – necessary if the crystalline precipitate was made of macromolecules rather than just salt. Aaron sat down with the list of ring diameters and was able to work out the size and symmetry of the crystal. Although of no use for structure analysis, this was the first demonstration of a crystalline sample of tRNA[3]. It unleashed numerous competing attempts to obtain good single crystals of a tRNA[4].

The two main contenders in the race turned out to be Aaron's group (in collaboration with Brian Clark's group) at the LMB in Cambridge, England, and Alex Rich's group at the Massachusetts Institute of Technology (MIT) in Cambridge, Massachusetts. Rich had started out as a postdoctoral fellow with Linus Pauling at Caltech taking X-ray photos of DNA fibres. He became a professor at MIT at the tender age of 31. He worked on RNA and DNA–RNA three-stranded complexes and explored non-Watson–Crick base pairs. Rich was also a member of the RNA Tie Club (ARG – arginine). In 1955, he visited Crick in Cambridge for a few weeks[5] to avail himself of the MRC's powerful rotating-anode X-ray tubes. He planned to take X-ray diffraction photographs of fibres of polyadenylic acid, an RNA-type synthetic polymer, which was later also worked on by Aaron and John Finch. He stayed with Francis and Odile Crick in their house in Portugal Place ('The Golden Helix'). By chance, over breakfast, Crick and Rich read about a new form of poly-L-glycine

[3] Clark, B.F.C., Doctor, B.P., Holmes, K.C. *et al.* (1968) *Nature* **219**, 1222–1224.
[4] Chedd, G. (1968) *New Scientist* **40**, 606–607.
[5] http://library.cshl.edu/oralhistory/interview/james-d-watson/writer/conversation-francis-crick-nature-paper/

from the Courtauld laboratory, just published in the journal *Nature*. That same morning, they managed to work out its structure. Then they realised that this structure was a model for the structure of collagen, one of the secondary structures of proteins that had not already been predicted by Linus Pauling. These ramifications took time to work out, and Rich stayed in Crick's home for six months. Rich did finally take data from the polyadenylic acid fibres[6].

Rich was an RNA specialist. At the same time as the LMB group were producing micro-crystals, members of the MIT group were trying similar methods, also on initiator tRNA. At the end of 1968, together with Sung-Hou Kim, Rich reported single crystals of initiator tRNA. Unfortunately, the crystals were soft and not suitable for a structure determination. A couple of years later, the MIT group were able to report usable crystals of a bacterial phenylalanine tRNA (referred to as tRNA^{Phe}).

One prerequisite for success in crystallisation is to have large quantities of pure material available. At the LMB, Bhupendra Doctor was only able to produce small quantities of initiator tRNA. Thus Daniela Rhodes, a doctoral student with Aaron, and Ray Brown, who had recently joined Aaron's group, switched to column chromatography using DEAE Sephadex chromotography gel and the new Pharmacia fraction collectors. This allowed them to purify several different tRNAs in quantity. Their source of tRNA was the thermophile bacterium *Bacillus stearothermophilus*. They obtained large crystals of several purified tRNAs, but none diffracted X-rays to high resolution. Aaron then urged Jane Crawford (later Ladner), a new recruit to Aaron's group, to crystallise putative tRNA dimers formed by mixing *B. stearothermophilus* tRNA^{Val} and yeast tRNA^{Phe} purchased from Boehringer Ingeheim (which was the same tRNA as was being used by Rich). Jane Ladner obtained some terrible-looking crystals that resembled insects' wings. Several months later, Brown decided to take an X-ray diffraction photograph. The crystals diffracted to 2.5 Å resolution, which was the highest resolution yet obtained for tRNA crystals. Analysis showed that the crystals contained only yeast tRNA^{Phe}.

This result presented Aaron and his group with a moral dilemma. Should they compete with Alex Rich, knowing that they were working

---

[6] Rich, A., Davies, D.R., Crick, F.H.C. and Watson, J.D. (1961) *J. Mol. Biol.* **3**, 71–86.

on exactly the same molecule? They decided to enter the race on the grounds that in Cambridge (England) the crystals had a monoclinic unit cell, whereas in Cambridge (Massachusetts) they had an orthorhombic unit cell. The monoclinic crystals diffracted to higher resolution than the orthorhombic crystals. Aaron maintained that his group would get better (higher-resolution) data, and that the different crystal form justified the morality of the competition.

Daniela Rhodes and Brown managed to grow better crystals by switching to dialysis bags for concentrating the solution. Rhodes grew a huge number of crystals, all in the cold room at 6 °C. Brown and Finch took some 800 procession photographs to find derivatives. They were joined by Jon Robertus, a postdoctoral fellow from Joe Kraut's laboratory in San Diego. By the time Robertus joined them, Aaron's group had already found a number of heavy-atom derivatives. Robertus collected X-ray diffraction data on Perutz's four-circle diffractometer. Since the crystals had been grown in the cold room, Greg Petsko came over from Oxford, where he was a doctoral student with David Phillips, to fit a device that could blow cold air over the crystals on the diffractometer so that data could be collected at 6 °C. By the autumn of 1973, the Cambridge group was able to calculate a good electron density map. Nevertheless, initial attempts to fit the RNA chain into the map failed because all base pairs look rather alike. Using RNA sequencing methods, Rhodes[7] discovered which base carried a platinum ion that had been attached as a heavy atom derivative for obtaining phase information. This provided a landmark for fitting the tRNA nucleotide sequence into the electron density map, and by spring 1974 the group had built an accurate atomic model and were preparing their work for publication.

How do you get started on building an atomic model of tRNA? On the basis of his sequence data, Holley had proposed that there were four stacks of base pairs forming short DNA-like helices, giving rise to four arms (Figure 14.2). Furthermore, these helical fragments would form an 'A-type' helix[8]. Holley's sequence data seemed so compelling that the

---

[7] Rhodes, D., Piper, P.W. and Clark, B.F.C. (1974) *J. Mol. Biol.* **89**, 469–475.

[8] The ribose sugar forms a furanose (five-membered) ring, but this ring is not flat. It can take one of two 'chair' conformations. Either the C2′ or C3′ carbon sticks out of the plane to form the back of the chair (so-called C2′ endo or C3′ endo conformations). Flipping between these two chair conformations was thought to be the primary determinant of whether the helix takes the 'A form' or 'B form'. The A form of DNA is C3′ endo, and the more common

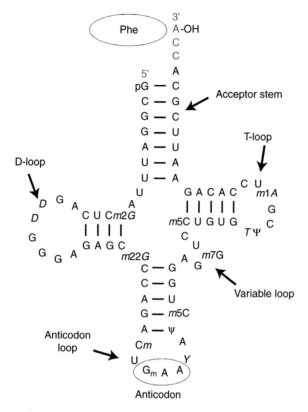

Figure 14.2 Secondary cloverleaf structure of phenylalanine tRNA from yeast (after Holley). Note the four stems that are common to all tRNAs. Each tRNA contains a specific anticodon triplet sequence (shown bold at bottom of diagram) that can base-pair to complementary codon sequences in the messenger RNA corresponding to one specific amino acid. For example, the codons for phenylalanine are UUU and UUC; the anticodon of a phenylalanine tRNA is G$_m$AA. The modified base G$_m$ can pair with both A and C, allowing both UUU and UUC as codons for phenylalanine. Each tRNA is amino-acylated (or *charged*) with a specific amino acid (in this case phenylalanine) by a specific enzyme (amino-acyl tRNA synthetase) at the 3' end. A number of the bases have been methylated or otherwise chemically modified; for example, D is dihydrouridine. Modified bases are shown in italic. (Diagram modified from Wikipedia ; image by Boumphreyfr. Licence: CC-BY-SA-3.0).

B form is C2' endo. The DNA (deoxyribose) sugar may alternate between 2' endo and 3' endo conformations, allowing DNA to switch between B form and A form. RNA has one more OH group on its ribose sugar than DNA. The extra 2' OH group locks the RNA into a 3' endo chair conformation so that it can only form an A-type helix.

Figure 14.3 John Finch, Aaron Klug and Brian Clark inspecting their atomic model of tRNA[phe].
(© MRC Laboratory of Molecular Biology)

problem appeared to reduce to finding out how to stack the four Holley helical arms on top of each other. In 1968, Mike Levitt came to Cambridge as a doctoral student with Bob Diamond. He was a theoretician and had just completed a two-year Royal Society Fellowship at the Weizmann Institute in Israel with Shneior Lifson. Levitt produced a structure for tRNA based on a computer modelling of the interactions between Holley's helical stems[9]. He predicted that the acceptor stem (see Figure 14.2) would stack onto the T-loop (also known as the Tψ-loop) and that the D-loop would stack with the anticodon stem. He also predicted some of the non-Watson–Crick base-pair interactions that indeed play an important role in holding tRNA together. He did not foresee the L-shaped molecule that Rich's group discovered.

How does one represent atomic models? For his work on myoglobin, John Kendrew had developed skeletal models consisting of rods and connectors. Each connector represented a chemical bond between two atoms. Each atom was represented by a polyhedron of rods joined at the atomic centre, some tetragonal, some trigonal and some binary, representing the appropriate symmetry for a particular atom. In the case of a

[9] Levitt, M. (1969) *Nature* **224**, 759–763.

polymer such as RNA, one ends up with a long chain of rods and connectors with the bases sticking out. The bases were pre-formed wire models of adenine, uracil, guanine and cytosine (A, U, G and C). X-ray crystallography yields 3D electron density maps, which in the days before computer modelling were presented as contours drawn by hand on stacks of transparent plastic sheets. The game was to fit the skeletal models into the electron density map so that the (known) sequence matched the density. This was carried out in an 'optical comparator', basically a half-silvered mirror that allowed one to look at the model together with the superimposed image of some specific region of the electron density, which was chosen by sliding the appropriate plastic sheets in or out. The arrangement had been developed by Fred Richards, the pioneer of American protein crystallography, and was colloquially known as 'Fred's Folly'. In the pre-computer-graphics age, model building was a labour-intensive activity that took lots of time and produced sore fingers.

In January 1973, Alex Rich's team published a description of the shape of the molecule[10]. Their electron density map was of low resolution, so they were unable to discern the base sequence, but they could make out the overall shape of the molecule and see stretches of A-type RNA double helix. Using the Holley cloverleaf structure as a basis, they were able to arrive at an approximate atomic model of tRNA: the acceptor stem stacks on the T-loop, and the D-loop stacks on the anticodon loop (as suggested by Levitt). The two stacks then form an 'L'. This was an unexpected result.

At about this time, Sung-Hou Kim left MIT to open his own laboratory at Duke University. Subsequently, the collaborating groups acquired better heavy-atom data that enabled a much more detailed electron density map to be produced. A description of an atomic model was published in *Nature* in March 1974. It turned out that the model built at MIT was flawed.

Aaron was disappointed that the MIT group had apparently got so far in front. Thus a certain *Schadenfreude* accompanied the receipt of Rich's March 1974 paper, because it became clear that the MIT group had made some mistakes, particularly concerning the structure of the anticodon loop and the join region (shown black in Plate 3). They had

[10] Kim, S.H, Quigley, G.J., Suddath, F.L. *et al.* (1973) *Science* **179**, 285–288.

missed the fact that the join region is a triple helix held together by a combination of Watson–Crick base pairs and other non-standard base pairs. This part of the molecule is structurally important because it stabilises the characteristic 'L' shape of tRNA.

In the spring of 1974, Alex Rich paid LMB Cambridge a visit. Aaron became very nervous: Rich might somehow try to make use of the Cambridge result before it was published. Aaron took the precaution of having the model room boarded up. Moreover, for safety, Aaron sent the LMB group's paper describing the atomic structure of tRNA off to *Nature*.

The Harry Steenboek conference takes place annually in Madison, Wisconsin. In June 1974 the subject was tRNA. Although this was an important conference, Aaron could not attend because Liebe's father Alter, who was now resident with the Klugs in Cambridge, was struggling with impending blindness and needed Aaron's presence. Thus, at the Madison meeting, Cambridge (England) was represented by Brian Clarke and Jon Robertus. Cambridge (Massachusetts) was represented by Alex Rich. Rich showed a slide with his published structure, whereas Robertus presented what turned out to be the correct structure. Robertus and Clark now felt secure because their work had already been accepted for publication by *Nature*. Even so, in discussion they were cagey. Notwithstanding their reticence, Rich realised what was afoot and moved quickly to rectify the situation. A corrected structure from the MIT group was accepted by the journal *Science* on 4th July and was officially published on 2nd August[11]. The LMB structure was not published until 16th August[12].

Two aspects of this story disturbed Aaron: the first was the extraordinarily short publication time of four weeks achieved by Rich, where two to three months would be normal; the second was that Rich apparently could rebuild his model in two or three days. Model building in a 'Fred's Folly' took weeks rather than days. The explanation was that the correct model already existed. Communication between the Rich group and the Sung-Hou Kim group was intermittent, and by the summer of 1974 there were in fact two models: one with errors at MIT and a more or less correct one at Duke University. Robertus's talk

[11] Kim, S.H., Suddath, F.L., Quigley, G.J. *et al.* (1974) *Science* **185**, 435–440.
[12] Robertus, J.D., Ladner, J.E., Finch, J.T. *et al.* (1974) *Nature* **250**, 546–551.

convinced Rich that the MIT structure was flawed. There was neither the time nor the need for a rebuild: the Duke University version had to be published quickly before the LMB group could take all the credit. Rich rapidly wrote a new paper based on the Duke structure, which he submitted to *Science*. Although it had been submitted a week after the Madison meeting, it still managed to beat the LMB paper into print.

When Alex Rich's *Science* publication reached Cambridge (England) at the end of July, it provoked an embittered 'I told you so' response from Aaron. Crick wrote a very strong letter of complaint to Rich, pointing out that in his *Science* paper he did not refer to the source of information that led him to correct his structure (Robertus's talk at Madison). After a four-week delay, a long letter arrived from Rich, which included a complaint about Brian Clark and Jan Robertus's reticence at Madison. Nevertheless, as with most disputes over priority, the issue soon became of minor interest to the broad scientific community. It was generally accepted that the MIT group had done laudable pioneering work and had published the first 'L' model of tRNA. The LMB group had published the first accurate atomic model. When all the dust had settled, Aaron remarked laconically: 'I spoke to Alex on the phone; he told me he was no saint.'

These studies of tRNA were important for establishing that RNA is capable of building structures rivalling proteins in complexity. RNA has one more OH group on its sugar ring than DNA. This extra OH group is chemically important: it allows RNA to function as an enzyme. Well before Thomas Cech and Sid Altman's demonstration of the enzymatic role of RNA, Aaron with Francis Crick, Sydney Brenner and George Pieczenik[13] proposed, in a hypothetical pre-ribosome world, a mechanism whereby the small soluble RNA molecules such as tRNA might have been able to synthesise proteins by an enzymatic process that involved binding to the coding RNA. The intricacies and complexities of the tRNA structure nonetheless led Aaron to remark that it showed why evolution had needed to 'invent' proteins in order to create a much more diverse library of structures and catalysts.

---

[13] Crick, F.H.C., Brenner, S., Klug, A. and Pieczenik, G. (1976) *Origin Life* 7, 389–397.

15

Chromatin and Nucleosomes

Isle de Port Gros is a magical island off the coast of Provence. On the west side of the island a few buildings cluster round a natural harbour; elsewhere there are just evergreen oak forests and a few palm trees. In the morning not much happens before the arrival of the ferry from Le Lavandou bringing the daily supply of baguettes. There is one Hotel, Le Manoir. Here, in May 1971, Giorgio Bernardi from the CNRS laboratory in Strasbourg organised a successful meeting[1] of around 50 scientists (including Francis Crick), devoted to a discussion of the structure of eukaryotic DNA. Large regions of repetitive sequences were being found. Did they determine the structure of the chromosome? Two conflicting theories were being propounded: both could not be right. Playing the elder statesman, at the end of the meeting Francis Crick choreographed the duelling parties to a grudging acceptance of each other's existence.

If the entire DNA[2] in a human cell were laid out end to end, it would extend about 2 metres. Since the cell nucleus is only 5–6 micrometres across, wrapping up the DNA to fit into the nucleus is a serious packing problem. The DNA in the nucleus interacts with histone proteins to

---

[1] See *Nature New Biol.* (1971) **231**, 68.
[2] Human DNA is organised in 23 paired chromosomes. At cell division the chromosomes are replicated. The nuclear membrane is dispersed and the DNA is compacted by a further factor of 10,000 into chromosomes that are visible in the microscope.

form the complex known as chromatin. Chromatin both packages DNA into a smaller volume and controls gene expression.

By 1969, Roy Britten and David Kohne at the Carnegie Institution in Washington DC had established that most of mammalian DNA consisted of repetitive sequences that have nothing to do with coding proteins. Work from Bernardi's lab led to the same conclusion. Francis Crick came back from the meeting on the Riviera full of ideas about chromosome structure. He summarised these in a paper in *Nature* that appeared in November 1971. It turned out this was not his best paper. Crick correctly assumed that most eukaryotic DNA is there for controlling gene expression rather than for coding proteins. However, he proposed that the controlling sequences would be exposed as single strands so as to enable a controlling protein, perhaps via a nucleic acid intermediary, to recognise a specific sequence[3]. This was wrong: in eukaryotes, proteins that recognise DNA sequences contain structural motifs that specifically bind to double-stranded DNA. Aaron was later to discover one of these DNA binding motifs. He showed that in some cases the recognition of specific sequences in double-stranded DNA involved the concatenation of structural motifs called 'zinc fingers', which were to become Aaron's swan song.

Early hypotheses about chromatin structure often envisaged the histones wrapping round the DNA to shield it from replicating enzymes. Structural data were minimal; they consisted mostly of an X-ray diffraction ring from chromosome gels showing a 110 Å spacing that had been measured some years ago by Vittorio Luzzatti and Maurice Wilkins. Moreover, electron microscopy was getting nowhere. At the end of the Cold Spring Harbor Laboratory Symposium devoted to chromatin in 1973, Hewson Swift felt compelled to comment: 'Spread whole chromosomes under the electron microscope look even at their best something like a bad day at a macaroni factory.' Apparently, specimen preparation was wrecking whatever structures might be there. Fortunately, in 1973 there was a notable improvement: using a new spreading technique, Ada and Don Olins at the Oak Ridge National Laboratory in Tennessee managed to produce electron micrographs of chromatin fibres spilling out of ruptured nuclei, showing arrays resembling beads on a string.

---

[3] See Olby, R. *Francis Crick, Hunter of Life's Secrets*. Cold Spring Habor Laboratory Press (2009) pp. 351–359.

The beads were spherical particles about 70 Å in diameter[4]. Likewise, the work of Dean Hewish and Leigh Burgoyne in Flinders University in Australia had shown that the nucleases in cells cut the chromosomal DNA in a regular way: the DNA fragments each contained about 200 base pairs[5]. These two short papers turned out to provide important clues in unravelling the structure of condensed DNA fibres.

Roger Kornberg came to the LMB in Cambridge in the spring of 1972 to work as a postdoctoral fellow with Aaron. He particularly wanted to get experience in X-ray diffraction. At this time Aaron was preoccupied with tRNA and with TMV (tobacco mosaic virus) assembly. Neither theme appealed to Kornberg, but a chance conversation with Mark Bretscher drew his attention to Crick's *Nature* paper and the general problem of the structure of chromosomes. He discussed with Aaron the idea of starting a research project in this area. Aaron pointed out that chromosome structure was Crick's project and they should first consult with him. Crick was enthusiastic, so Aaron duly established a chromatin research group. Crick and Aaron spent a lot of time trying to guess how DNA might fold to satisfy the X-ray diffraction repeat of 110 Å. Kornberg quickly became disillusioned with this approach and decided to go for biochemistry in the hope that this might lead to an understanding of the DNA folding.

The biochemistry of chromatin is messy. The constituent proteins are known as histones. There are five types of histone, designated H1, H2A, H2B, H3 and H4, but individual histones proved to be extraordinarily sticky, binding to DNA and to one another promiscuously. Histones appeared to be a kind of glue for coating the chromosomal DNA. However, drawing on some experiments that Aaron was carrying out, Kornberg became aware of the work of two biochemists from Cape Town who had established a method of purifying the histones from chromatin fibres by a mild treatment that preserved their integrity[6]. These gently extracted histones associated with each other in two groups, H2A/H2B and H3/H4.

Kornberg measured the molecular weight of the purified H3/H4 preparation by equilibrium ultracentrifugation, which showed that H3

[4] Odins, A.L. and Odins. D.E. (1974) *Science* **183**, 330–332.
[5] Hewish, D.R. and Burgoyne, L.A. (1973) *Biochem. Biophys. Res. Commun.* **52**, 504–510.
[6] van der Westhuyzen, D.R. and von Holt, C. (1971) *FEBS Lett.* **14**, 333–337.

and H4 were in the form of a double dimer, an $(H3)_2(H4)_2$ tetramer. Through a conversation with Richard Perham, who was a lecturer in the Biochemistry Department and also worked at the LMB, Kornberg was introduced to Jean Thomas, who was a recently appointed lecturer in the Biochemistry Department. Jean Thomas examined the structure of the H3/H4 complex by an independent method, chemical cross-linking, and produced the same result as Kornberg. This led to a very fruitful collaboration[7].

After some deliberation, Kornberg came to the conclusion that the structure of the unit histone particle must be an octamer: $(H2A)_2(H2B)_2(H3)_2(H4)_2$. A histone octamer would be associated with about 200 base pairs of DNA. The $(H3)_2(H4)_2$ tetramer reminded him of haemoglobin, also an $\alpha_2\beta_2$ tetramer. Such proteins were compact, with no holes for DNA to pass through. Thus it seemed likely that the DNA in chromatin must be wrapped around the outside of the histone octamer with about 200 base pairs of double-stranded DNA per octamer. These ideas were published[8] in a *Science* paper that accompanied Kornberg's paper with Jean Thomas. In collaboration with Thomas, Aaron and John Finch spent the next few years characterizing the histone octamer with its attendant DNA, which became known as the nucleosome.

Pre-dating Kornberg's breakthrough, Aaron had used the rather meagre X-ray diffraction pattern given by solutions of histone–DNA complexes to follow the reconstitution of histones with DNA[9]. These X-ray studies showed that almost 90% reconstitution could be achieved when the DNA was simply mixed with an unfractionated total histone preparation, but all attempts to reconstitute chromatin by mixing DNA with a set of all four purified single species of histone failed. However, using the milder methods of histone extraction described by van der Westhuyzen and von Holt, he found that the native structure could be reformed readily if the four histones were kept together in two pairs, H3 and H4 together, and H2A and H2B together, but not once

[7] Kornberg, R.D. and Thomas, J.O. (1974) *Science* **184**, 865–868.
[8] Kornberg, R.D. (1974) *Science* **184**, 868–874.
[9] The following account is taken in part from Aaron Klug's Nobel lecture in 1982. A popular account is given in Kornberg, R.D. and Klug, A. (1981) *Scient. Am.* **244**, 52–64. See also Kornberg, R.D. (1977) *Ann. Rev. Biochem.* **46**, 931–954.

they had been taken apart. It was this work that started Roger Kornberg on his journey to the discovery of the histone tetramer $(H3)_2(H4)_2$, which in turn led him to the model of the nucleosome described above.

In the summer of 1975, Aaron and his colleagues set about trying to prepare nucleosomes in forms suitable for crystallisation. Nucleosomes purified from the products of micrococcal nuclease digestion contain about 200 base pairs of DNA, but such preparations were not homogeneous enough to crystallise. However, this variability can be eliminated by further digestion with micrococcal nuclease. While the action of micrococcal nuclease on chromatin is first to cleave between nucleosomes, it subsequently acts on the excised nucleosome, shortening the DNA to about 146 base pairs. During this last stage the histone Hl is released, leaving a particle containing about 146 base pairs of DNA complexed with a set of eight histone molecules. This reduced form of the nucleosome is called the core particle, and its DNA content was found to be constant over many different species. The DNA removed by the prolonged digestion, which had previously joined one nucleosome to the next, is called the linker DNA. A core particle contains a well-defined length of DNA and is homogeneous in its protein composition.

After some time, Aaron's co-worker Leonard Lutter found a way to produce homogeneous preparations of nucleosome core particles. Using the nuclease DNAse1 to cut the DNA at exposed positions, Lutter also accumulated a wealth of data that was consistent with the DNA wrapping in two superhelical turns around the histone octamer. In the summer of 1976, Daniela Rhodes obtained the first nucleosome core crystals. These gave a diffraction pattern out to about 20 Å resolution. With Finch and co-workers, Aaron examined the crystals by a combination of X-ray diffraction and electron microscopy[10]. This work is remarkable for its methodology – the use of electron micrographs of small crystals to obtain phases for the X-ray reflections (see Chapter 9) – as well as the use of every possible experimental detail to derive a model. The proposed structure remained the definitive model of the nucleosome for several years, and later work confirmed all its key conclusions. The nucleosome core particle turned out to be a flat disk-shaped object, about 110 Å by 110 Å by 57 Å, somewhat wedge-shaped, and strongly

---

[10] Finch, J.T., Lutter, L.C., Rhodes, D. *et al.* (1977) *Nature* **269**, 29–36.

divided into two layers. Aaron said it should be called a 'platysome' to reflect its flattened shape – until then, all the diagrams had shown a more-or-less spherical ball. Needless to say, the name did not stick. In the model, the DNA was wound into about 1¾ turns of a shallow super-helix of pitch about 27 Å around the histone octamer (see Figure 15.1). There are about 80 nucleotides in each turn of the super-helix. Diffuse X-ray scattering from the crystals showed that the DNA of the core particle is in the B form.

The position of the DNA in the core particle could also be revealed by the method of contrast variation using a beam of neutrons. Neutron scattering can distinguish between nucleic acid and proteins. The scattering power of the solution in which the nucleosomes are immersed can be adjusted to match the protein or nucleic acid scattering by varying the proportion of heavy water (D_2O) to ordinary water (H_2O). If the scattering power of the solution is the same as the protein (or nucleic acid) then the latter becomes invisible (no scattering). Neutrons are strongly scattered by hydrogen atoms. Proteins contain relatively more hydrogen than nucleic acids. Heavy water, which

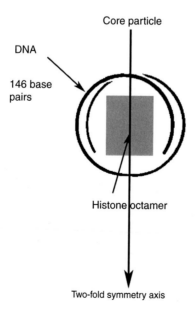

Figure 15.1 Diagram of the histone core particle showing the possible role of the linker histone H1. (© MRC Laboratory of Molecular Biology.)

contains deuterium instead of hydrogen but is otherwise just like water, scatters neutrons much less strongly than normal water. Thus by mixing water and heavy water in various proportions, one can match the scattering of either the nucleic acid or the protein to the scattering power of the solution and thereby find out where each one is.

Morton Bradbury and his collaborators pioneered the study of nucleosomes by neutron scattering. They showed that the DNA was exterior to the protein. However, their studies were with solutions of nucleosomes. John Finch went off to the research reactor at Grenoble with large crystals. He managed to collect neutron diffraction data from crystals of core particles immersed in various mixes of water and D_2O. These data confirmed what the X-ray work had already made clear, namely that the nucleic acid really was wrapping round the outside of the protein core.

An alternative to separating the contributions of the DNA and the protein by neutron diffraction is to study the histone protein octamer directly. Kornberg and Jean Thomas discovered that adding lots of salt to a solution of nucleosomes displaced the DNA to leave the histone octamer intact. Aaron's group attempted to crystallise the octamer, but they obtained hollow tubular structures rather than crystals: these were investigated by electron microscopy. To a resolution of 20 Å they showed that the histone octamer, just like the nucleosome core particle, is a wedge-shaped particle showing a system of surface ridges forming a helical ramp of external diameter 70 Å and pitch about 27 Å – just right to act as a spool on which about 1¾ turns of a super-helix of DNA could be wound. The resolution of the octamer map was too low to define individual histone molecules, but, using published results on the chemical cross-linking of histones to nucleosomal DNA, Aaron, Finch and their collaborators were able to propose a spatial arrangement for the eight histones in the octamer[11].

The model of the histone core particle revealed a problem that has become known as the 'linking number paradox'. Structural analysis[10] showed that the double helix of DNA in a nucleosome is wound into nearly two turns of a shallow super-helix – the 'linking number' change caused by the supercoiling would be nearly 2 – whereas measurements

[11] Klug, A., Rhodes, D., Smith, J., Finch, J.T. and Thomas, J.O. (1980) *Nature* **287**, 509–516.

on closed circular DNA extracted from SV40 chromatin indicated on binding to the histone core a linking number change of 1 per nucleosome, not 2. Finch *et al.*[10] and Klug and Lutter[12] proposed that the explanation of the paradox lay in a difference in the screw of the DNA (the number of bases per turn) when wound round the nucleosome and when free in solution. Marcus Noll had estimated the periodicity of the DNA when wound round the histone core by digesting the DNA with DNAse1, an enzyme that can cleave the minor groove[13] of a DNA double helix if it is exposed. The size of the fragments produced (the cutting number) gives an estimate of the size of a single turn. The number he obtained was 10 base pairs per turn. Based on this result, and taking account of the supercoiling, the estimated number in solution should be about 10.4. Apparently the commonly accepted number 10 (obtained from DNA fibres) is not correct for DNA in solution.

Daniela Rhodes and Klug[14] tried to define the geometry of the DNA in solution by using DNAse1 to digest DNA lying on a flat surface. The cutting number (and hence the helical periodicity) they obtained was 10.6, which is about halfway between the canonical A form (11 per turn) and B form (10 per turn). From an energetic analysis of the stability of DNA structures, Levitt[15] had also predicted 10.6 for DNA in solution. A further study by Rhodes and Klug of a synthetic DNA[16] in which one chain was made purely of A (adenine) and the other of T (thymine) showed that in this very special case the number of bases per turn was 10, but this DNA was apparently too stiff to wrap around the histone octamer. The final number for the helical periodicity *on the nucleosome* settled down at 10.2 base pairs per turn (see the discussion of work by Drew and Travers below). Even after taking account of the supercoiling, this is significantly tighter than the 10.6 measured in solution. Thus the linking number paradox appeared to be solved: DNA takes on an extra twist when it binds to the nucleosome[17].

[12] Klug, A. and Lutter, L. (1981) *Nucleic Acids Res.* **9**, 4267–4283.
[13] The two chains of DNA are not symmetrically grouped round the helix axis, with the result that of the two grooves formed by the sugar-phosphate chains, one is bigger than the other. These are referred to as the major and minor grooves.
[14] Rhodes, D. and Klug, A. (1980) *Nature* **286**, 573–578.
[15] Levitt, M. (1978) *Proc. Natl Acad. Sci. USA* **75**, 640–644.
[16] Rhodes, D. and Klug, A. (1981) *Nature* **292**, 378–380.
[17] Travers, A.A. and Klug, A. (1987) *Phil. Trans. Roy. Soc. Lond.* **B317**, 537–561.

Nevertheless, there were now problems with what defined the A and B form of DNA. The textbook concept was that the sugar pucker determined the DNA type: C2′ endo was B form, C3′ endo was A form. This was known as the 'rigid nucleotide rule'. In 1979, Aaron and co-workers put forward the idea that the sequence played a role in determining the DNA geometry[18]. Just a year later, new structural data from X-ray diffraction studies of crystals of a 12-base-pair fragment in the B form transformed our understanding of DNA[19]. This first crystal structure of a DNA fragment in the B form came from Dick Dickerson's lab at CalTech in Pasadena. The new structure demonstrated that the 'rigid nucleotide rule' was a bad rule. Analysis[20] showed, among many things, that the classical base pair was not planar, as had originally been assumed. There was a propeller-like twist between the two base pairs of up to 15–20°. The authors also noted wide fluctuations in the geometry of the sugar ranging from C2′ endo to C3′ endo. The conformations of sugars in paired bases (1/24, 2/23, 3/22, etc.) tended to be anti-correlated: if one base-pair conformation is A type (C2′ endo) then the following base pair is a B type (C3′ endo). Aaron brought this to the attention of his Peterhouse engineering friend Chris Calladine, who, using standard engineering concepts, suggested what might be going on[21]. Because of the propeller-like twist in the B form, the large purine residues of one base collide with the neighbouring base on the opposite strand if it is also a purine. This 'steric clash' has important consequences for the structure of the DNA that Calladine was able to predict. Moreover, whether a purine stacks parallel to a neighbouring purine in its own chain or parallel to the purine in the neighbouring chain makes a large contribution to the difference between the A and B forms of DNA.

In Dickerson's laboratory, Horace Drew had crystallised the 12 base-pair DNA fragment in the B form. In 1982, Drew came to the LMB as a postdoctoral fellow with Aaron. Drew was most impressed by Calladine's insight and sought him out. An active collaboration on DNA structure and mechanics ensued, culminating in a text-book[22]. Later, on

[18] Klug, A., Jack, A., Viswamitra, M.A. et al. (1979) *J. Mol. Biol.* **131**, 669–680.

[19] Wing, R., Drew, H.R., Takano, T. et al. (1980) *Nature* **287**, 755–758.

[20] Drew, H.R., Wing, R.M., Takano, T. et al. (1981) *Proc. Natl Acad. Sci. USA.* **78**, 2179–2183.

[21] Calladine, C.R. (1982) *J. Mol. Biol.* **161**, 343–352

[22] Calladine, C.R. and Drew, H.R. *Understanding DNA: The Molecule and How It Works.* Academic Press (1997).

the occasion of Calladine's retirement, Drew helped to organise a *Festschrift* for Calladine, to which Aaron contributed.

Drew addressed the question of whether there was a structural code in DNA (within the sequence) to help in wrapping around the histone octamer. At this time, it was known that certain sequences such as AA steps had a narrow minor groove and GG steps a wide minor groove, which would affect the ability of the DNA to bend. Drew collaborated with Andrew Travers (a group leader at LMB) and Sandra Satchwell to sequence about 150 DNA fragments extracted from nucleosome cores[23] using a method they called 'statistical sequencing'. They found a sequence periodicity of 10.17 base pairs for DNA wound round the nucleosome.

Daniela Rhodes, Ray Brown and Barbara Rushton grew crystals of core particles prepared from seven different organisms: all gave essentially identical X-ray patterns, testifying to the universality of nucleosomes. In 1978, Tim Richmond from Yale joined Aaron's group as a postdoctoral fellow. He developed an interest in the nucleosome work and joined Rhodes and Brown in their attempts to obtain better crystals of the histone core particle. The problems were caused by heterogeneity: no two nucleosomes were quite the same because each carried a different DNA sequence. Thus in detail each particle in the crystal was different, and such a hotchpotch crystal of nucleosome cores produced from native sources would not diffract to high resolution. Daniela Rhodes[24] made synthetic nucleosomes of defined DNA sequence by assembling histone core particles from histone octamers and commercially available synthetic DNA made of alternating adenine and thymine bases. This did indeed produce crystals, but too small for use in a structure determination.

Furthermore, on account of the heterogeneity of the DNA, different batches of crystals had slightly different unit cell dimensions, which made the collection of X-ray diffraction data very difficult. Richmond tried shrinking the unit cells to a standard size by putting the crystals in MPD (2-methyl-2,4-pentanediol) solutions, as this compresses the macromolecules. After some years, in 1984, Aaron's group were finally able to calculate a density map of the nucleosome at about 7 Å

---

[23] Satchwell, S.C., Drew, H.R. and Travers, A.A. (1986) *J. Mol. Biol.* **191**, 659–675.

[24] Rhodes, D. (1979) *Nucleic Acids Res.* **6**, 1805–1816.

resolution[25], enough resolution to see the DNA backbone clearly and to see some elements of the protein structure. This result was in good agreement with the earlier low-resolution model. It showed that the DNA curved smoothly but not uniformly round the histone core (no obvious kinks). However, in the neighbourhood of the dyad, it made a couple of sharper bends, giving the double ring of DNA approximately the shape of a key ring.

Other groups were now busy in the field. In particular, Evangelos (Van) Moudrianakis and his co-workers at Johns Hopkins University and Medical School in Baltimore had managed to crystallise the histone octamer (the core particle without DNA). In 1985, they reported their X-ray structure at 3.1 Å resolution[26]. The polypepetide chain could be traced though large regions of the H2A and H2B histone molecules. Unfortunately, their model was nearly twice as large as the Cambridge model. They maintained that the Cambridge histone octamer was an artefact. Aaron was very bothered by his work being questioned and insisted that *Science* publish his commentary on the Moudrianakis paper in the same issue. Aaron in turn maintained that the Cambridge structure was right, and that the Johns Hopkins group had made a mistake with their phasing procedures. By this time, Gerry Bunick and his group at the Oak Ridge National Laboratories (Tennessee) had produced crystals of the nucleosome (with DNA) and had successfully analysed their X-ray diffraction data using the Cambridge low-resolution octamer model as a starting point. Bunick was also of the opinion that the Johns Hopkins group had somehow got things wrong.

In 1988, Moudrianakis and Aaron were invited to a structure meeting in Spandau, Berlin, to discuss their two versions of the histone octamer structure. Aaron took the opportunity of crossing into East Berlin via the notorious Friedrichstraße crossing to visit the Pergamon Museum. From the Friedrichstraße U-Bahn station, a short walk along Dorotheenstraße brings you to the Museum Insel. Aaron's interest was not just the impressive Pergamon Altar but also the reconstruction of the Ishtar Gate and Processional Way of Babylon. Dedicated to Ishtar, the goddess of love, the gate was constructed using blue glazed

[25] Richmond, T.J., Finch, J.T., Rushton, B.J., Rhodes, D. and Klug, A. (1984) *Nature* **311**, 532–537.

[26] Burlinghame, R.W., Love, W.E., Wang, B-C. *et al.* (1985) *Science* **228**, 546–553.

bricks decorated with alternating rows of dragons and aurochs (bulls) in gold. The Ishtar Gate was once considered one of the Seven Wonders of the World. The amazing Berlin reconstruction was built from material excavated in Babylon by Robert Koldewey and finished in the 1930s. Its sheer size is awesome. Moreover, it is sobering to think that the construction of the Ishtar Gate by King Nebuchadnezzar was contemporaneous with the Jewish captivity in Babylon.

Back in Spandau, the organisers had allocated a whole morning for the discussion of the structure of the histone octamer in a rather gladiatorial setting. Neither combatant relished the idea of a public confrontation, and Aaron withdrew. Moreover, conversations with Moudrianakis shed light on an episode that had disturbed Aaron deeply. At the time of the publication of the Moudrianakis article in *Science*, a student newspaper at Johns Hopkins had published a triumphal piece claiming that a Hopkins student had shown that Nobel Prize winner Klug was wrong. Persons unknown but of malicious intent had mailed this to Aaron. Aaron had assumed it came from Moudrianakis and was naturally most perturbed. His reaction had been to insist that *Science* publish his refutation. The realisation that Moudrianakis was not involved in the newspaper incident took much of the heat out of the dispute, and the two scientists became more friendly.

Aaron had also been invited to give the keynote evening lecture at Spandau. He did not mention histones. Indeed, his attention had turned elsewhere, to 'zinc fingers', a theme that preoccupied him for the rest of his life.

The confidence of Moudrianakis's group was being eroded because they were unable to fit atomic models to parts of their electron density maps. Something was wrong. They decided to re-determine the position of the heavy atom used for phasing and found this to be 2.7 Å wrong. Because the heavy atom was close to a symmetry axis in the crystal, this small error led to a very large effect. It had produced a secondary image of the core particle rotated through 120° and superimposed on the correct solution[27]. The secondary image made the core histone octamer appear much larger than it really is. After correcting the heavy atom position and using the same analytical procedures employed in the 1985

[27] Wang, B-C., Rose, J., Arents, G. and Moudrianakis, E.N. (1994) *J. Mol. Biol.* **236**, 179–188.

paper, they were able to determine the complete atomic structure of the core histone octamer[28]. Now that the structure was canonical, the dispute with Aaron was resolved. The results from the Johns Hopkins group showed that the histone proteins were elongated, mostly α-helix, and interpenetrated in a way that could not be seen in the earlier low-resolution work. At the places where the histones contacted the DNA, the 'helix–turn–helix' protein structure motif was found. This motif is common in DNA binding proteins. Thus, 15 years after its discovery, the atomic structure of the core histone octamer finally emerged.

In 1987, Tim Richmond, his wife Robin and his student Song Tan left Aaron's group for appointments at the Institute of Molecular Biology and Biophysics at the Eidgenössische Technische Hochschule (ETH) in Zürich. Richmond took the structural investigation of the nucleosome core with him. He and his co-workers took the approach to the hetero-geneity problem that had been tried out by Rhodes, by tailor-making DNA fragments all with the same 146 base-pair DNA sequence[29]. They then added these back to purified histone octamers to make identical nucleosome particles. Meanwhile, the Oak Ridge National Laboratory team led by Gerry Bunick was making good progress. In 1993, at a meeting of the American Crystallographic Association, Bunick's group reported crystals scattering to high resolution from nucleosomes con-taining a palindromic DNA sequence derived from repetitive human satellite DNA[30]. Subsequently, both groups produced high-resolution structures of the nucleosome using this palindromic sequence[31].

There was still no clear idea of the relation of one nucleosome to another along the basic chromatin fibre (the coiled-up nucleosome chain), nor of higher levels of organisation. The thickness of fibres observed in EM studies of whole chromosome specimens varied from about 100 to 300 Å in diameter, depending on ionic conditions. Taking this as a clue, Finch and Aaron carried out some EM experiments in which magnesium ions were added. The nucleosome chain coiled up

[28] Arents, G., Burlingham, R.W., Wang, B-C., Love, W. and Moudrianakis, E.N. (1991) *Proc. Natl Acad. Sci.* **88**, 10148–10152.

[29] Richmond, T.J., Searles, M.A. and Simpson, R.T. (1988) *J. Mol. Biol.* **199**, 161–170.

[30] Harp, J.M., Ueberbacher, E.C., Roberson, A.E. *et al.* (1996) *Acta Cryst.* **D52**, 283–288.

[31] Luger, K., Mäder, A.W., Richmond, R.K., Sargent, D.F. and Richmond, T.J. (1997) *Nature* **389**, 251–260.

    Harp, J.M., Hanson, B.L., Timm, D.E. and Bunick, G.J. (2000) *Acta Cryst.* **D56**, 1513–1534.

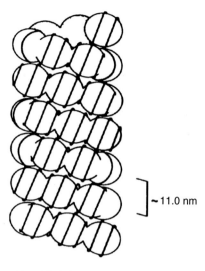

~11.0 nm

Figure 15.2 Solenoid model of chromatin fibres: each of the ball-shaped particles represents a histone core particle. (Figure 7 of Finch, J.T. and Klug, A. (1976) *Proc. Natl Acad. Sci.* 73, 1897–1901, reproduced with permission.)

into thicker, rather knobbly fibres about 250–300 Å in diameter, corresponding apparently to the turns of an ordered but not perfectly regular helix or supercoil. Since the term 'supercoil' had already been used in a different context, they called this a 'solenoid'[32]. On the basis of these micrographs and companion X-ray studies, they suggested that the second level of folding of chromatin was achieved by the winding of the string of nucleosome particles into a helical fibre with about six nucleosomes per turn. The linker DNA (that DNA between nucleosomes not incorporated into the nucleosome) would be on the inside of the cylinder.

Moreover, they found that when the same experiments were carried out on H1-depleted chromatin, only irregular clumps were formed, showing that the fifth histone H1 is needed for the formation and stabilisation of the ordered solenoid structure.

Although the solenoid structure made it into the textbooks, even within the LMB it did not find unanimous support. An alternative

[32] Finch, J.T. and Klug, A. (1976) *Proc. Natl Acad. Sci.* 73, 1897–1901.

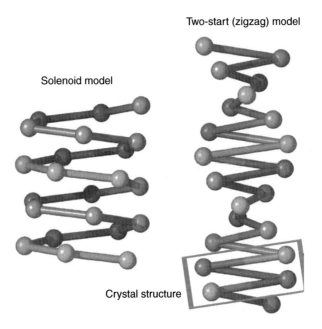

Two-start (zigzag) model

Solenoid model

Crystal structure

Figure 15.3 Diagrammatic representation of the two contending models for the chromatin fibre: the solenoid model (left) and the two-start or 'zigzag' model (right). The balls represent nucleosome core particles. The rods represent the 'linker DNA' that is found between the nucleosome particles. In the two-start model, this DNA is fully extended B form so that the diameter of the chromatin fibre reflects the lengths of the linkers (which are variable). In the solenoid model, the nucleosomes form a flat left-handed helix. In the zigzag model, the nucleosomes form a right-handed helix, but because each turn is nearly 180° this appears as a two-stranded left-handed helix. X-ray crystallography of a tetramer of nucleosome particles[27] shows the geometry found in the box marked 'crystal structure'. (Reprinted from Robinson, P.J.J. and Rhodes, D. *Curr. Opin. Struct. Biol.*16, 336–343 © 2006, with permission from Elsevier.)

'two-start' or 'zigzag' structure had been proposed that could fit John Finch's micrographs.

Somewhat ironically, those who were to become the chief proponents of the two contending theories were both originally co-workers of Aaron: Daniela Rhodes, group leader at LMB and later professor in Singapore, with her co-worker Phillip Robinson produced EM evidence in support of the solenoid model; Tim Richmond, now professor at the ETH Zürich, produced a X-ray structure at 9 Å resolution from crystals of four histone core particles arranged in the zig-zag geometry[33], just as would be expected from the two-start model. The arguments for and against the two models are summarised in two review articles: from Rhodes and Robinson[34] and from Andrew Travers and co-workers[35].

A little later, a detailed study by Rhodes and her co-workers showed that both structures can occur[36]. The solenoid model occurs for longer linker DNA and incorporates the fifth histone H1. The zig-zag model occurs for short DNA linkers and does not require the histone H1. Whether either structure is present in major proportions in cell nuclei remains unresolved.

[33] Schalch T., Duda S., Sargent D. and Richmond T.J. (2005) *Nature* **436**, 138–141
[34] Robinson, P.J.J. and Rhodes, D. (2006) *Curr. Opin. Struct. Biol.* **16**, 336–343.
[35] Wu, C., Bassett, A. and Travers, A. (2007) *EMBO Reports* **8**, 1129–1134.
[36] Routh, A., Sandin, S. and Rhodes, D. (2008) *Proc. Natl Acad. Sci. USA* **105**, 8872–8877.

16

Zinc Fingers

From his work on nucleosomes, Aaron became intrigued by 'active chromatin', chromatin that is involved in transcription of a gene, or in other words the making of an RNA copy from genetic DNA. He became aware of the work of Don Brown, at the Carnegie Institute in Baltimore, on the 5S RNA gene from the South African claw-toed frog *Xenopus laevis*. The 5S RNA gene is an important constituent of the ribosome. The enzyme RNA polymerase III transcribes 5S RNA, but the initiation of transcription requires the binding of a protein 'transcription factor IIIA' (TFIIIA, or TF-three-A). This factor interacts with a region of 50 nucleotides in the gene called the internal control region, thereby allowing transcription. It also binds to the product of transcription, the 5S RNA, to form the 7S particle. This property makes the transcription factor relatively easy to purify. TFIIIA was the first eukaryotic transcription factor to be recognised. It is present in large quantities in the immature eggs of the South African claw-toed frog.

In 1985, Aaron entrusted a new graduate student, Jonathan Miller, with the task of purifying TFIIIA from frogs' eggs. Initially, Miller had little success. Then Aaron noted that the protein was reported to contain an abnormally high proportion of cysteine and histidine residues. The side chains of both of these residues can be involved in binding metal ions, such as zinc. Normal purification procedures for proteins contain chelating agents for removing metal ions, including zinc. Working on Aaron's hunch that TFIIIA might be a zinc binder,

Jonathan Miller found that if he purified the protein without chelating agents the situation altered markedly and he could get good yields.

TFIIIA contains between seven and eleven tightly bound zinc ions. Extending work that was started by Don Brown and his group, who had shown that the TFIIIA protein could be broken into smaller parts by proteolysis, Miller found that proteolysis gave a ladder of bands on a gel suggesting that TFIIIA was made up of a repeating unit: the protein might be an assembly of a number of compact zinc binding domains perhaps held together by flexible linkers, which could be digested away by proteolysis. About this time, the amino acid sequence of TFIIIA was published. Aaron quickly spotted that there was indeed a pattern of nine very similar sequences, each about 30 amino acids long and each containing two pairs of histidines and cysteines, which were ideally suited for binding a zinc ion. A computer analysis of the sequence of TFIIIA by Andrew McLachlan was fully consistent with Aaron's model[1]. At the same time, a group that included Aaron's former co-worker Ray Brown at the European Molecular Biology Laboratory in Heidelberg came up with essentially the same result based purely on a computer analysis of the sequence[2]. Somehow, this caused Aaron considerable unease. He stressed that *his* result was based on biochemical insight.

In his paper, Aaron proposed that each of the zinc binding domains could recognise a particular sequence of bases in intact DNA. Then the concatenation of zinc binding domains would bind to a number of adjacent sites along a DNA double-stranded molecule, thereby resulting in a very specific binding to a certain DNA sequence. On account of their putative gripping of the DNA, Aaron and Daniela Rhodes referred to these domains as 'zinc fingers'[3] (see Figure 16.1, from a more recent paper). The name stuck.

By looking at the places that showed variations among the nine fingers, Aaron tried to guess where the chemical distinctiveness would lie. Moreover, Aaron thought it was unlikely that such a good system would be wasted on a single gene in a frog and was not surprised when Herbert Jäckle in Tübingen discovered a second such motif in the fruit fly *Drosophila*. With a feeling of elation, Aaron realised that you might

---

[1] Miller, J., McLachlan, A.D. and Klug, A. (1986) *EMBO J.* **4**, 4170–4174.
[2] Brown, R.S., Sander, C. and Argos, P. (1986) *FEBS Lett.* **186**, 27–274.
[3] Klug, A. and Rhodes, D. (1987) *Trends Biochem. Sci.* **12**, 464–469.

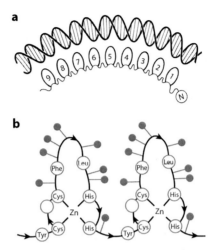

Figure 16.1 The zinc finger domains in TFIIIA, and their proposed structure.
a, TFIIIA contains a concatenation of nine zinc fingers that may bind to a sequence of adjacent sites on a DNA double-stranded molecule with high specificity. The region between the zinc fingers is unstructured and flexible.
b, A diagrammatic representation of the proposed structure of two adjacent zinc fingers. Pairs of cysteine residues (Cys) and histidine residues (His) coordinate a zinc ion (Zn), which constrains the connecting protein chain into a loop or 'finger'. The grey circles represent amino acid residues that differ between domains. The differences would allow each domain to bind specifically to a certain sequence of DNA. (From Klug, A. *Quart. Rev. Biophys.* 43, 1–21 © 2010, reproduced with permission.)

be able to string together a number of specific zinc fingers to recognise any DNA sequence you wanted. The more such fingers in the string, the more specific would be the binding. Such a string would be the ideal tool for modifying DNA sequences at the required place – the gateway to specific genetic engineering and gene therapy.

Subsequently, numerous laboratories confirmed the correctness of Aaron's concept. Gradually, more and more examples came to light. Now that the sequence of the human genome has been determined, it appears that up to 2% of protein encoded in the human genome might be of this type[4].

How do the zinc fingers recognise their target? To look at this in detail, the crystal structure of the complex was necessary. Initially,

[4] Tupler, R., Perini, G. and Green, M.R. (2001) *Nature* **409**, 832–833.

crystallisation of zinc fingers proved difficult. However, they were small and were therefore ideal objects for structure determination by nuclear magnetic resonance (NMR).

The crux of the NMR method is that the nuclei of hydrogen (and many other atoms) have an intrinsic magnetic dipole. On application of an intense magnetic field, the nuclei will line up with their magnetic dipole either in the same direction as the field, or the opposite direction. They can flip between these two states, and the flips can be detected by radiowave spectroscopy. If two hydrogen atoms are close together, then their nuclear spins tend to flip together. By identifying pairs of hydrogen nucleii with correlated flips, one can build up a list of distances between hydrogens in the molecule. With a lot of ingenuity, this information can be turned into an atomic model of the molecule. The method works best on small proteins. There is a great deal of enjoyable physics in the method, but until now NMR had not commanded Aaron's attention. Realising the method's relevance to zinc fingers, Aaron initiated NMR studies at LMB on two fingers joined together (Figure 16.2), since Aaron suspected this would be the minimum binding unit. In point of fact, the in-house NMR efforts were overtaken by Peter Wright's group in California, who were working with a simpler single zinc finger, which is easier to solve. Wright's results showed that the zinc finger consisted of a β-hairpin alongside an α-helix, all glued together by the tetrahedrally coordinated zinc and a hydrophobic core, a structure that had already been proposed on general structural considerations by Jeremy Berg at Johns Hopkins University in Baltimore. In bacteria, it was known that for DNA-binding proteins the binding is mediated by α-helices sitting in the major groove of the DNA. Thus it seemed very likely that the α-helix of the zinc finger would bind to the DNA.

In the absence of structural information about how the zinc fingers of TFIIIA bound to their target DNA, Louise Fairall, Daniela Rhodes and Aaron probed the accessibility of the DNA in the protein–DNA complex to chemical reagents and enzymes[5] (the contact points would be masked and unavailable to the enzymes). They found a series of contacts about five base pairs apart, or half a double-helical turn, in the major groove of the DNA along its entire length. They presented two

---

[5] Fairall, L., Rhodes, D. and Klug, A. (1986) *J. Mol. Biol.* **192**, 577–591.

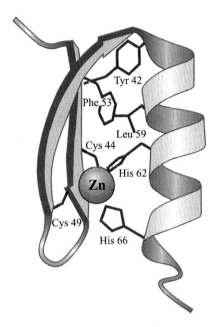

Figure 16.2 The structure of a zinc finger domain as revealed by NMR spectroscopy. The second finger of TFIIIA is shown, starting at residue 41. The domain consists of a β-bend (left) and an α-helix (right), held together by a zinc ion and a small hydrophobic core (tyrosine Tyr 42, phenylalanine Phe 53 and leucine Leu 59). The zinc ion is tetrahedrally coordinated by two cysteine side chains (Cys 44 and Cys 49) and two histidine side chains (His 62 and His 66). (From Klug, A. *Quart. Rev. Biophys.* 43, 1–21 © 2010, reproduced with permission.)

models for the overall geometry of the interaction: in model 1 the string of zinc fingers follows the DNA helix; in model 2 the zinc fingers stay on one side of the DNA. Aaron strongly favoured model 2. Subsequent crystal structures have only shown model 1.

Rhodes and her co-workers tried to crystallise a complex of a zinc finger protein bound to a piece of DNA, but in 1991 it was Nikola Pavletich and Carl Pabo in the Biophysics Department at Johns Hopkins who published the first crystal structure of a DNA oligo-nucleotide bound to zinc fingers[6]. The DNA was specifically manufactured to bind to a DNA-binding domain containing three zinc fingers from the mouse

---

[6] Pavletich, N.P. and Pabo, C.O. (1991) *Science* **252**, 809–817.

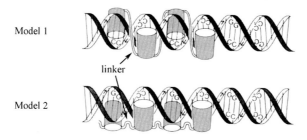

Model 1

linker

Model 2

Figure 16.3 Two possible modes of binding of TFIIIA to its target DNA. DNA is shown as two intertwined helices; the zinc fingers are depicted as cylinders. In model 1, the chain of zinc fingers follows round the DNA helix. In model 2, the zinc fingers all lie on one side of the helix. (Reprinted from Fairall, L., Rhodes, D. and Klug, A. *J. Mol. Biol.* 192, 577–591 © 1986, with permission from Elsevier.)

transcription factor Zif268. This showed, as anticipated, that the binding sites are on the α-helices of the zinc fingers. They bind in the major groove of the DNA. Moreover, three adjacent amino acids on each helix form hydrogen bonds with three adjacent bases on the DNA along the major groove, wrapping round the DNA as in model 1 in Figure 16.3. Thus each zinc finger binds a DNA triplet – a neat result consistent with the modularity of zinc finger proteins. The result suggested the possibility of a 'code': one particular zinc finger could bind to one particular triplet base sequence. Zinc finger motifs appeared to be the ideal natural building blocks for the design of proteins for binding to any particular sequence of DNA. However, Rhodes' group produced a structure of a two-zinc-finger peptide bound to a cognate DNA that revealed further complexity[7]. Not only did each zinc finger contact three bases with three amino acid side chains, but also a fourth side chain contacted the opposite DNA strand. This finding made life much more complicated because it introduced an interaction between neighbouring zinc fingers: the simple idea that one zinc finger equals one base triplet was modulated by cross-talk arising from this fourth side chain. The code for binding to DNA, if any, was not simple.

[7] Fairall, l., Schwabe, J.W., Chapman, L., Finch, J.T. and Rhodes, D. (1993) *Nature* **366**, 483–487.

Yen Choo, a graduate student and later co-worker with Aaron, following the method of Greg Winter, had the idea of producing random zinc finger sequences by using combinatorial chemistry and then using phage display (explained further in Chapter 18) to discover which DNA sequences they happened to bind to. In this way he planned to set up a zinc finger library[8]. In 1999, Yen Choo and Aaron founded a small company called Gendaq (dactyl is Greek for finger) in the MRC Technology building at Mill Hill, North London, for 'exploiting gene regulation technology'. Timothy Brears, a cofounder, became the CEO. Yen Choo decided that he would be happy to be a part-time scientific adviser. In fact, he soon moved to Mill Hill full time, together with Aaron's student Mark Isalan, who had just completed his doctorate on finding methods for improving the DNA binding specificity of zinc fingers. At this time, Aaron was President of the Royal Society. Furthermore, his son Adam was terminally ill in Beer-Sheva, with the result that Aaron was often in Israel. Thus he was not really able to take part in the day-to-day running of Gendaq.

Edward Lanphier, who was a business development manager, early recognised the therapeutic potential of zinc fingers. He founded Sangamo – a company aimed at the exploitation of the zinc finger technology – at Point Richmond near Berkeley, California. Carl Pabo, whose laboratory had first determined the structure of a zinc-finger: DNA complex, was associated with Sangamo. He and Lanphier wanted Yen Choo and Aaron to be on the scientific advisory board. Choo and Aaron were at first reluctant and thought they would see how far they could get with Gendaq; but Lanphier had steered Sangamo into a dominant position in the zinc finger field by acquiring all the relevant patents. Gendaq was small and underfinanced but had lots of useful know-how. It was inevitable that, in 2001, Sangamo should buy out Gendaq. Aaron and Yen Choo then joined the Sangamo board. Gendaq still continued doing research in London with a group of about 20 people.

During his presidency of the Royal Society and afterwards, Aaron kept a small research group at LMB with two postdoctoral fellows and a student working on correcting mitochondrial DNA with zinc finger

---

[8] Choo, Y. and Klug, A. (1994) *Proc. Natl Acad. Sci. USA* **91**, 11168–11172.

constructs. Monica Papworth demonstrated that you could switch off the growth of herpes simplex virus in a cell culture using a zinc finger designed to bind specifically to the virus DNA. The efficiency was too low to be of therapeutic significance, but it was a very interesting proof of principle. Moreover, Frank Lu, a postdoctoral fellow working with Barbara Searles (Aaron's technician), solved the crystal structure of a zinc-finger:RNA complex that revealed two modes of molecular recognition[9]. It showed that the RNA binding geometry was significantly different from DNA binding.

One way to find out what a particular gene does in a cell is to disable the gene and then see what happens to the cell. This is known as gene knockout, and is the most powerful tool for determining the function of a gene and also for modifying the phenotype of a cell. The problem is, how do you target some specific gene? Classically, you hit the cell with radiation that makes many DNA breaks at random and use genetic selection to find the rare cell in which just the gene of interest has been knocked out – which can be a lot of work. A very similar problem comes up in gene therapy. One needs to put the healthy gene into the correct site in the genome. Only then will it be properly integrated and properly expressed under control of the appropriate transcription signals. Could one really intervene in gene expression? After several years of diligent work, Yen Choo showed that it was possible to switch off a specific oncogene in a rat cell line using designed zinc fingers[10]. It was now clear that the system really worked. One could in principle target any DNA sequence, and therefore one should be able to regulate gene control, by attaching the zinc finger to an appropriate activation domain or a repression domain.

An important advance was made by Srinivasan Chandrasegaran and his co-workers[11] at the Johns Hopkins School of Public Health in Baltimore. The basis of their method is to couple a string of three zinc fingers to a DNA cleaving enzyme, FokI-EL, which then cleaves the DNA at a point close to the zinc finger binding site. This construct was known as a zinc finger nuclease (ZFN). To cut both the strands of the DNA at the same place you need a specific dimer, one molecule specific

[9] Lu, D., Searles, M.A. and Klug, A. (2003) *Nature* **426**, 96–100.
[10] Choo, Y., Sánchez-García, I. and Klug, A. (1994) *Nature* **372**, 642–645.
[11] Kim, Y.G., Cha, J. and Chandrasegaran, S. (1996) *Proc. Natl Acad. Sci. USA* **93**, 1156–1160.

for each strand. This requirement means that six zinc fingers have to recognise the target DNA sequence at the same time, increasing the specificity to a point where the ZFN dimer complex will bind and cut at only one specific sequence even when confronted with the whole of the human genome. On account of this high specificity, ZFNs did indeed appear to provide a way forward for gene therapy.

The use of ZFNs makes it possible to introduce genes at uniquely defined sites in the genome by cutting the DNA. The cell has enzymes for repairing double-stranded breaks in DNA. Researchers can use the cell's own repair mechanisms to incorporate the required new sequence of DNA at the break point. Confronted with a DNA double-stranded break, a set of repair enzymes uses the sequence of the sister chromatid (the other identical copy of the chromosome) to substitute for the damaged DNA sequence (homologous recombination). However, it does not have to be the sister chromatid: any DNA with the correct flanking sequences will do. Thus if one introduces the specific ZFN together with a modified DNA with the correct flanking sequences, it will be incorporated with quite high efficiency at the site of the double-stranded break, and the existing damaged sequences will be edited out.

In 2000, Aaron stepped down as President of the Royal Society and had more time to devote to Sangamo. He visited Richmond, California, quite frequently in the ensuing decade. This collaboration led to Aaron's last scientific publication. Engineering gene knockouts using zinc finger nucleases involves a lot of zinc finger design and genetic manipulation, and delivery of the ZFN into the target cell is always a problem. Thus it was late in the decade that the Sangamo team were able to demonstrate the successful knockout with high efficiency of a specific gene in a Chinese hamster ovary cell line (CHO)[12].

Aaron summarised the work at Sangamo in a review he wrote in 2010[13]. He reported that Sangamo had promoted zinc finger technology in a therapeutic context, thus fulfilling his vision. Unfortunately, it turns out that making zinc fingers is expensive, because their binding specificity is difficult to predict. Thus while the nuclease method of marking the DNA for incision and insertion of desired sequences remains very

[12] Santiago, Y., Chan, E., Liu, P-Q. *et al.* (2008) *Proc. Natl Acad. Sci. USA* **105**, 5809–5814.
[13] Klug, A. (2010) *Quart. Rev. Biophys.* **43**, 1–21.

important[14], alternative genome editing schemes have since been discovered (TALEN, standing for Transcription Activator-Like Effector Nucleases; and CRISPR, for Clustered, Regularly Interspaced, Short Palindromic Repeats, which uses the enzyme Cas9[15]). In particular, because it is based on a DNA/RNA hybrid recognition system, CRISPR/Cas9 (the system comes from bacteria) is much cheaper to use than zinc fingers and has become the method of choice for attempted gene therapy[16]. Neverthelelss, in an interview, Aaron said[17]:

> And I have to say that we don't work on zinc fingers because basically
> I discovered them, it's because they are the best system. I thought that in
> time other systems might arise but since the human genome is now known
> one sees there is no other candidate... The beauty of the zinc finger is that
> it has the combinatorial system and each finger, for the first
> approximation interacts only with three successive bases. And that is
> about as good as you can get and so that's the reason, it must be the
> biological reason, why 2% of the human genome uses zinc fingers.

[14] Baker, M. (2012) *Nature Methods* **9**, 23–26.
[15] see e.g. Gaj, T., Gersbach, C.A. and Barbas, C.F. (2013) *Trends Biotechnol.* **7**, 397–405.
[16] see e.g. Ledford, H. (2015) *Nature* **522**, 20–23 and www.nature.com/crispr
[17] Klug, A. (2001) interviewed by Tony Crowther and John Finch, 13–14 December 2001 at CUMIS Studio Cambridge, Tape 5 (© MRC Laboratory of Molecular Biology, Cambridge)

Part III

Duties and Rewards

17

Peterhouse

In the autumn of 1961, John Kendrew dined with the Klugs in Sutcliffe Close. Aaron had already signed his contract with the MRC and from the beginning of 1962 was often in Cambridge. Kendrew wished to interest Aaron in taking a teaching fellowship at his college, Peterhouse. Kendrew was Director of Studies in Natural Sciences in Peterhouse, but, alongside his job as Head of Structural Studies in the new Laboratory of Molecular Biology, he was also working for the Ministry of Defence and was finding it difficult to carry out his commitments. Aaron agreed to Kendrew's suggestion: he liked teaching, and the extra cash would be most welcome. Thus at the start of the Michaelmas term 1962, on John Kendrew's recommendation, Aaron was appointed a fellow of Peterhouse, the most ancient and smallest college in the University of Cambridge. It was founded in 1284 by Hugo de Balsham, bishop of Ely and granted its charter by Edward I. The land on which the college was built was purchased from a member of the flourishing community of Cambridge Jews who were thrown out of England in 1290.

Peterhouse had a traditional strength in history and engineering. Aaron's election to the fellowship strengthened the Natural Sciences. Aaron taught for another 23 years, and for most of this time he was Director of Studies in Natural Sciences. Being the Director of Studies involved organising the Entrance Scholarship exam, monitoring changes in Tripos regulations and a host of other tasks. He became in effect 'chief scientist', giving advice to the Master and the Governing

Body about all manner of problems. Aaron gave supervisions (tutorials for small groups of undergraduates) twice a week, Tuesdays and Thursdays. He taught second-year physics and became rather good at solving the 'Cavendish Problems in Classical Physics'.

Peterhouse is renowned for the quality of its dinners and for its excellent wine cellar. As a junior Fellow, Aaron was asked to be the 'Auditor of Wine'. This is part of the annual Audit of college finances, which takes place in November. The Wine Auditor goes down into the capacious wine-cellar with the Cellarman, and counts the number of bottles in each of the 70-or-so 'bins'. The task takes at least a whole day, but, having done the job, a new Fellow would have a very good overview of what was in the cellar; and could buy some fine wines from the cellar at the price the college had paid for them.

Aaron had a college 'set' (a bedroom plus living or 'keeping' room) at C4 Old Court, on the first floor. A round table that Aaron used for teaching stood in the middle of the room. On the wall to the left of the entrance are two windows that look south into Old Court at the Combination Room and Dining Hall. In summer, the window ledges were graced with geraniums. Between the windows was a bookcase holding the *Proceedings of the Royal Society* that Aaron received every two weeks. During the course of three decades, this amassed knowledge became more than the bookcase could comfortably sustain, causing it to take on a trapezoidal form. On the wall opposite the door was a bookcase holding textbooks, including a very annotated version of R. W. James's *Optical Principles of the Diffraction of X-rays* which Aaron had proofread way back in his Cape Town days. Adjacent was a door leading to a small bedroom containing a simple iron bed. Aaron never slept on it, but Liebe did borrow it once as a prop in a play she was directing.

The north wall has a fireplace in the centre. To the right of the fireplace hung a large and rather menacing portrait of Peter Guthrie Tait, which had previously hung in the college Hall. It was evidently provided by John Kendrew, who was keeper of the College's pictures; his set C3 was just across the hallway[1]. Tait had been the Senior

---

[1] Kendrew founded the *Journal of Molecular Biology*, and in its early days the journal was run from C3 Old Court. The juxtaposition resulted in Aaron getting many manuscripts to edit. Aaron was an editor of the *Journal* for 50 years.

Wrangler – the top mathematics undergraduate – at Peterhouse in 1852 and was elected to the Fellowship. He left to be the professor of mathematics in Belfast where he became acquainted with quaternions, invented by the great Irish mathematician William Hamilton. Quaternions are an extension of complex numbers into four dimensions that can be used to represent 3D rotations. Tait became a proselytiser for quaternions. Although they were superseded by vector algebra, they are nowadays back in fashion and are much used in computer graphics to represent 3D rotations. Tait had an impressive bald head and a stern visage; Aaron thought Tait's mien would steady the minds of his students as they grappled with problems in analytical geometry.

A window to the left of the fireplace looks out over the west end of Little St Mary's Church, which had earlier been dedicated to St Peter and had lent its name to the college. Rosalind Franklin's sister, Jenifer Glynn, lives in Cambridge. Her husband, Ian Glynn, was Professor of Physiology and is a fellow of Trinity College. Their daughter Sarah, then aged 16, was impressed by the view out of Aaron's north-facing window: she captured the essence of the view from his desk in a watercolour (Plate 4).

At the time of Aaron's appointment, the Master of Peterhouse was Herbert Butterfield, a historian who had made his name by criticising the Whig Interpretation of History. Butterfield, who was Regius Professor of Modern History and was a Methodist of strictly Nonconformist leanings, took a very jaundiced view of any movement that might be termed liberal. He wrote a book on the history of science and was responsible for getting History of Science accepted as a half-subject in the Natural Sciences Tripos. One of his appointments to the Fellowship was the historian Maurice Cowling, who took Butterfield's right-wing philosophy to new heights. Cowling's avowed mission appeared to be the rolling back of the Enlightenment. Another notable appointment was Kingsley Amis, who did not stay in Cambridge for very long: his was only a college appointment and he discovered that he had little to say in the English Department.

Initially Aaron's social associations with the college Fellowship were limited. However, in the Easter Term of 1963, Aaron invited Huw Wheldon to join him at a college feast. Kingsley Amis knew Wheldon, and this – after the feast – resulted in their repairing to Amis's house in Barton Road for an exchange of views and a jolly alcoholic evening.

Aaron was not bothered by Amis's iconoclastic views, but Amis's failed marriage led to his departure later that summer before their friendship could deepen.

Aaron's firm friend in the early days at Peterhouse was Sir Michael (Munia) Postan – a medieval economic historian of international standing and for many years the Professor of Economic History in Cambridge. Postan was a Moldavian Jew who, in the turbulent years following the First World War, moved between various countries in search of stability and scholarship; he settled in London before becoming a Fellow of Peterhouse in 1935. Somewhat to the envy of his colleagues, he could speak all European languages fluently and express himself in wonderful lucid English. Aaron admired his consummate scholarship. Perhaps more important, he and Aaron could swap Yiddish jokes for which Aaron had a special weakness. Much to Aaron's regret, Postan died in 1981.

Butterfield was succeeded as Master in 1968 by J. Charles Burkhill, a mathematician of note and Fellow of the Royal Society, who had long been Senior Tutor of the College. His main mathematical subjects of interest were the theory of integration, and Fourier series, which should have interested Aaron. Burkhill was actually older than Butterfield, and his appointment caused Butterfield some consternation. Burkhill came to office at a time of student protest including the infamous Garden House riots. He managed rather astutely to defuse the situation by setting up machinery for consultation with the students without altering the statutes of the college. Brevity and conciseness were characteristic of Burkhill's interchange: his silences were eloquent. Burkhill's formidable wife Greta was famous for aiding refugees. Moreover, she furthered the Cambridge Graduate Society and found them premises next to Peterhouse.

After Burkhill's retirement in 1973, Grahame Clark was elected Master. Clark, who had been an undergraduate and later Fellow of the college since 1927, was the Disney Professor of Archaeology from 1952 to 1974. An authority on the Mesolithic settlement of northern Europe, he was an austere scholar often regarded with trepidation by his colleagues. Another elderly Master, he was cautious and made few new appointments, with the result that the Fellowship shrank. Political power in the Fellowship was increasingly wielded by a group of bachelor Fellows who dined together every evening. This group included the historian

Maurice Cowling, the mathematician Hallard Croft, the art historian David Watkin and the Dean Edward Norman. They were of a misogynous bent and were particularly concerned that the college should not admit women. The Governing Board had come within a whisker of changing Statute 1 to admit women in 1970, after which the misogynists took steps to afforce their numbers. Cowling, the great manipulator, recruited Norman in 1971. Moreover, Ed Shils, also against women, was elected in 1971.

On Clark's retirement, Cowling, playing the *éminence grise*, proposed Hugh Trevor-Roper for the Mastership. Trevor-Roper appeared to be a safe conservative pair of hands. In 1980, he retired from being Regius Professor of History in Oxford, and he thought that a term as Head of House might give him time to complete his book on the Puritan Revolution that had lain unfinished for a decade. Also, the Master's Lodge, a Queen Anne residence thought to be the best Lodge in Cambridge, did have its attractions.

Peterhouse was not unknown to Trevor-Roper[2]. In 1965 he had given the Trevelyan Lectures in Cambridge and on two occasions had stayed with Butterfield in the Master's Lodge. Even as far away as Oxford, Trevor-Roper had heard of the 'hom. party' (gay party) for which the college had become notorious. Stories were told of cross-dressing parties, of undergraduates and dons using girls' names. Armed with a certain Oxford arrogance, Trevor-Roper approached the job in the spirit of a colonial governor who had been sent out into the Fens to clean the place up. Despite Postan's recommendation of the Fellowship as a friendly bunch, the college was clearly a rum place. Trevor-Roper, whose middle name was not 'Tact', soon earned the disapprobation of the bachelor fellows. This induced Postan to comment: 'They are such fools, they thought they were electing a Tory and never realised they were electing a Whig.' Trevor-Roper was a great admirer of Edward Gibbon and his enlightened historicism. He felt no kinship with Cowling's mantra: *tackle liberals with irony and malice*. Without teaching and faced with a hostile mafia, he felt isolated. Slowly Trevor-Roper realised that there were some interesting and convivial members of the Fellowship; but they came not from the Historians but rather from

---

[2] Anecdotes pertaining to Hugh Trevor-Roper's Peterhouse adventures have been gleaned from Adam Sisman's *Hugh Trevor-Roper, the Biography*. Weidelfeld & Nicolson (2010).

Engineering and Natural Sciences. Indeed, there were two distinct groups: the lunch-time Fellows, and a half-dozen arch-conservative High Table dinner Fellows. Although Aaron was clearly a member of the former, the Laboratory of Molecular Biology was too far out of town for him to make lunch in college possible on a daily basis. Moreover, he did not dine in college regularly. Thus it took a little time for Aaron and Trevor-Roper to establish a rapport. Aaron's breadth of knowledge commended itself to Trevor-Roper: here was someone with whom you could really discuss things. Aaron admired Trevor-Roper's skill in formulating and writing elegant prose. The pair became friends. Moreover, Trevor-Roper struck up an enduring friendship with Max Perutz that continued until Perutz's demise.

About this time, Aaron recruited Tony Crowther to the Fellowship to teach 'Mathematics for Natural Scientists'. Physics students needed a working knowledge of matrix algebra, variational calculus, statistics and Fourier transforms just to understand what was being talked about in their lectures. Until this time, the 'pure mathematicians' – particularly Adrian Mathias – had taught this subject. They regarded it as 'easy', but Mathias was a specialist in mathematical logic and set theory, and Aaron found that the students were not being taught well. Crowther was a professional in mathematics applied to physics. He proved to be a great asset.

One of Aaron's firm friends at Peterhouse was Chris Calladine, who became Professor of Engineering. Calladine's primary interest was in the stability of thin-shelled structures and why they buckled, which found plenty of practical application in pipelines. Aaron interested him in biological phenomena, particularly at the molecular level. As mentioned in Chapter 15, Calladine wrote authoritative papers on how bacteria swim[3]. Later, he became interested in the mechanics and stability of the DNA molecule and contributed to an influential textbook[4]. Aaron and Calladine noted that over the years there had been a serious drop in the quality of student applications to Peterhouse. They decided to put this problem to the Governing Body.

The Governing Body of the college is the Master and full Fellowship excluding research fellows. For most of Aaron's time there, this was

[3] Calladine, C.R. (1974) *Nature* **249**, 385; and Calladine, C.R. (1975) *Nature*, **255**, 121–124.

[4] Calladine, C.R., Drew, H.R., Luisi, B.F. and Travers, A.A. *Understanding DNA: The Molecule and How It Works*. Third edition. Elsevier Academic Press (2004)

about 25 people. They met about four times a term on Mondays in the elegant Combination Room situated on the south side of the Old Court. The room is oak panelled and has stained-glass windows and tiles around the fireplace designed by William Morris and colleagues. Lord Kelvin, an alumnus of the college, had electric lighting installed in the college in 1884. The lighting in the Combination Room comes from small electric pseudo-candle lamps on the walls, which emit realistic candle-like illumination. Real candles grace the tables. Despite Lord Kelvin's magnanimity, the illumination is not overwhelming; Aaron wondered why they bothered to publish an agenda, since he was never able to read it. Meetings of the Governing Body played an important role in the social life of the College. After the meeting one dined in Hall. There was pre-prandial champagne, good food, excellent wine; everyone had an enjoyable time. Moreover, the Governing Body discussions, held under the chairmanship of the Master, were usually at a civilised level, which was not always the case at High Table.

At a meeting of the Governing Body in February 1982, Aaron and Calladine proposed that women should be admitted to Peterhouse. They noted the decline in the number of good students showing up in the Entrance Scholarship competition in the previous few years. Calladine produced a graph that showed the linear reduction over time of the number of entrance scholarships that were awarded – a line that would hit zero in about three years' time. Aaron and Calladine argued that this reflected the fact that Peterhouse was no longer able to compete with mixed-intake colleges. The proposal was not well received by the Governing Body and failed to attain the necessary two-thirds majority at the ensuing Statutory Meeting of the full Fellowship.

The Fellows' concern was apparently shared by the college staff. Three times a year, the Natural Science Fellows organised dinners in the Combination Room for their students and the external supervisors. Since these were private dinners, wives could be invited. On one such occasion, Liebe dared to ask Mr Moffett, the butler:

'What do you think about accepting women?'

He stopped serving, looked Liebe straight in the eye, and said:

'Begging your pardon, Ma'am. I don't think much to it.'

One year later, the same proposal was put to the Fellowship by the Master, this time with the support of Cowling, and it passed with but one dissenting voice. Now Statute 1 of Peterhouse admits that even women would benefit from an education at the college.

On Monday 11th October 1982, Aaron heard that he had won the Nobel Prize for Chemistry. Tony Crowther, newly elected Fellow, was in the Laboratory of Molecular Biology at the Annual Lab Talks as Aaron received the news. Crowther hurried off to college to a meeting of the Governing Body and passed the information to the Master. The Master's announcement was greeted with applause, a serious departure from tradition for this august body. Two days later, Aaron organised a pre-prandial champagne party, and everyone agreed that Aaron was a great guy and that the champagne was excellent. Later in the Michaelmas term, a celebratory dinner was held in Aaron's honour.

A few months later, Hugh Trevor-Roper became involved in a fiasco that seriously compromised his reputation. As a young man he had made his name with the book *The Last Days of Hitler*, which remains the authoritative work on the subject. As a consequence, he was often consulted about the authenticity of material ostensibly related to Hitler. In April 1983, the *Sunday Times* was offered the chance of serialising what purported to be 'Hitler's Diaries', which had been uncovered by Gerd Heidemann[5], a reporter from Hamburg, and which were to be serialised in the German weekly magazine *Stern*. Rupert Murdoch, owner of *The Times* and *Sunday Times*, got wind of the deal, and Trevor-Roper was hustled off in a hurry to inspect samples held in a Zurich bank vault (Trevor-Roper was a Director of Times Newspapers). *Stern* informed him that the paper, ink and handwriting had all been authenticated by acknowledged experts, but in point of fact none of this had happened. Although he had serious doubts about some of the content, Trevor-Roper had to admit that the handwriting looked genuine and that most of the content was as banal as he would expect. He was inhibited from consulting with other experts by a pledge of secrecy. Moreover, he was under pressure to give a quick telephone appraisal, so he reported that the Diaries *could* be genuine. In an ensuing visit to Hamburg, he met Heidemann and saw a large collection of Hitler

---

[5] Heidemann had acquired the 60 volumes of the Diaries on behalf of *Stern* magazine for $5 million from Konrad Kujau, an antiques dealer and painter.

memorabilia in his apartment, which he said had all come from 'the aeroplane' carrying Hitler's belongings and the diaries, which was purported to have crashed in Bavaria.

The *Sunday Times* then got wind of another newspaper's plan to publish first. Trevor-Roper was prevailed upon to write a piece for the Saturday edition of *The Times*, the day before the publication in the *Sunday Times*. Early on the Friday, he wrote his piece and sent it off to London by motor-cycle courier. While the motor-cycle was en route, he began to have serious doubts. It dawned on him that the aeroplane would have needed to be enormous to hold all the memorabilia he had seen in Hamburg: it must all be a fake. So he telephoned *The Times* to urge caution, but it was too late to stop the presses rolling[6]. The following week, *Stern* organised a televised press conference in Hamburg, where Trevor-Roper appeared self-contradictory and confused. After that, the whole episode began to unravel. By the time Trevor-Roper returned to Cambridge, it was clear that he had been duped.

As soon as the press got an inkling of this, Trevor-Roper was besieged by paparazzi. He had to exercise considerable ingenuity in getting from the Master's Lodge across Trumpington Street to the college office. On one occasion, Aaron happened to have his car in the car park at the back of the Lodge, and there was a tapping at the window. He was prevailed upon to drive round to the former Deer Park at the back of the college with Trevor-Roper hiding on the back seat.

The Hitler's Diaries episode did nothing to enhance Trevor-Roper's authority with the Fellowship. Cowling, who on more than one occasion had suffered from Trevor-Roper's biting criticism, gloated.

In 1984, Peterhouse had existed for 700 years. To mark the occasion, there was a meeting of the alumni, and the Master's wife, Lady Alexandra Howard-Johnston (Xandra), organised a concert. At a meeting of the Governing Body, Trevor-Roper proposed a series of five commemorative lectures to be held in the Lecture Theatre of the New Library, which had recently been refurbished. The general title of the series was 'Aspects of university culture in seven centuries'.

---

[6] Some of this account comes from Chris Calladine. After the event, over lunch in college, Trevor-Roper would talk freely with the Fellows about the episode. Adam Sisman's biography of Trevor-Roper gives a very comprehensive account that differs in detail from some of Calladine's recollections.

He suggested that the college historians might like to contribute a lecture, but they declined. Aaron, who spoke about the cultural background contributing to the rise of Natural Sciences in Cambridge, was the only College Fellow to contribute. Trevor-Roper himself contributed two lectures.

When Trevor-Roper retired he commented, 'Seven wasted years.' In 1987, the church historian Henry Chadwick was appointed as his successor. Chadwick had been Dean of Christ Church Oxford; it is believed that he is the only person ever to have been Head of both an Oxford and a Cambridge college, and also a Regius Professor (of Divinity) in both universities. Chadwick had translated works of Origen and had written a book on Augustine. Thus he commended himself to Aaron, who had an interest in the early church dating from his Cape Town interchange with the ethics professor, Martin Versfeld. Chadwick's Mastership in Cambridge seems to have been for him a more positive experience than his Deanship in Oxford. He continued the reforms initiated by Trevor-Roper. He insisted on civility in the college and removed one dissenting Fellow. The atmosphere at High Table improved.

Aaron stopped teaching after 1985 when he became Director of the Laboratory of Molecular Biology. Nevertheless, he continued his association with Peterhouse as an Honorary Fellow. The college threw a garden party to mark his appointment to the Order of Merit. When Chadwick retired in 1993, Aaron was sounded out by the Fellows to see if he would consider letting his name go forward for election as Master of Peterhouse. Aaron felt he was already overcommitted and sadly declined. John Meurig Thomas, a surface chemist of international renown and formerly Director of the Royal Institution in London, was elected to the Mastership instead.

Edward Shils had been a Fellow of Peterhouse since 1971 and was often in Cambridge, although his official appointment was at the University of Chicago. He was a sociologist of note and an authority on Max Weber, one of the founders of the subject and originator of the phrase 'Protestant Work Ethic'. Shils started out in Cambridge as visiting Professor and Fellow of King's College under the Provostship of Noel Annan. In 1963, he founded the journal *Minerva*, which is devoted to the study of ideas, traditions and cultures in science. *Minerva* was partially financed by the CIA. Edmund Leach, who succeeded Annan as Provost of King's, found the CIA contact incompatible with his view of the King's Fellowship and urged Shils to move on. He came to rest three blocks down the road at Peterhouse. Shils found the relatively anarchic Fellowship of Peterhouse more to his taste and became an enthusiastic member and benefactor of the college. Aaron and Shils became friends. Shils died in January 1995, and Peterhouse organised a memorial service in Little St Mary's. This might have been thought unusual because Shils was Jewish and Little St Mary's is decidedly High Church with incense and bells. To redress the balance, Aaron recited the Kaddish. Hugh Trevor-Roper was an admirer of Shils, and he gave the laudation. One paragraph from the laudation[7] says much about the college:

> Surely Peterhouse was ideally suited to Shils. A small college, a "primary group" held together not by the distant authority of its Master but by tribal cohesion at a lower level, a society tenacious of tradition and yet containing within itself forward-looking scientists of the greatest distinction. Surely they were made for each other – at least if Peterhouse was viewed from outside, from Provost Leach's Abbaye de Thélème down the road. Internal experience may not have entirely confirmed external views: it seldom does; but certainly Shils became devoted to Peterhouse. He showed his devotion by staying there after his retirement and by great practical generosity. He also inspired devotion in others. In his last illness, the former college butler, Mr Moffett, his neighbour in Tennis Court Terrace, insisted on flying to Chicago to look after him, as he had done so often before.

Peterhouse prospered under John Meurig Thomas and his successors. The return to successful normality did not please everyone: Roger Scruton, a research fellow in 1970 and later editor of the *Salisbury Review*, maintained that 'Peterhouse was a great institution, but probably is no longer – it was gratuitously destroyed by the admission of women.'[8] Even so, it is now generally agreed that the Peterhouse Fellows are a friendly bunch and that the food at High Table is still excellent. The rancour induced by Cowling and his acolytes is no more.

---

[7] Trevor-Roper, H.R. (1995) *New Criterion* **14**, 77–80.

[8] Mullan, J. and Foden, G. Peterhouse blues. *The Guardian* (10 September 1999); http://www.theguardian.com/theguardian/1999/sep/10/features11.g27

Figure 17.1 A composite photo of the speakers at Aaron Klug's 70th Birthday Meeting. Shown are: Ken Holmes, Tim Richmond, Uli Laemmli, David De Rosier, Tim Baker, Aaron Klug, Steve Harrison, Roger Kornberg, Michael Levitt, Alfonso Mondragon and Hillary Nelson. (© MRC Laboratory of Molecular Biology.)

In 1996, on 9th September, Tony Crowther, Daniella Rhodes, John Finch and Jo Butler organised a one-day meeting at Peterhouse to celebrate Aaron's 70th birthday. Speakers were selected from attendees not coming from Cambridge. The theme of the meeting was 'Understanding Macromolecular Structures', which was an apposite summary of Aaron's life. The day culminated in a splendid meal in the Combination Room presided over by the Master, John Meurig Thomas. Not only was this a very successful scientific meeting, it also made a deep impression on Aaron.

18

Director of the Laboratory of Molecular Biology

In 1970, Ted Heath's newly elected government asked Victor Rothschild, then Research Director of Shell, to investigate the basis of government research funding. Rothschild's report, *A Framework for Government Research and Development*, issued in 1972, recommended a 25% cut in the basic funding of the Research Councils. This money should be transferred to government departments and allocated on the basis of a customer–contractor relationship between the Research Councils and the government.

While this might be a reasonable strategy for applied research, there were fundamental reasons why it was a bad idea for basic research. These had been formulated half a century earlier as the Haldane Principle[1]: decisions about how to spend research funds should be made by research scientists rather than politicians. Haldane felt this rule (among other reasons) was necessary to avoid politicians indulging in funding for their own hobby-horses. Alan Hodgkin, in 1972 President of the Royal Society, took issue with Rothschild's report, arguing strongly that for pure research, scientists were the best judges of what should be done in science. Nevertheless, the Rothschild report left its mark. It determined how Margaret Thatcher's government thought about

---

[1] Richard Burdon Haldane, a senior minister in Asquith's Government, in 1904 and from 1909 to 1918 chaired committees and commissions that recommended this policy and founded the Research Councils. He was uncle to J. S. Haldane and Naomi Mitchison.

science funding[2]. It impinged on the Laboratory of Molecular Biology in two ways: money became more difficult to obtain; and the *laissez faire* attitude to scientific direction so successful during Perutz's period as chairman was no long accepted as the norm.

In 1974, Max Perutz was 60, the official retiring age for directors of Medical Research Council Laboratories. On the occasion of Perutz's reaching the official retirement age, the MRC commissioned a report on the LMB from a committee chaired by David Phillips. In 1976, David Phillips's committee produced a hostile report on the LMB, demanding a 25% cut in funding and staffing, and recommending a reorganisation of the management structure by strengthening the role of the Director (Perutz had always eschewed this title and had referred to himself as chairman of the board). Furthermore, it decreed that the laboratory should work more closely with the University departments. Now was scarcely the right time to relieve the helm. Perutz petitioned the MRC to extend his directorship for five years. With some relief at having avoided a difficult decision, the Council agreed to extend Perutz's term of office until 1979 when he would be 65. The Council added that in the intervening years, Sydney Brenner should prepare to take over. He immediately took over the financial management of the laboratory and spent several years in trying to get the finances on a proper basis. Money was tight, and not enough attention had been given to the rising costs of research. From 1977, Brenner was to carry the title Proleptic Director (Brenner himself suggested that the title epileptic director[3] would be more appropriate).

Brenner officially took over from Perutz on 3rd September 1979. Most unfortunately, a couple of days earlier he had been knocked off his motorbike, resulting in a badly damaged leg. Thus he started his directorship from a hospital bed at Addenbrooke's Hospital, on the other side of the road from the LMB building, and this was where he spent most of his first year as Director. Brenner did not allow his immobility and considerable pain to stand in the way of administering the laboratory. Frances Taylor, the administrator, and Margaret Brown, the Director's secretary, made frequent visits. Michael Fuller, the laboratory manager,

---

[2] Agar, J. (2011) *Notes Rec. Roy. Soc.* **65**, 215–232.

[3] I have gleaned much of the following from Errol Friedberg's informative book, *Sydney Brenner, A Biography*. Cold Spring Harbor Laboratory Press (2010)

could be called for audiences as appropriate. Brenner was able to check with his co-workers on the progress of his nematode project.

But there were problems. The lab had a serious overdraft[4]. These were times of hyperinflation in Britain. Nor was the mood at the MRC Head Office quite so benevolent as in Harold Himsworth's golden days. Sir James Gowans was now secretary of the MRC. As a result of the Rothschild report, the Research Council Chiefs found themselves fighting for funds in competition with the University Grants Committee. Added to all this, as an exception to the MRC rules, Perutz still had a very active research group in the lab and was omnipresent. It appears that even Gowans could not quite bring himself to tell Perutz[5] that it was time to go and tend his garden.

Thus Brenner's directorship got off to an inauspicious start, and it seems to have headed downhill from there. In contrast to Perutz's apparently *laissez faire* management, Brenner took the title of Director seriously. While generally supportive of young people, his manner with senior colleagues left something to be desired. He had a depressing effect on the lab morale. His relationship with Hugh Huxley, the deputy director, was not good. They had a long-standing dispute over who had discovered 'negative staining' in electron microscopy, which neither had forgotten. In 1977, Francis Crick left Cambridge lured by the warmth, comfort and remuneration of Southern California, leaving Brenner feeling rather alone with no one to parry his ideas. On occasions when Huxley and Brenner needed to meet to discuss lab matters, Brenner could pontificate, which Huxley found tiresome. However, alongside such tribulations, apparently Brenner's most heinous crime was to reserve a large chunk of space in the overcrowded lab for a 'Director's Section'.

Brenner's directorship nonetheless saw two major scientific developments, both of which posed questions for Aaron to face when he succeeded Brenner in 1986. From 1962 on, it had seemed to Brenner

---

[4] Brenner explained to the lab that the financial problems were serious but that he was not yet recommending the use of both sides of toilet paper.

[5] In this period, the LMB benefited considerably from a donation of $500,000 that Perutz had negotiated with Thomas C. Usher, a US philanthropist who had made his fortune through the development of dextran-based blood substitutes and was indebted to Perutz for advice he had received many years previously. See: http://www.legacy.com/obituaries/vancouversun/obituary.aspx?n=thomas-c-usher&pid=3235846&fhid=5845

and Francis Crick that the ramifications of the genetic code work were fast being exploited by *hoi polloi*. What should be the proper pursuit of *hoi oligoi*? The nervous system? Crick wanted to go for the brain. In contrast, Brenner, bearing in mind the great success of the genetic analysis of bacteriophages, sought an animal in which one might correlate induced genetic abnormalities with behaviour. What Brenner wanted was a simple organism in which one could could watch development cell by cell. It should be small and transparent, and it should reproduce fast enough to allow genetic analysis. After much deliberation, he hit on the nematode roundworm *Caenorhabditis elegans*.

Brenner hoped that he would be able to see direct links between the worm's genes and its development from egg to adult. First, the wiring diagram of the worm's nervous system was to be established by identifying every nerve cell and plotting its connections (at the cellular level, all the worms are identical: they contain 959 cells, excluding the gonads). This was done by serial section electron microscopy. John White was hired to computerise the whole process and ended up designing a confocal microscope (more on this later). Assisted by a series of mostly American postdocs, Brenner was initially very successful in isolating mutants and mapping them onto the six chromosomes of the worm. Mapping mutants into behaviour did not turn out to be as straightforward as hoped, but *C. elegans* was a great system for studying development. Brenner was joined by John Sulston and Bob Horwitz. Together they mapped out the whole developmental process from the egg to adult animal. They showed how important specific programmed cell death (*apoptosis*) was on the developmental program. In the unfolding growth of the animal, at certain precise moments certain cells had to die. If this went wrong (mutants inhibiting cell death were found) then the development went wrong. This work earned John Sulston, Bob Horwitz and Sydney Brenner the Nobel Prize in 2002.

In 1975, César Milstein, then a senior staff member at the LMB, together with Georges Köhler, a visiting scientist from Freiburg, showed that one could fuse a mouse B-lymphocyte cell secreting a certain antibody with a human cancer cell to produce an immortal hybrid cell that would secrete the antibody. These hybridoma cells grew *in vitro* to yield the antibody (a protein molecule) in good yield. This notable discovery was honoured with the Nobel Prize in 1984. It was the beginning of a series of developments made at the LMB involving the production of antibodies that led to a sea change in the attitude of the MRC to patenting and the commercial exploitation of discoveries made within MRC-funded laboratories[6]. Finally, it led to the LMB generating, for a period of over 20 years, more income in licence revenue than the lab cost. However, Milstein and Köhler did not patent their discovery and were somewhat appalled when, two years later, the Wistar Institute in Philadelphia was granted an American patent on their method. It was thought that the patent would not be approved because it depended on known facts (*prior art*), and indeed it was not granted in Britain. Patenting and licensing was handled (or in this case mishandled) by a government body, the National Research Development Corporation (NRDC). The MRC was required to channel commercial activities through the NRDC.

Clearly something had gone wrong. Criminations and recriminations reverberated through the Research Council and Ministries. A Government enquiry led to the Spinks report that unfairly exonerated the NRDC and laid the blame at the feet of the scientists. In point of fact, most research scientists in the LMB were not enthusiastic about patents, since they felt that public-sponsored research belonged in the public domain (through publication). This began to change.

Brenner was involved in wide-ranging discussions about how to improve the LMB's income and its control over patents and licences, and spent a considerable amount of his life in Whitehall. On account of a 25% cut in MRC funding, a search for alternative sources of funding including possible contract work for external drug companies was becoming imperative. In line with the Thatcher government philosophy, the NRDC lost its exclusive rights to inventions arising in the public sector. With the active support of the MRC, including Brenner and the LMB, a new firm, Celltech, was founded in 1980 as the first European biotechnology firm. Celltech was initially endowed with exclusive rights to exploit and market findings arising in MRC-funded laboratories on condition that part of the royalties would flow back into research[7]. Gowan appears to have been the architect of this agreement. This was something new. On account of hybridoma technology, Celltech was successful right from the start.

[6] For a very informative history and analysis see: de Chaderevian, S. (2011) *Isis* **102**, 601–633.
[7] The NRDC pocketed all royalties, which flowed back to the Treasury.

In Perutz's time, the Director's (Chairman's) office was on the first floor in the middle of the Structural Studies division. Next door were the offices of the administrator and the Director's Secretary, Margaret Brown. When Brenner took over in 1979, he created a Director's Office Suite (rather like the Chairman's Suite in some American Universities) on level 5 of the new adjoining MRC building (Block 7)[8] and moved Margaret Brown up there. Here in 1983 he was joined by Bronwen Loder, as Assistant Director (administration), seconded from MRC Head Office in London. Loder held a PhD from Melbourne University and was a skilled and experienced scientific administrator. She was also put in charge of the MRC centre in Block 7, an administrative group that provided services to all the MRC units in Cambridge. Brenner and Loder got along very well. Brenner at last had someone to talk at, and spent considerable periods up on level 5 bouncing ideas off Loder, which led to the perceptive comment from Jenny Brightwell, the secretary of the Structural Studies Division, 'Oh, they're upstairs playing Kings and Queens.'

Loder's executive efficiency may have led to her downfall. Brenner threw out ideas without much filtering: one needed to sort the wheat from the chaff. Loder was perhaps over-zealous in implementation. Whatever the case, in September 1985, something that Bronwen Loder said or did upset the Secretary of the MRC, Sir James Gowans, to such an extent that he summarily dismissed her from MRC service[9]. Brenner was already irritated with Gowans over not getting his agreement to opening a new neurobiology division at the LMB. After the Loder episode, Brenner told Gowans he would not seek an extension of the directorship beyond his 60th birthday in 1987. Without Loder, the administrative load was becoming tiresome. Moreover, he did not like the job. Apparently the feelings of irritation were mutual. Dai Rees, Gowans' appointed successor as Secretary of the MRC, indicated to

---

[8] In 1980, Block 7 was built by the MRC at right angles (pointing north) to the main LMB building near the east end. It was not primarily intended for the LMB although the LMB gradually leaked into it.

[9] Cambridge gossip has it that Brenner had been negotiating with an American pharmaceutical company to offer exclusive rights on LMB Intellectual Property in return for a considerable quantity of cash. In anticipation of this cash flow, Brenner ordered a much needed main-frame computer for the lab that was authorized by Bronwen Loder. This far exceeded her authority, and when Gowans became aware of the episode he was not amused.

Brenner that he should resign forthwith. Nevertheless, Brenner had strong ideas about his successor, whom he decreed should be an external appointment. Despite Brenner's opinions, in June 1986 the MRC recommended that Aaron Klug should apply for the post. Thus, in August 1986, Aaron became the third Director of the LMB.

Sydney Brenner and Aaron Klug share many attributes. They are of similar age and height, and both grew up in South Africa. Both came from Jewish immigrant families originating in Lithuania. Both were child prodigies able to read practically before they could walk. Both attended Medical School at Witwatersrand as 16-year-olds. Both are brilliant intellects with encyclopaedic memories. As scientists they are both right at the top of their fields. But there the similarity ends.

Brenner is a consummate lecturer and performer with a wonderful sense of timing. He enjoys playing the *enfant terrible*. With the sharpest of wits, he is renowned for his one-liners, usually fiercely accurate and generally not flattering, but he needs the limelight. One of Oscar Wilde's epigrams seems appropriate:

> *There is only one thing in the world worse than being talked about, and that is not being talked about.*

In contrast, Aaron appears self-effacing and eschews displays of emotion. He is not a brilliant lecturer, nor is he great with anecdotes. But he is a talented teacher and manager who can direct and inspire his co-workers to perform at a very high level. After seven strenuous years with Brenner's nervous energy, the LMB welcomed a more even-handed boss. In August 1986, Aaron marked the occasion with a garden party on the Fellows' Lawn at Peterhouse. Although Francis Crick now lived and worked in California, he turned up nonetheless. The event was also a goodbye party for Hugh Huxley, who was moving to Brandeis University in Massachusetts. In 1984, Huxley reached the age of 60. An extension of his contract beyond 65 would have needed Brenner's approval, which was not forthcoming, so Huxley had taken a position at Brandeis to which he moved in October 1987.

Aaron reintroduced a Perutzian management scheme – with an open office door. As soon as possible, he got rid of the Director's Suite on the fifth floor of Block 7 and re-established the Chairman's office on the first floor. This was an essential move for Aaron because he was still leading a very active research group working on chromatin structure and zinc fingers in the Structural Studies Division. Richard Henderson and Nigel Unwin took over the direction of the Structural Studies from Aaron and Hugh Huxley, with an intervening year in which Henderson and Huxley were joint heads. Unwin arrived in 1987 as Huxley departed for Brandeis.

On his retirement from the directorship, the MRC granted Brenner a small Unit for Molecular Genetics in the Department of Medicine across the road from LMB. Here Brenner started to do the genomic DNA sequence of the pufferfish, chosen because its genome was only a tenth of the size of the human genome. This work was also being carried out in parallel at Brenner's lab in the Institute of Molecular and Cell Biology in Singapore, an Institute that was founded by Brenner while he was still director of the LMB. For his Cambridge work, Brenner applied to use the LMB stores and services. Bearing in mind the high cost of reagents used in genome sequencing, Aaron refused this request. Brenner treated this refusal as an affront, and the relationship became difficult. Brenner remained peripatetic.

By the time Aaron took over as director, problems with LMB's exclusive association with Celltech had become evident. Firstly, Celltech was located in Slough[10] rather than Cambridge, which diminished LMB's influence. Secondly, the management did not seem particularly interested in exploiting and marketing antibody technology; and thirdly, none of the income from patents or licences accrued to the inventor. Moreover, Celltech management tended to be dismissive of scientists' abilities to understand the commercial world. Strongly supported by César Milstein, Aaron negotiated for a break with Celltech[11]. The exclusive agreement with Celltech was coming up for renewal, and Aaron did not think it should be renewed. Gowans, who had been actively involved in setting up Celltech, was adamant that Celltech should retain its exclusive rights to MRC discoveries. The negotiations were tense.

Greg Winter had just achieved a breakthrough in the production of 'humanised' monoclonal antibodies (antibodies made by identical

---

[10] The Slough Trading Estate is located near London Heathrow Airport. Its aesthetic appeal may be gauged from John Betjeman's infamous poem that starts: 'Come friendly bombs and fall on Slough'.

[11] Ed Lennox and David Secher, two distinguished LMB scientists, went to work at Celltech and did not return to LMB, so the cultural split was not total.

cloned immune cells). In the method developed by Georges Köhler and César Milstein, monoclonal antibodies against specific antigens are produced by injecting the antigen (for instance, a virus) into an animal, often a rat or mouse, to provoke an immune response. In the course of the immune response, B-cells are produced that secrete antibody molecules binding with high affinity to the injected antigen. The B-cells giving rise to the required antibody are then isolated and fused with a human cancer cell to give a hybridoma that secretes just one kind of antibody. But it secretes a rat or mouse antibody, and if this is injected into a human it is recognised as 'foreign' and often provokes a strong immune response or even anaphylactic shock. Such antibodies are not useful for therapeutic purposes. Many labs were trying to get round this problem.

Building on the work of Michael Neuberger, Greg Winter and his collaborators used genetic engineering to transplant the antigen binding regions (the hypervariable 'complementarity-determining regions' or CDRs) of a rat antibody into the framework of a human antibody[12], which was then expressed in suitable eukaryotic cells. This avoided the problems of provoking an immune reaction when injected into a human. They chose to make a humanised antibody to the CAMPATH-1 antigen, which is a surface marker (CD52) found on mature T-lymphocytes. If the antibody binds to this marker, the cell is destroyed. A subsequent mini-trial of a humanised rat antibody against CAMPATH-1 on two patients in Addenbrooke's suffering from a non-Hodgkin lymphoma (a T-lymphocyte cancer) showed complete remission for both patients[13]. Things looked interesting. LMB patented this technique for 'humanising' antibodies[14].

Winter did not think that the exploitation of the humanising antibody patent should be handed over to any one firm and made his views known to Gowans, which did nothing to increase Winter's popularity. Gowans was due to retire, and the incoming Secretary of the MRC, Dai Rees, who came from industry, was much more open to the idea

---

[12] Riechmann, L., Clark, M., Waldmann, H. and Winter, G. (1988) *Nature* **332**, 323–327

[13] Hale, G., Clark, M.R., Marcus, R. *et al.* (1988) *The Lancet* **332**, 1394–1399. This antibody is now marketed as a drug for the treatment of T-cell lymphoma under the name of Alemtuzumab.

[14] The details of the ensuing negotiations with Celltech and the setting up of Cambridge Antibody Technology (CAT) can be found in de Chaderevian, S. (2011) *Isis* **102**, 620–625.

of non-exclusive contracts. He apparently told Aaron to 'hang in there'[15] until he took over. When he did so in 1987, a limited and non-exclusive contract was negotiated with Celltech for just the hybridoma applications, and Celltech did not gain control over Greg Winter's important humanising patent. The humanisation technology was licensed non-exclusively to many companies.

Winter soon developed plans for human antibody production that he thought could be best served by forming his own company. His concept was to mimic nature's method by generating a very large number of the DNA sequences that would potentially code for millions of antibodies. One proceeds by joining antibody-coding DNA sequences to the ends of the DNA of a suitable bacterial virus (such as the fd filamentous bacteriophage). On infecting a bacterial host, bacteriophage particles are produced with the antibody expressed on the surface. A bacteriophage carrying an antibody that binds to some desired antigen can be isolated by mixing it with antigen molecules fixed to a substrate. Only bacteriophages that carry antibodies with high affinity for the chosen antigen will be stuck to the substrate. Since the bacteriophage also carries the DNA coding for the high-affinity antibody, the required DNA can be recovered from the immobilised bacteriophage. This method is termed 'phage display'.

The method will not work well with whole antibody molecules (they are too big) but is efficient with the soluble fragments of the antibody termed scFv that actually carry the antigen binding sites. Each scFv carries two independently folded domains known as VH and VL, each of which contains three surface loops called complementarity-determining regions (CDRs) that can have practically any sequence. The possible number of combinations of these six CDR sequences is astronomical. Winter's plan was to use combinatorial chemistry to make a very large library of scFv molecules and then scan these by phage display against pharmacologically interesting antigens. The successful scFv-carrying bacteriophage would be isolated to yield the DNA coding for the high-affinity antibody. This DNA could be built into the DNA coding for a human antibody (IgG) molecule, which could be

---

[15] Much of the information in the following section has been taken from Aaron Klug's extensive interview with John Finch, available under Web of Stories 86 (http://www.webofstories.com/play/aaron.klug/86) and surrounding sections.

expressed in eukaryotic cells to yield human antibodies that had never seen a mouse or a hybridoma.

Winter presented this plan to Aaron, who responded enthusiastically. However, the plan was too big for the LMB. They needed an external firm. Where should they raise money? Aaron arranged meetings with Glaxo, Unilever and Wellcome Biotech, all giants in the field of drug marketing[15]. They showed polite interest but did nothing. They did not think an antibody could be a drug. Aaron also approached Victor Rothschild, who had now given up science to concentrate on running the family bank. Unfortunately, he also showed no interest, possibly because Brenner, who was his scientific adviser, had similar plans using the bacteriophage lambda. These refusals cost them hundreds of millions of dollars.

In the end, Geoffrey Grigg, a large, genial Australian entrepreneur who earlier had been a student of Fred Sanger, offered to provide start-up capital for Winter's new firm. Aaron and Winter were of the opinion that the MRC should hold equity in the new firm, but negotiations with head office ran into a snag. Aaron spoke with the Executive Director of the MRC, Norma Morris, who pointed out that the MRC was not allowed to own equity. Aaron pointed out that they already did! This had come about because an American biotech firm, Somatogen, had financed a postdoc working in Kiyoshi Nagai's laboratory. Nagai had managed to express the protein *globin* in bacteria (the first sizeable eukaryotic protein to be made in this way). When combined with a chemically synthesised haem group, this assembled into haemoglobin, the red oxygen-carrying pigment of the blood. There was hope that this method might lead to a blood substitute; but Somatogen ran out of cash to pay the postdoc. Brenner had left behind some short-term positions that he had negotiated for the Director's Section. Aaron agreed with Somatogen to pay the postdoc out of one of the Director's positions in exchange for stock in Somatogen. Thus the LMB acquired 50,000 shares in Somatogen for the MRC (which were to prove quite valuable). Somewhat shocked by these revelations, the MRC entered into negotiations with the Treasury who retrospectively sanctioned the deal. This development in turn led to a permanent change in Treasury policy: MRC was permitted to own small amounts of equity in firms exploiting MRC discoveries.

Winter's new firm was called Cambridge Antibody Technology (CAT). It was closely tied to the LMB. Milstein and Winter were on the scientific advisory board, and Aaron became a director. The MRC agreed to take equity in CAT. The local newspaper showed photographs of César Milstein and Aaron talking about the creation of CAT. Although neither Milstein nor Aaron received remuneration for their services[16], some unconvinced member of the lab posted the newspaper article in the elevator emblazoned with the comment: 'Who are you working for?'

In Aaron's second year as Director, in a Lab Symposium talk, Aaron said that if people were working on problems that had different facets, they should choose the facet that might have some application. This provoked a letter from the head of one of the Divisions saying it appeared that Aaron's intention was to turn the laboratory into a laboratory of *applied* molecular biology. Many scientists were of the conviction that Science and Mammon do not make connubial bedfellows. Nevertheless, Aaron remained proud of his role in helping to re-orientate the MRC towards shareholding and founding CAT. Since the whole lab has recently moved without protest into a futuristic new building financed from the Greg Winter patents, it appears that Mammon has gained some acceptance.

Winter's scientific plan was successful. In 1996, CAT produced the first therapeutic antibody, Adalimumab, to be made from the phage display technology. Adalimumab (now known as Humira) was approved by the FDA (the US Food and Drug Administration) in 2002. It acts as an anti-inflammatory drug by binding to tumour necrosis factor (TNFα – a protein that can cause death of tumour cells), preventing it from activating TNF receptors; TNFα *inactivation* is important in down-regulating – that is, decreasing – the inflammatory reactions associated with autoimmune diseases including rheumatoid arthritis. Humira, now marketed by Abbott Laboratories, was a blockbuster. It has produced an income of over $350,000,000 for the MRC (although a protracted court case was needed to extract the money). At the time of writing, Humira is the world number one drug by annual sales value.

Another earlier money-spinner for the lab was confocal microscopy. The original inventor of the confocal microscope was the polymath

---

[16] They received token amounts of stock in the firm.

Marvin Minsky who filed a patent in 1957 but did not get round to publishing the idea in any other way. The basic idea is to illuminate the specimen with a single point of light (as can be produced in the focal plane of a microscope from a point source) and to observe the scattered light with another microscope focused onto the same point. By inserting an aperture into the focal plane of the observing microscope, all scattered light that does not originate at the focal point of the observing microscope can be eliminated, cleaning up the image hugely and giving 3D resolution. Scanning the beam across the samples yields a complete image.

John White, who had been in Brenner's group, was using a fluorescence microscope on nematode specimens with a laser illumination for the ablation of cells. He discussed with Aaron the idea of inserting an appropriate aperture in the optics he was using to produce a confocal beam. He realised that the controlled tilting of two small mirrors would be the way to scan. White, who was by training an engineer, collaborated with Brad Amos, who knew about microscopes and optics. Richard Durbin wrote user-friendly programs to assemble the scanning signals into a 3D image that could be viewed on a computer display from any angle.

Aaron could see that the confocal microscope was going to be a winner. In fact, it became the most powerful instrument in cell biology. He gave it top priority in the workshop, and also arranged with Ken Smith in the Electrical Engineering Department of the University to supply a suitable frame store to hold all the scanned information. Together with Mick Fordham in the workshop, Brad Amos, John White and Richard Durbin built the prototype of the world's first commercial confocal microscope. Aaron persuaded the firm BIORAD to market the microscope. BIORAD had a good sales network and placed the microscope in well-respected labs. Brad Amos provided after-sales help and instruction. BIORAD was later bought out by Zeiss, who wanted the two-photon microscope patents that BIORAD owned. Zeiss did not continue manufacturing the BIORAD instrument, but in the early years the LMB derived considerable royalties from the world's first commercial confocal microscope[17].

---

[17] Later, a badly drafted patent offset clause caused the royalties effectively to drop to zero.

One problem that Aaron had to face was how to share out payments accruing from patents and licences. According to a MRC rule this income was shared three ways: one-third to the MRC, one-third to the LMB and one-third to the inventors. For large earnings, there was a sliding scale to stop obscene quantities of money accruing to employees. As director, Aaron had to decide on the distribution. He often went beyond the names listed on the patent to include technical staff whose input had been of great value.

As head of Structural Studies, Aaron wanted to start a nuclear magnetic resonance (NMR) group. NMR can be used to solve the structure of small protein molecules in solution (without crystallisation). Many large proteins are built of small domains strung together. Often, small domains could be isolated from larger proteins by biochemical methods or genetic engineering. The structures of these small domains were amenable to analysis by NMR. Zinc fingers were an obvious application for NMR, but until he became director, Aaron had not been able to raise sufficient funds. After a week in Zürich with Kurt Wüthrich, Aaron became convinced that NMR spectroscopy and its application to structure determination of small protein molecules was a manageable technique. In September 1988 he recruited David Neuhaus from Oxford to run an NMR group at the LMB. Neuhaus's first job was to determine the structure of zinc fingers. At first a few spectra were run on a 300 MHz machine at the Parke-Davis pharmaceutical company and on Alan Fersht's 500 MHz NMR instrument that he had left behind at Imperial College; in December 1988, the LMB acquired its own Brucker AMX 500 machine. Thus the LMB, the founding laboratory for X-ray crystallographic structure determination of proteins, belatedly embraced the complementary culture of NMR for structure determination in solution. In 1994, when Gabriele Varani was recruited to work on protein–RNA complexes, the LMB acquired a 600 MHz machine and then in 2001 an even more expensive 800 MHz instrument, which was shared with Alan Fersht (see below) and an LMB start-up company, RiboTargets[18].

The University Department of Pharmacology was housed in a building adjacent to the LMB that had been built in 1973. In 1989, the

---

[18] RiboTargets collaborated with Jonathan Karn on drug design arising from control proteins. Karn was a group leader working on the control of gene expression in HIV.

Department of Pharmacology moved into a new building in Tennis Court Road. In 1988, Alan Fersht and the Secretary of the MRC Dai Rees came up with the idea of opening an MRC Centre for Protein Engineering in half of the now vacant Old Pharmacology building. The director would be Fersht, with Winter as deputy director. Aaron welcomed the move because it would reduce the number of people in the LMB working on applied problems, which was beginning to become an embarrassment. The motivation for the founding of the new centre was that in 1988 Fersht, who was originally from the LMB, moved back to Cambridge from Imperial College London to become Professor of Organic Chemistry. His research on protein folding depended on protein engineering of which he was one of the luminaries. Margaret Thatcher opened the new centre in 1991. Two of Winters's co-workers took appointments there. Fersht needed powerful NMR machines, which were housed in the Department of Chemistry, but later they were moved to a special building that was constructed on the Addenbrooke's site. This finally housed all the NMR machines from the two labs[19]. When Fersht retired in 2010, his research group and spectrometers were fully incorporated into the LMB.

The University was generally indifferent to the LMB. Nevertheless, Aaron's prominence ultimately led to an initiative in neurobiology at the LMB. Aaron's Nobel award brought him to the attention of Martin Roth, Professor of Psychiatry and an expert on Alzheimer's disease. Roth had shown that Alzheimer's was a disease of the elderly and that the manifestations were two lesions: neurofibrillary tangles inside the cells and amyloid deposits outside the cells. Roth spoke to Aaron and Tony Crowther over coffee and asked if one could perhaps characterise these lesions and work out where they came from. Aaron consulted the literature and found that the lesions were fairly specific and initially occurred in very definite locations in the brain. Otherwise, not much was known.

Electron microscopy had shown the neurofibrillary tangles to comprise two intertwined filaments, termed paired helical filaments (PHF).

---

[19] From 2001, the two labs shared the 800 MHz machine, which together with the original LMB 500 MHz housed in Medicinal Chemistry, the LMB 600 MHz machine, Fersht's two 500 MHz and later a 600 MHz machine from the Chemistry Department, gave *in toto* six NMR spectrometers on the Addenbrooke's site.

Crowther thought it was an interesting project but needed someone from the medical side to do the preparation and biochemistry. Roth suggested Claude Wischik, a graduate student in his Psychology Department. Working with Murray Stewart, Wischik began using sucrose gradients to try to purify the PHFs. Crowther did image reconstruction on PHFs[20]. Now that more was known, Aaron became interested, and he provided a bridge to John Walker for the protein chemistry and César Milstein for making monoclonal antibodies. As a result, a small protein sequence was derived from the neurofibrillary tangles. At this stage, Aaron involved Michel Goedert, who had been recruited by Brenner but left at a loose end when Brenner retired. Aaron suggested that he should attempt to clone a gene containing the DNA sequence corresponding to the PHF fragment. He succeeded quite quickly. Comparison of the sequence of the cloned gene with that of a mouse protein called tau[21] showed that both contained the unusual PGGG repeats. Goedert had cloned human tau. Thus he, together with Wischik, Crowther and John Walker, was able to show that the neurofibrillary tangles were abnormal aggregates of tau, a brain-protein that binds to microtubules and controls their stability in the axons of nerve cells. Crowther then used electron microscopy to work out the structure of the paired helical filaments. The LMB was into neurosciences[22].

When in 1992 Eric Barnard retired as Director of the MRC Neurobiology Unit, Aaron reopened negotiations with Dai Rees, Secretary of the MRC, to establish a Neurobiology Division within the LMB. A decade earlier, negotiations between Brenner and Gowans had stalled. Relationships had been thoroughly soured when a couple of months later Gowans appointed Barnard as Director of an MRC Molecular Neurobiology Unit housed adjacent to the LMB. This time Aaron obtained Rees's backing and a fourth division (Neurobiology) was added to Structural Studies, Cell Biology, and Protein and Nucleic Acid Chemistry. The first head of division, appointed in 1992, was Nigel

[20] Crowther, R.A. and Wischik, C.M. (1985) *EMBO J.* **4**, 3661–3665.
Wischik, C.M., Crowther, R.A., Stewart, M. and Roth, M. (1985) *J. Cell Biol.* **100**, 1905–1912.
[21] Marc Kirschner had just cloned and identified the tau protein from the mouse.
[22] Goedert, M., Wischik, C.M., Crowther, R.A., Walker, J.E. and Klug, A. (1988) *Proc. Natl Acad. Sci. USA* **85**, 4051–4055; Wischik, C. M., Novak, M., Thøgersen, H.C. *et al.* (1988) *Proc. Natl Acad. Sci. USA* **85**, 4506–4510; Wischik, C. M., Novak, M. Edwards, P.C. *et al.* (1988) *Proc. Natl Acad. Sci. USA* **85**, 4884–4888.

Unwin, who was working on the structure of the acetylcholine receptor by electron microscopy. Initially the division remained small. Unwin carried out much of his experimental work in Kyoto in Yoshinori Fujiyoshi's lab, where electron microscopes with specimen stages cooled to liquid helium temperatures were available. Goedert's group joined the new Division, and later Goedert became its head.

As mentioned above, John Sulston had been working with Brenner on the development of nematode worms. After establishing the embryonic lineage of the worm, John Sulston wondered about mapping the whole genome of *C. elegans* (putting all the DNA markers that had been collected associated with various genes in order). Brenner (who at this time was still Director) gave Sulston space in the Director's Section in Block 7. Under Aaron's management, Sulston effectively became head of the Director's Section, which was renamed the Genome Section[23]. On Fred Sanger's retirement in 1983, Sulston was joined by the DNA sequencing expert Alan Coulson. Then Bob Waterston joined them: Waterston, who was acknowledged as a leading member of the *C. elegans* community, had come to Brenner's lab as a postdoc from Chicago in 1972. Subsequently he opened a Division in Washington University, St Louis, using the nematode to study the genetics of muscle. In 1985 he came back to Cambridge for a sabbatical year. Coulson, Waterston and Sulston together put the following idea to Aaron: given the difficulties one had in finding and mapping the genes of an animal, would it not be simpler to determine the total sequence of the animal's genomic DNA? Then one would have all the genes and could see where the markers were. Thus the idea of the nematode genome-sequencing project was born. Sequencing projects became a very serious part of Aaron's directorship. The nematode genome project was the harbinger of the human genome-sequencing project.

In 1988, Jim Watson became head of the Office of Genome Research in Washington[24], a newly founded office entrusted with planning the Human Genome Project. The project was officially started in 1990 with the aim of sequencing the human genome by 2005, but Government

---

[23] In 1993 the Director's Section was phased out following the creation of the Sanger Centre at Hinxton (now the Sanger Institute), to which John Sulston moved as its first Director.

[24] The ensuing story has been extracted from *The Common Thread* by John Sulston and Georgina Ferry, Bantam Press (2002).

funds had already become available by 1989. The initial aim was to develop technology through smaller-scale projects such as the sequence of the fruit fly or the worm. Noting that Sulston at the LMB and Waterston in Washington University Missouri were jointly by far the world's most productive sequencing group, in the spring of 1989 Watson encouraged them to apply for funds from the National Institutes of Health to do the sequence of the worm. This would be a big project with at least 20–30 people in each lab. Sulston returned to England in some excitement and immediately went to talk things over with Aaron. Aaron was initially doubtful about the worm – there was so much more interest in fruit flies – but John Sulston's convictions swayed him: 'John was the standard bearer and that's why it had to be the worm.' In September 1989, Waterston visited Cambridge to draft a joint application to the MRC and the National Institutes of Health. They made an application for £1,000,000 over three years for the Cambridge part and an equal amount for the Washington University lab. With Aaron and Watson's enthusiastic support, this was funded.

In Cambridge, new space had to be found. Aaron negotiated to take over the remaining half of the building that had recently housed the University Department of Pharmacology group. Known as the 'Old Pharm', this space was rapidly transformed by Mike Fuller (and money) into the most modern sequencing lab.

The joint efforts of the Sulston and Waterston labs were most successful, and two years later it was clear that they could finish the worm DNA sequence for a reasonable outlay. By extrapolation, the human genome could be financed. Now Mammon raised its ugly head. Enormous quantities of money could be extracted from the financial system on the basis of possible drug discoveries. The human genome was going to reveal a plethora of drug targets: clearly, you'd better patent the lot. But published data cannot be patented. Thus the sequencing was a tempting prospect for a private consortium that did not need to publish the results.

In January 1992, Waterston and Sulston were approached by a private investor, Frederick Bourke, who lured them with a new institute in Seattle, very adequate funding and princely remuneration. Sulston and Waterston were basically the only game in town, but they were completely dedicated to making all sequence results available immediately (even before publication) and abhorred the idea of patenting DNA sequences. This was in the tradition of Fred Sanger, who had never patented even his method of sequencing DNA. Nevertheless, the news that Sulston and Waterston might accept the King's shilling reached Jim Watson and caused him great unease. Watson was also dedicated to free exchange of information, and Watson considered Sulston and Waterston his protégés. He called Aaron to express his 'annoyance and rage' at the Bourke approach. He phoned Waterston, who quickly put his mind at rest but pointed out that they had no funding for their work after the initial three years were up.

Watson decided to exploit the situation. He promptly called on Rees at the MRC Headquarters to make him aware how close the MRC was to losing one of its flagship enterprises. Rees quickly organised long-term funding to finish the worm sequence. Aaron and Rees discussed a possible human genome initiative. Realising that this would be too big for the MRC alone, Rees immediately agreed to a request from Aaron that he might contact the Wellcome Foundation.

Just at this moment, on account of profits made from sales of the anti-HIV drug AZT, the Wellcome Trust was awash with money. The newly appointed chief of the Trust, Bridget Ogilvie, was initially enthusiastic about a partnership funding with the MRC. However, she quickly realised that much more was possible. In the spring of 1992, Aaron and Sulston attended a meeting of the Wellcome Trust's Genetics Advisory Group. At the meeting Aaron argued forcefully for large-scale sequencing of the human genome. He took as an example the gene for Huntingdon's disease, which had been located on a particular chromosome but had resisted all efforts from conventional genetics to locate it within the chromosome. With the sequence at hand, one could just read it out. The Wellcome Trust was convinced, and decided to set up a large-scale sequencing centre with John Sulston as director. A big question was where to put it. As with so many things affecting the LMB, Mike Fuller played a pivotal role. He pointed out that Hinxton Hall, a country house and grounds 8 miles south of Cambridge, was available for rent. This proved to be an ideal site and was soon purchased outright by the Wellcome Trust, becoming the Wellcome Trust Genome Campus. Somewhat to Rees's annoyance, the MRC contribution was limited to the worm genome sequence. In recognition of Fred Sanger's pioneering work in setting up DNA sequencing methods, John Sulston's lab was entitled the 'Sanger Centre'. Sanger agreed to the use of his name on condition that the lab did good work. Sulston turned out to be a very talented manager: to put Sanger's mind at rest, this lab sequenced one-third of the human genome.

Aaron was appointed Director of the LMB in 1986 for seven years, until 1993, which would take him to 67. The Medical Research Council approached Richard Henderson and asked whether he would be interested in taking over the Directorship in 1993. He replied that on account of his relative youth and his research being in a very productive phase he thought someone else would be better. The MRC then persuaded Aaron to stay on for another three years. In the meantime, César Milstein (then deputy Director) persuaded Henderson to be willing to take over in 1996, when he was 50. Towards the end of his extended appointment, Aaron was sought out to be President of the Royal Society and was duly elected President, starting in December 1995. Thus, for nine months, Aaron overlapped the job of Director of the LMB with President of the Royal Society.

By this time Aaron was on the scientific boards of many institutes. Three of these, the Salk Institute in California, the Weizmann Institute and the fledgling Institute at the Ben Gurion University, he took very seriously – and thus spent rather a lot of his life in aeroplanes.

19

Presidency of the Royal Society

Aaron's son David Klug, having graduated in physics from University College in 1984, was accepted as a PhD student by George Porter at the Royal Institution and was awarded his doctorate in 1987. This successful arrangement served to deepen the friendship between Aaron and George Porter. George married Stella Jean Brooke in 1949, the year he obtained his doctorate in Cambridge with work that would later earn him a Nobel Prize. Stella was diminutive and beautiful. As recorded earlier, she had first attracted Aaron's attention when he saw her from the window of his shared apartment in Chesterton Road, as she hung out washing in the next-door garden. Stella had a forceful personality and was a teacher of ballet. Indeed, George and Stella were a lively pair: George was an enthusiastic Gilbert and Sullivan performer. He was appointed President of the Royal Society in 1985, taking over from Andrew Huxley. For a couple of years, he combined the presidency with being Director of the Royal Institution, but after 20 years at the Royal Institution he handed it over to John Meurig Thomas. When George and Stella moved their London residence from Albemarle Street to Carlton House Terrace, Stella was most irritated to find that the President's flat was smaller than the flat occupied by the Executive Secretary of the Royal Society, Peter Warren. Moreover, George Porter was put out by being expected to share a secretary with the Executive Secretary.

In the greater scheme of things, Porter walked into a much more serious problem: the hostile attitude of the then-current government to science. Porter well understood the importance of science to the community, and he argued that without a strong science base Britain would quickly become 'well prepared to join the Third World of science'. In contrast, some senior members of Margaret Thatcher's government such as Michael Heseltine basically considered science and education a waste of money that might be better invested in something useful – such as banking. Although Margaret Thatcher was supportive of environmental science and European ventures such as CERN (the European Organization for Nuclear Research) and EMBL (the European Molecular Biology Laboratory), her attitude to the University Grants Committee was iconoclastic. In February 1985, in protest against her cuts in funding for higher education, the Congregation of the University of Oxford refused to bestow on her an honorary degree. In 1983, during Andrew Huxley's presidency, she had only been grudgingly elected to the Royal Society, an honour accorded to many prime ministers. Thus, a continuous conflict with Whitehall over funding levels for science determined Porter's presidency. Nevertheless, Porter ensured that the Royal Society became more engaged in British public life. He drew attention to the ever-widening ozone-hole produced by the enthusiastic commercial use of chlorofluorocarbons (CFCs) as refrigerants, which led to an important international ban on their use. He also made global warming and greenhouse gases a mainstream issue.

Half way through his five-year term, Porter started planning his successor. His natural choice was Aaron, as a Nobel Laureate and, moreover, a scholar of note whose interests embraced the 'Two Cultures'. Porter arranged that Aaron should become a member of the Royal Society Council. In December 1989, Aaron and Liebe had planned to go to South Africa for a few weeks. Cambridge at the Winter Solstice can be rather cold and depressing, and there were matters awaiting in South Africa. Liebe waited patiently at Heathrow while Aaron attended a Royal Society Council Meeting in London. Somewhat later, an ashen-faced Aaron appeared at Heathrow: Council had offered him the presidency and he had turned it down[1].

---

[1] Part of the content of this chapter, including the text of interviews conducted with Aaron and Liebe Klug, has been made available to me by Peter Collins, Emeritus Director at the

This was an unusual step. The only previously recorded case of an offer of the presidency being refused had been by Michael Faraday. Normally Council is careful to check that its approaches will not fall on barren ground, but this time they had slipped up. Aaron reasoned that his commitment to the LMB at this moment was crucial. He had just set up a new department of neurobiology; he had got the John Sulston nematode genome-sequencing project off the ground, and complicated negotiations were under way about transferring the patent rights on human cloned antibodies to Celltech, a firm partly owned by the MRC. Moreover, Aaron had recently discovered zinc fingers and wanted time to explore the exciting implications of this discovery.

This left Council in something of a dilemma. They first asked Porter if he would extend a year, but he and Stella were not prepared to stay for a minute over the allotted time. After some deliberations, Council offered the presidency to Michael Atiyah, a most distinguished mathematician and Master of Trinity College Cambridge, who accepted. Michael Atiyah and his wife Lily seemed unbothered by the size of the presidential apartment; the Master's Lodge at Trinity was already palatial enough, and Lily intended to stay in Cambridge. Traditionally the presidency alternates between the two 'sides' of the Royal Society: the physical side (physicists, chemists and engineers) and the biological side (biologists, biochemists, physiologists and medics – social scientists generally end up in the British Academy, just across the Duke of York steps). Porter was firmly on the physical side, and Aaron could be construed as being biological. Atiyah was the first mathematician appointed to the presidency for over a century. There isn't really a mathematical side, so Atiyah was something of a wild card. Thus, five years later, in December 1994, Council felt emboldened to offer Aaron the presidency from the B side. Atiyah is reported to have said to Aaron, 'I'm keeping the chair warm for you.' As retiring president, Atiyah negotiated with Peter Warren to ensure that the Klugs would be offered the larger apartment. In the meantime, Aaron's status had been further enhanced by his election to the exclusive Order of Merit. This time Council selected Tony Epstein, who had been Foreign

Royal Society. I am most grateful for his help and criticism of the whole chapter. In addition, Liebe has told me many things, and I have also drawn on the Aaron Klug correspondence in the Churchill College Archive.

Secretary of the Society, as their emissary. The approach was well prepared: Tony Epstein made a strong appeal to Aaron's sense of duty. Any boy who had been through the Durban High School for Boys knew the imperative nature of a call of duty. Aaron, now 69, acquiesced.

Every year, fellows of the Royal Society celebrate the founding of the Society on 30th November 1660 with an Anniversary Meeting. Every five years or so, this is also the occasion for the installation of the new president. But Aaron had agreed to address a festive session of the Senate of Ben Gurion University in Beer-Sheva on 28th November to celebrate 25 years since the founding of the University. Very appropriately, Aaron's lecture was 'Some Reflections on Science and Science Policy', a theme that would dominate the first year of his presidency of the Royal Society. The lecture in Beer-Sheva was necessarily followed by a precipitate journey back to London to be in time for the Anniversary meeting on 30th November at which he would be installed as President.

At the meeting, the current president presents his address, which includes a summary of the Society's year. On Thursday 30th November 1995, Michael Atiyah presented his valedictory address. Such an address can afford to be more personal since it does not need to consider future presidential policy. Atiyah chose to present a strongly argued attack on post-war UK atomic weapons policy, pointing out that the Governments misuse of vast quantities of Research and Development funds for defence needs was responsible for the relative decline of the British economy compared with the similarly funded but nuclear-free German economy. Atiyah then vacated the president's chair in favour of Aaron. At that time the president's chair, which carried the coat of arms and motto of the Society, *Nullius in Verba* (roughly, 'Take no one's word for it') was rather large (it has since been replaced by a chair more commensurate with the average president), and Aaron could only huddle in a corner. At the end of the meeting, the officers of the Society processed out of the room following the newly installed president, who was preceded by a bearer carrying the Ceremonial Mace that had been presented to the Royal Society by Charles II.

The following Monday was Aaron's first working day with Peter Warren, which showed up some important cultural differences. Before his appointment as Executive Secretary, Warren had spent three years as a senior civil servant in the Cabinet Office. Rather naturally, he assumed the manners of a Permanent Secretary of State who was required to deal with presidents (or ministers) who come and go every few years. The proper role of presidents was to define policy but not to execute. Since the Executive Secretary's office dealt with presidential correspondence it was not clear to Warren why presidents might require their own secretary (this had already been a bone of contention with Porter). It is true that most presidents were academics with no executive experience. However, Aaron, as director of the LMB, knew how to handle an organisation that was bigger than the administration of the Royal Society. Moreover, he had lots of experience in negotiating with government departments and commercial organisations. He and Warren soon agreed that he would have a secretary of his own.

Meanwhile, Liebe's reforming zeal was rampant. She was determined to make the Royal Society more attractive and more open. She improved the standards of catering quite remarkably, and introduced music to the reception after the Anniversary Meeting. In time, she reformed the lunchtime cafeteria and redecorated the presidential apartment to make it more suitable for entertaining and interchange. Aware of the 'Two Cultures', Aaron initiated a series of informal dinners inviting both scientists and representatives of the humanities to get to know each other in a relaxed setting. It transpired that many quite sophisticated and learned people across society had never even heard of the Royal Society and certainly had no idea what it was there for. During the latter part of his presidency, Aaron set up a one-day conference aimed at bridging the gap between the two cultures with the idea of making this an annual event. Unfortunately, the stresses surrounding the illness and death of their son Adam made it impossible to continue the programme, but it appears to have had a nucleating effect: under succeeding presidents the Outreach programmes became important and sustained activities of the Society.

Liebe herself was concerned with what she saw as the split in the Royal Society: apparently on the one hand were the professionals – the paid staff – and on the other the amateurs – the scientists. She felt that this gap led to a permeating feeling of deadness. In her view, the culture of the RS was out of keeping with the way that better scientific institutions functioned. Liebe herself attempted to loosen these barriers by establishing friendships with staff members. She and Ling Thompson, the retiring Under Secretary of the Society, formed an enduring friendship. The terminal illness of Aaron and Liebe's son Adam during the last part of Aaron's presidency drew staff and president together.

The year of 1996 was tough. Aaron was still Director of the MRC Laboratory of Molecular Biology and at the same time was trying to get to grips with his new job. It was necessary to spend at least two full working days in each week in London. The scope of activities was more extensive than he had anticipated. The Royal Society had a permanent staff of over 100. Raising money was a headache. Peter Warren was heavily involved with Project Science, a fund-raising exercise that had been inaugurated by Aaron's predecessor Michael Atiyah with the aim of safeguarding the financial independence of the Society. The Society had reached the stage when nearly 80% of its annual expenditure came from government sources, a proportion that appeared to compromise the independence of the Society. In fact, the financial imbalance of the Society was a product of its own success. The University Research Fellowship scheme was working very well. This with the Royal Society Professorships accounted for over half of the Grant-in-Aid from the government. Another extensive activity was offering the government advice on science policy, an activity that had been expanded considerably while Lord Todd was president. The Project Science was particularly important because it enabled the Society to maintain its independence when offering opinion and advice to the government. The Duke of Edinburgh was its patron, and committee meetings of Project Science were often held in Buckingham Palace. Aaron found the Duke's down to earth approach refreshing.

It was also a year of travelling. International scientific exchanges are an important part of the Society's work, and in 1996 Aaron and Liebe made two official visits to South Africa, a visit to Malaya and a visit to India. At the end of January, Aaron and Liebe spent two weeks in South Africa at the invitation of the Foundation for Research Development (FRD) in Pretoria. The President of the FRD, Dr Rein Arndt, was a personal friend of Aaron's. As he had done six years previously, when he refused the presidency, Aaron left after a Thursday Council meeting to join Liebe at Heathrow for the overnight flight to Johannesburg. They flew on to Cape Town and stayed at the Portswood Hotel close to their youthful stamping ground. On the Monday, Aaron gave the opening lecture at a Symposium on Cell Growth Control, and on Tuesday he gave the plenary lecture at a symposium at the University to commemorate the 100-year anniversary of R. W. James' birth. Aaron spoke about X-ray structure analysis in biochemistry and biology. John Juritz told the story of R. W. James and how he had trained two Nobel Prize winners. The following day, Aaron visited the University of the Western Cape where he met up with Rein Arndt, followed by a visit to Pretoria and excursion to the University of the North[2] to meet students: the University of the North was set up under Apartheid for black students and rather naturally became a centre of the anti-Apartheid movement.

In May, Aaron represented the Society at the inauguration of the new South African Academy of Science. For about a century, the national science 'academy' had comprised two separate institutions – the Royal Society of South Africa based in Cape Town and the *Suid-Afrikaanse Akademie van Wetenskap en Kuns* (SAAWEK), which had an Afrikaans language focus, based in Pretoria. It was the official national academy until 1994. In post-apartheid South Africa, a new model was required. The Academy of Science of South Africa (ASSAf) was inaugurated in Pretoria in May 1996 by Nelson Mandela, winner of the Nobel Peace Prize, who sought the backing of other Nobel Prize winners. Thus, for the celebration, Sherwood Roland, Mario Molina, Aaron and their wives were invited. Sherry Rowland and Mario Molina had developed the 'CFC-ozone depletion theory[3]' for which they had just been awarded the Nobel Prize in Chemistry. Mandela was in surprisingly good spirits, considering the fact that his divorce from Winnie Mandela had become finalised that day: he had been freed from 27 years of incarceration to find that he didn't really have a wife any more. After his address to the assembled company, he came back to the Nobel Prize winners' table and remarked rather wistfully that they were fortunate to have wives who stood behind them and supported them in their work. He also told Aaron that South Africa needed him and that he ought to return.

Aaron, in his Anniversary Day Address to the Royal Society, empha- sised the importance of the Royal Society in offering unbiased advice to

---

[2] 350 km northeast of Johannesburg, now called the University of Limpopo.

[3] CFC = chlorofluorocarbons. CFCs at high altitudes are destroyed by solar radiation: the chlorine atoms produced by the decomposition of the CFCs catalytically destroy ozone, leading to severe ozone depletion and allowing through more ultraviolet radiation, particularly in the southern hemisphere. CFCs were a common ingredient in refrigerators. Their industrial use is now banned.

the government and its increasingly important role in offering independent judgements on contentious public issues. He discussed the Prior Options Study[4] as a recent and ongoing example of its involvement in science policy. It illustrated that an important role of the Royal Society was to ensure the healthy funding of basic science at a time when the prevailing philosophy held that public sector activities should be subjected to market forces. An extreme form of this view was that basic research should not be funded by government at all but that all research could be more efficiently done in the private sector. Aaron reiterated what many knew, namely that such a strategy was shortsighted and would inevitably lead to a serious decline in the quality of the UK science base. The yield of basic science in terms of industrial or medical application is often measured in decades. No private firm could afford to take such a long view. In his discussions with the Ministry of Trade, Aaron was particularly emphatic that the LMB should not be sold off to a predatory drug company. The Prior Options Study of research council laboratories was based on two questions: 'Is the work being done in the research establishment significantly important?' and 'If so, would it best be continued in the public sector or would it best be done by a privatised institute?' Chosen institutes were scrutinised by senior civil servants who were singularly lacking in an appreciation of the way that creative science was organised, and expected each institute to be managed by a senior administrator and certainly not by a scientist. They would have had no understanding of the way the LMB functioned, and fortunately they did not attempt to find out. As Aaron put it: 'The best labs are not working at the frontiers of science, they're creating the frontiers of science.'

Aaron was at this time on the Council for Science and Technology, which was chaired by David Hunt. Hunt confided to Aaron that basically the whole idea had been shelved. Thus, at the end of his Anniversary Address, Aaron was happy to be able to report that doctrinaire privatisation was apparently no longer a central issue in John Major's government. In his Anniversary Address a year later, Aaron

---

[4] On 3rd February 1997, John Major's Government finally backed down on plans to privatise research council laboratories. The Prior Options Study, a Government privatisation probe of more than 40 public sector research facilities, concluded that most should retain their present status.

summarised why he thought basic research should be funded. He referred to the role of delegated curiosity. In his Annual Report[5] for 1998, Aaron once again emphasised this aspect of basic research, which is central to the philosophy of the Royal Society:

> *Science is possible only because 'society' tolerates it. This relation has several components. Taxpayers, represented by the Government that collects and spends the taxes, seek practical benefits. But the so-called 'general public' often also takes great interest in the ideas that science throws up – from the heliocentric system and evolution to black holes, chaos theory and continental drift. Scientists are, certainly, motivated by the possibility that their work may have beneficial outcomes, but for many of us the overriding driver is fascination with the workings of nature. That fascination is the most powerful source of new knowledge, and it is the fruits of that new knowledge, which eventually benefit society. For science policy, then, the challenge is to understand how these components can coexist and to create an environment – an ecosystem as it has been put – in which researchers can eventually deliver various advances that society eventually welcomes. The key characteristic of this ecosystem is freedom: freedom to set the research agenda and freedom to change it in the light of unexpected discoveries. There is very extensive evidence that, at the highest levels of research, it is this freedom which produces the breakthroughs that, literally, change the world.*

In August 1996, Aaron gave up his Directorship of the LMB and passed the sceptre to Richard Henderson.

On 8th January 1997, Don Caspar was 70. A couple of years earlier, Don and Gladys Caspar had moved from Brandeis to the Florida State University in Tallahassee and sold their nice house in Brookline. Caspar wished to work with Lee Makowski, his one-time student, who was now Chair of Biophysics at Florida State. In January 1997, Makowski organised a two-day seminar to commemorate Caspar's seniority and science. Aaron was chosen as the Keynote Speaker. In spite of Tallahassee's relative remoteness, Aaron felt he had to go. His talk was on 'Protein Designs for Manipulating DNA'.

---

[5] Klug, A. *Notes Rec. Roy. Soc. Lond.* (1999) **53**, 157–167.

In June 1997, Peter Warren, the Executive Secretary, retired. The event was marked by a number of receptions at the Royal Society; indeed, because of great interest, the farewell reception given by former members of Council was repeated on two successive evenings. Aaron, in his warm appreciation of Peter Warren's 20 years of service, emphasised his selfless dedication to the Society. It was Peter Warren who had got the Society involved in defining science policy and who in 1981 had appointed Peter Collins to run the Science Advice Section. Aaron added that there was hardly anything that the Society had done in the past 20 years that did not bear Peter Warren's creative imprint. Warren's successor was Stephen Cox. A former employee of the Society, he returned from the Commonwealth Institute, where he had been Director-General.

The year of 1997 brought an extended trip to China. It had been pending for some years: Zhang Youshang, who had been a postdoc with Aaron in 1962, had invited Aaron in 1990. Finally, in 1996, Aaron was formally invited as President of the Royal Society by Zhou Guangzhao, the President of the Chinese Academy of Sciences (CAS)[6]. The visit took place in September 1997. The delegation included Brian Heap, the Foreign Secretary of the Royal Society, and Ling Thompson, now Head of International Affairs, who fortunately was Mandarin-speaking. The visit was broadened to include South Korea where Aaron had been invited to give the Hallim Distinguished Lecture at KAST, the Korean Academy of Science and Technology. He spoke on the regulation of gene expression on this and other occasions during the trip, including a day at the CAS Institute of Biophysics in Beijing.

The stay in Beijing required a number of official visits, including a grand meeting with the President and Vice Presidents of CAS. Aaron was impressed by the Institute of Atmospheric Physics, which is a regional research centre of the Global Change System for Research and Training and had close ties with senior UK scientists working on global warming. He hoped that the Chinese Government would listen to the warnings emanating from this well-informed institute. At the express wish of Lu Shengdong, the Vice President of the Chinese

---

[6] The Chinese Academy of Sciences (CAS), formally known as the Academia Sinica, is the National Academy for Natural Sciences of the People's Republic of China. The name Academia Sinica is now used in Taiwan.

Academy of Medical Sciences (CAMS), Aaron gave a lecture at CAMS entitled 'The Development of Structural Molecular Biology in Cambridge'.

After the lecture, the party was invited to a very adequate lunch at the Peking Duck, which provided a suitable prelude for Brian Heap's afternoon lecture, 'Feeding a Population of 8 Billion by the Year 2020. Biotechnology – Will It Help?', at the China Association for Science and Technology (CAST). The next day, a meeting with the Chairman of the National Natural Science Foundation of China, Zhang Cunhau (Plate 5), terminated their visit to Beijing.

After Beijing, Aaron, Liebe and Ling Thompson flew to Shanghai. Aaron visited the CAS Institute of Biochemistry where he was able to greet Zhang Youshang. A visit to the Shanghai Institute of Metallurgy stuck in Aaron's mind. He had an illuminating conversation with the director and Vice President of CAS, Jiang Mianheng[7], about the problems of managing CAS, which had been set up on the Russian model.

From Shanghai, the party made a stop in Fuzhou, where the Klugs had friends, and then home via Hong Kong. The transfer of Hong Kong to Chinese sovereignty had happened a few weeks before the Klug's arrival. A feeling of uncertainty pervaded the ex-colony. The Klugs and Ling Thompson met Lee Quo-wei (known as Sir Q. W.), the Director of the Council of the Chinese University and of the Hang Seng Bank. Sir Q. W. was one of the co-founders of the Ho Leung Ho Lee Foundation, which promotes science and technology in China. Aaron and Sir Q. W. had met in 1985 at Hull University, where both had been awarded honorary doctorates. Sir Q. W. was sanguine about the future of Hong Kong but unfortunately appeared uninterested in supporting 'Project Science'.

Aaron had given eight lectures during the trip, and he had been inundated with information, people, and experiences. The visit did indeed foster relationships between CAS and the Royal Society. The pragmatic search for knowledge, as embodied in the Royal Society motto, transcends cultural differences. Moreover, CAS was hotly debating the same problems that were bothering British science policy: how do you make research relevant; how do you commercialise research?

---

[7] Jiang Mianheng was the son of the President of China, Jiang Zemin.

Aaron was also relieved to note that some intelligent people in China were very concerned about the problems of global warming, to which China itself was contributing on a massive scale.

Aaron had been offered an Honorary Doctorate by the University of Cape Town. The ceremony would take place in December. There could scarcely be a more opportune time of year to exchange grey London for sunny Cape Town. Liebe was always thrilled to get back to Cape Town. Thus, directly after the Anniversary Meeting, they were off to South Africa for the third time in the year. Woefully, after the degree-giving at the University of Cape Town, when they were looking forward to a couple of weeks in the sun, the heavens fell in. Adam's wife Debbie phoned from Israel to say that Adam was seriously ill. Moreover, the Israel medical services were on strike! A precipitate return to England ensued in time to meet Debbie and Adam from Israel. Adam was admitted to Addenbrooke's Hospital in Cambridge for tests. It took until February to get a definitive diagnosis: pancreatic adenocarcinoma. Debbie, a medical doctor, had access to the pathology tests and kept the awful reality from Aaron and Liebe for two long weeks until the diagnosis was made known.

Adam was in and out of Addenbrooke's for pre-op medication. Adam and Debbie's sons Yoel and Omri, who had been parked with their Israeli grandparents at Hod Hasharon, 100 km to the north of Omer, joined them in Cambridge. During the eight-hour operation, Debbie and Liebe spent the time cooking. Adam came through the operation well; the boys returned to Israel, and Aaron and Liebe moved with Debbie and Adam to the flat at the Royal Society so that Liebe could indulge Adam's enthusiasm for operas and concerts. The whole family even went to see the Coen brothers film *The Big Lebowski*. Then Adam and Debbie went back to Omer, where Adam received a hero's welcome when he rejoined his department at the Ben Gurion University.

The prognosis for pancreatic cancer is very poor. Soon the liver was involved. There followed two years of crisis, chemotherapy and operations. Adam weathered each setback and remained extraordinarily productive in his field. Adam and Debbie decided to live life to the full. They even managed a holiday in Venice where Adam could sit within sight of the Grand Canal. Aaron and Liebe's commitment to the Royal Society and the support of the staff were very important factors in helping them to come to terms with Adam's terminal illness. Adam died in Beer-Sheva in August 2000.

In 1996, the President and Council instituted a medal, the King Charles II Medal: *For foreign Heads of State or Government who have made an outstanding contribution to furthering scientific research in their country.* The King Charles II medal was in fact created for Emperor Akihito[8] of Japan. In May 1998, the Emperor Akihito of Japan paid an official visit to the United Kingdom. Akihito is a serious scientist who maintains an electron microscope in the cellar of the imperial residence in Tokyo. The main theme of his research is the taxonomy of goby fishes, and he has also published papers on the history of science. On his visit, Akihito was accompanied by the Empress Michiko. On the third day of their visit the Royal Couple came to the Royal Society for the presentation of the King Charles II Medal. At the ensuing reception, Aaron and Liebe established a warm relationship with Akihito and Michiko, who extended an invitation to the Klugs to visit them in Tokyo.

The explosive progress of biology in the second half of the twentieth century brought with it hotly debated issues of policy and ethics. In 1978, the first human baby resulting from *in vitro* (external) fertilisation or IVF was born. This raised a host of ethical issues that were addressed carefully by a Committee of Enquiry set up by the UK Department of Health and Social Security, and this led to the Human Fertilisation and Embryology Act of 1990. The Committee of Enquiry was chaired by Dame Mary Warnock, Mistress of Girton College and a philosopher of note. One of the most active members of the Committee was Anne McLaren, later to be Foreign Secretary of the Royal Society at the time Aaron started his presidency. Anne McLaren was a leading figure in developmental biology. Together with John Biggers, she had shown that early mouse embryos could be cultured for a day or two *in vitro* and still develop into adult animals after transplantation into the uteri of surrogate females. This study provided the essential backdrop for the development of *in vitro* fertilisation.

The Act of 1990 provided a legal basis for embryo research and recognised that the issues were too complicated to be dealt with purely

---

[8] The King Charles II Medal has since been awarded to Abdul Kalam, President of India; Angela Merkel, Chancellor of Germany; and Wen Jiabao, Premier of the State Council of China.

by legislation. It empowered the Human Fertilisation and Embryology Authority to regulate the practice of human *in vitro* fertilisation in Britain. McLaren served with the Authority for ten years. One stipulation of the Act was to forbid experimentation on embryos that were more than 14 days old. This cut off was derived from the age at which an embryo first develops the neural groove, the antecedent of the spinal cord. When Aaron heard about this stipulation he was reminded of the teachings of Thomas Aquinas, who had maintained that the soul entered the foetus on the fortieth day after conception (somewhat later for women).

On 22 February 1997, Dolly the Sheep was presented to the public. 'Dolly' had been created using the technique of somatic cell nuclear transfer (SCNT), in which a cell nucleus from an adult cell is transferred into an egg cell that has had its cell nucleus removed. The hybrid cell is then stimulated to develop into a blastocyst[9], which is implanted in a surrogate mother. The birth of Dolly showed that the genes in the nucleus of a mature differentiated somatic cell can still revert to an embryonic cell, which can then develop into a normal foetus and healthy animal. This discovery opened the door to therapeutic cloning, for example creating healthy nerve cells to treat Parkinson's disease from embryonic blastocyst stem cells. Culling stem cells from an embryo requires the destruction of a latent human being. There was a strong public reaction against using embryonic cells in this way, with the result that all research involving embryonic cells was called into question. The UK's 1990 Act was relatively liberal (in Germany, embryo stem cell research is forbidden) but there was strong lobbying to forbid research on stem cells derived from human embryos. It was argued that the destruction of human embryos was proscribed on ethical grounds. Parliament was under pressure to amend the 1990 Act.

Aaron became involved in this debate and observed that the 14-day rule coupled with the Human Fertilisation and Embryology Authority was actually working well. Research was already tightly regulated. Any further restrictions would drive the research away from Britain. Aaron felt strongly about this issue and personally wrote letters to all

---

[9] The blastocyst is formed in the early development of mammals. In humans, its formation begins five days after fertilisation. It possesses an inner cell mass (ICM) that subsequently becomes the embryo.

650 Members of Parliament urging them not to restrict research any further. In 2001, the debate on stem cell research reached a crescendo because President Bush had to decide whether or not to forbid US Government funding. In the end, Bush permitted very limited funding of stem cell research[10]. Britain did not enact legislation to limit therapeutic cloning any further.

The issue of genetically modified (GM) crops also became highly polarised during Aaron's tenure as president. The scientific side, led by the Royal Society, portrayed itself as the embodiment of dispassionate truth, while the press attacks were entirely *ad hominem*. Broadly speaking, the press maintained that the whole exercise was a tremendous cover up with vested interests extending at least to the President of the United States. Conspiracy theories tend to be popular and help sell newspapers. In September 1998, the Royal Society felt that some issues should be clarified and issued its first report on the use of GM crops, entitled 'Genetically Modified Plants for Food Use'. The chairman of the group producing the report was Peter Lachmann, Biological Secretary of the Royal Society. The report emphasised the benefits of GM plants in agriculture, medicine, nutrition and health, especially in alleviating food shortage in developing countries. Lachmann quickly became a target of the anti-GM food lobby.

No peer-reviewed studies investigating the safety of GM food had been published before 1995, when the Scottish Office of Agriculture (SOAEFD) commissioned a three-year project, 'Genetic engineering of crop plants for resistance to insect and nematode pests: effects of transgene expression on animal nutrition and the environment'. The collaborating institutions were the University of Durham, the Rowett Research Institute (RRI) in Aberdeen and the Scottish Crop Research Institute (SCRI). The role of the RRI was to determine the level of expression of the inserted transgene products and to determine any effects on rats. Three genes were selected: snowdrop lectin (*Galanthus nivalis* agglutinin, GNA), jackbean lectin (concanavalin A or ConA, known to be toxic to higher animals) and the *Phaseolus vulgaris* or

---

[10] President George W. Bush decreed that federal funds might be awarded for research using human embryonic stem cells if the derivation process for those cells (which begins with the destruction of the embryo) had been initiated prior to 9:00 pm EDT on 9th August 2001. Many of the cell lines thus derived were not very useful.

common bean lectin (PHA)[11]. They were chosen for their effects on insect pests and differences in the severity of the effects of purified protein on the mammalian gastrointestinal tract. Target crops were potato, oilseed rape and strawberry. Árpád Pusztai, a member of the RRI team, stated in an interview on a 'World in Action' programme (Granada Television) that he had misgivings about GM crops: his group had observed damage to the intestines and immune systems of rats fed the genetically modified potatoes. He also said, 'If I had the choice I would certainly not eat it,' and 'I find it's very unfair to use our fellow citizens as guinea pigs.'[12] For the anti-GM movement, this looked like the Holy Grail.

The results that Pusztai quoted in his interview were a comparison of rats fed ordinary potatoes and rats fed potatoes with genetically inserted GNA. He maintained that the rats on the GM diet grew less well and had immune problems, even though the lectin itself caused no adverse effects at high concentrations. His conclusion was that the GM process had somehow made the potatoes less nutritious. But the GM potatoes were not a commercial variety and were never intended for human consumption: nobody was being used as a guinea pig. Furthermore, newspaper stories generated confusion over the nature of the genetic modification. The data Pusztai cited were concerned with GNA, but the press articles refer to potatoes modified with the lectin from jackbean (ConA), known to be poisonous to mammals. Subsequent work has shown that even the ConA-containing potatoes are not detrimental at the levels of expression achieved in potatoes. However, this does little to convince the anti-GM lobby.

James suspended Pusztai for speaking in public about unpublished collaborative work. He also set up a committee to re-evaluate the data. Although that committee concluded that there was no statistically significant support for Pusztai's conclusions, 23 European and American scientists released a memo supporting Pusztai, who acquired a reputation as a victimised 'whistleblower'.

Because the controversy impinged on the conclusions of the 1998 Royal Society Report, in April 1999 the Society convened a Working Group to examine Pusztai's evidence that genetically modified

---

[11] Lectins are proteins that bind selectively to certain sugars. GNA binds the sugar mannose.
[12] Quoted by James Randerson, *The Guardian*, 15th January 2008.

potatoes adversely affected the health and growth of rats. The group was chaired by Noreen Murray, known for helping to develop a vaccine against hepatitis B. Other members were Brian Heap (Foreign Secretary of the Royal Society), William Hill, Jim Smith, Michael Waterfield and Rebecca Bowden (Secretary). In June the Working Group concluded[13]:

> ...it appears that the reported work from the Rowett [Institute] is flawed in many aspects of design, execution and analysis and that no conclusions should be drawn from it. We found no convincing evidence of adverse effects from GM potatoes. Where the data seemed to show slight differences between rats fed predominantly on GM and on non-GM potatoes, the differences were uninterpretable because of the technical limitations of the experiments and the incorrect use of statistical tests.

This took the wind out of the sails of the anti-GM lobby, albeit temporarily. There was little in the way of hard data to back up the contention that GM foods were harmful. Pusztai's experiments were eventually published as a letter in *The Lancet* in 1999. Because of the controversial nature of his research, the letter was reviewed by six reviewers – three times the usual number. Although two of the referees advised rejection, the Editor, Richard Horton, decided to publish anyway. The letter reported differences between the thickness of the gut of rats fed genetically modified potatoes and in those fed the control diet. Richard Horton, in his editorial, admitted that the paper was controversial. Peter Lachmann had phoned Horton to urge him in strong terms not to publish bad science (Lachmann was dismayed that the polemic of the anti-GM lobby was preventing the use of genetically modified crops beneficial in many developing-world situations[14]). At this time Lachmann was no longer an officer of the Royal Society and his phone call was in no way connected with the Society. Horton reported that Lachmann had threatened him with dire consequences if he dared to publish and implied that the Royal Society were out to get him. For the popular press, no holds were barred: GM was evil, and the defenders of GM, including the whole of Tony Blair's government and the Royal Society, were all in the pay of Monsanto. The papers strove to outdo

---

[13] *Royal Society Report* (June 1999) Ref: 11/99.
[14] Peter Lachmann's viewpoint is set out in a chapter entitled 'Genetically Modified Organisms' in *Panic Nation*, ed. S. Feldman and V. Marks, published by John Blake (2005).

each other in a crescendo of allegations and innuendo. Even the serious-minded *Guardian*[15] reported Peter Lachmann's alleged intimidation of Horton and then maintained that the Royal Society was actively intervening in support of the pro-GM lobby:

> *According to a source the Royal Society science policy division is being run as what appears to be a rebuttal unit. The senior manager of the division is Rebecca Bowden, who coordinated the highly critical peer review of Dr Pusztai's work... The rebuttal unit is said by the source to operate a database of like-minded Royal Society fellows who are updated by email on a daily basis about GM issues. The aim of the unit, according to the source, is to mould scientific and public opinion with a pro-biotech line. Dr Bowden confirmed that her main role is to coordinate biotech policy for the society, reporting to the president, Sir Aaron Klug. However, she and Sir Aaron denied it was a spin-doctoring operation.*

At the Royal Society, the idea of Beccy Bowden spin-doctoring seemed quite droll. A month later, in his address at the Anniversary Meeting[16], Aaron attempted an even-handed appraisal:

> *There is one art whose usefulness has been much debated – if that is the right word! – over the past year or two. I have got this far without explicitly mentioning genetically modified plants, but I cannot avoid the issue. We have been accused of many things during the year, including 'breathtaking impertinence' for daring to review formally unpublished experimental data, but from which conclusions were drawn that were said to underpin anti-GM claims being vigorously pushed by certain pressure groups. It was also said that we had 'absolutely no remit' to get involved. These accusations entirely miss the point. We have been involved in the GM debate, not because we have a particular mission to defend the interests of biotechnology, still less because of vested financial interests (as some have tried to impute), but because of what I mentioned earlier: the Society's twin concerns with the values of science and with the useful arts. Where relevant, the Society is determined that public policy should be based on the best available science, rather than on propaganda or emotion; and, where a new technology has the potential to offer real*

---

[15] Flynne, L. and Gillard, M.S. *The Guardian* 1st November 1999.
[16] Klug, A. (2000) *Notes Rec. Roy. Soc. Lond.* **54**, 99–108.

*practical benefits, the Society is determined that that potential should be
fully examined along with the possible risks. In our work on GMs we are
following the tradition of our predecessors: it is of a piece with our history
over the last 340 years. And we are committed to maintaining that
tradition with increased vigour and effectiveness in the years to come.*

A strange aftermath was an article in the *Financial Times* on 3rd
March 2000 by Clive Cookson, maintaining that Aaron had been
cautious and inward-looking as president, and that he had frustrated
the Royal Society's attempt at a thorough review of GM foods.
Perhaps the comment reflected Aaron's attempts to keep the debate in
the realm of objective decision-making. The injustice of the *Financial
Times* article was galling. In point of fact, during Aaron's presidency
the Royal Society had been outstandingly active *pro bono publico* on a
range of contentious issues. Three vice presidents of the Society,
Patrick Bateson, John Enderby and Brian Heap, wrote a joint letter[17]
of protest to the editor of the *Financial Times*:

> . . .*He (Aaron Klug) brings to the Presidency intellectual rigor and
> integrity, penetrating insights and knowledge of a staggering array of
> fields, both scientific and cultural. He has been at the forefront in
> engaging the Society in matters of great importance. Far from frustrating
> the Royal Society's thorough review of the claims about GM foods he has
> insisted that it go ahead. . .*

– a fitting tribute to Aaron's presidency.

On 4th August 2000, Queen Elizabeth the Queen Mother was 100.
Since she was a Royal Fellow, Council determined that she should be
invited for tea. The Royal Society had apparently grossly under-
estimated the resilience of the elderly lady, since the invitation quickly
elicited the response that she would rather come for lunch. The date was
fixed for 1st June 2000. This was a remarkable occasion. First there was a
small reception during which the Queen Mother went round the room
and had a long chat with Miriam Rothschild; then lunch. Eight Fellows
whose age exceeded 100 years were invited, and three of them came. At
lunch the Queen Mother sat with Aaron on her right and an ancient
Fellow on her left. Aaron asked how she was getting along with her

---

[17] Aaron Klug collected papers in the Churchill Archive Centre, Cambridge

elderly neighbour, who was obviously hard of hearing, to which the Queen Mother responded robustly: 'It's OK. I just shout!'

In the Society, there was a tradition that a quinquennial dinner should mark the end of a presidency. In recent years this has fallen into abeyance, but in 2000 it was thought appropriate for Aaron and Liebe to hold a President's Reception and Dinner. Furthermore, the event was honoured by the presence of the Queen and the Duke of Edinburgh. The dinner on 22nd November 2000 was attended by the vice presidents, the president-elect Robert (Bob) May and the Executive Secretary and their wives (Plate 6).

The guests included Max Perutz and seven other Nobel Prize winners. George Porter came as an ex-president with a particularly close friendship with Aaron. The guest speaker was Lord John Browne, the Chief Executive Officer of British Petroleum. John Browne was a Cambridge-trained chemist who lived in Madingley, a village adjoining Cambridge, and was friendly with the Klugs. Three Royal Society Fellowship holders from China were also present. The Queen received a posy before touring the exhibits, then she left. The Duke was pleased to stay for dinner. In his speech John Browne summarised the trials and tribulations of Aaron's presidency and pointed out how fortunate they all had been to have Aaron's concentrated intelligence to help them steer through a difficult five years from which the Society had emerged stronger and more effective.

Aaron's presidency ended with the Anniversary Meeting on 30th November 2000. His valedictory address included a wide-ranging discussion of the sequence of the human genome, which had been released in a preliminary form in June 2000 and was published (99.9% complete) in *Nature* on the 15th February 2001. As recounted, the LMB had been a prime mover in getting this project set up. Aaron summarised the development[18]:

> *The biology of the nematode worm, C. elegans, had been a subject of study in the Laboratory [of Molecular Biology] as a model organism. To facilitate understanding of its genetic programme, John Sulston undertook a mapping of its genome. . . On completion of the genome map, in 1990, Sulston, in collaboration with Robert Waterston's laboratory at*

---

[18] Klug, A. (2001)*Notes Rec. Roy. Soc. Lond.* **55**, 165–177.

Washington University in St Louis, Missouri, began sequencing the C. elegans genome. This consists of about 100 million DNA bases, and was thus a formidable objective. I was then Head of the LMB and encouraged Sulston to go ahead. Sulston was supported by the Medical Research Council and Waterston by the US National Institutes of Health, the NIH. This was a bold step, undertaken at a time when people worldwide were still talking about the problems of genome sequencing of complex organisms, debating whether to wait for better biochemical techniques and more advanced automation. Sulston and Waterston simply got on with it, using whatever techniques were at hand, improving them, and incorporating advances. Moreover, there was concern about the possible cost. . . The nematode project therefore came to be looked on as the touchstone, or indeed pilot project, for human DNA sequencing. I well remember Jim Watson, who by then had become head of the potential US effort at the NIH, telling us that, if the cost of DNA sequencing could be reduced to 50 cents a base or less, he could get the go-ahead for human genome sequencing to begin in the USA.

By 1992, not only had the cost of nematode sequencing come down, but long continuous tracts of DNA sequence, over a hundred thousand bases long, had been obtained. Genomic sequencing had been demonstrated. This allowed Sulston and myself to make an approach to the Wellcome Trust, with a proposal to begin sequencing the human genome, based on the experience with the nematode. This led to a joint MRC–Wellcome initiative, in which the MRC continued to support nematode sequencing, while training people and setting up the methodologies for human sequencing, the latter to be supported by the Wellcome Trust. It was in this way that eventually the Sanger Centre was formed, with John Sulston and his colleagues from the LMB forming its core. We, in Britain, owe the Wellcome Trust a special debt for deciding to support human genome sequencing in this country. In 1998, the nematode C. elegans became the first multicellular organism to have its complete genome sequenced, the work having earlier illuminated the way forward.

The International Human Genome Sequencing Consortium was truly international, but the coordination (and much of the work) was largely a US–UK joint venture. In his address, Aaron stressed the importance of making scientific results public at the earliest possible moment. He expressed his distaste for the attitude of the Celera private human genome initiative, which published in *Science* at the same time as the *Nature* articles from the International Consortium: the Celera researchers kept their results to themselves but at the same time availed themselves of the public work of the Human Genome Project to order their DNA fragments.

Later in his address, Aaron returned to the threat of global warming. Indeed, he had brought up this topic in each of his Annual Addresses, which led Aaron to liken himself to Cato the Elder, a frequent speaker in the Roman Senate. In every speech, whatever the topic, Cato would end with the comment: *Carthago delenda est*[19].

In 2000, Aaron was elected to honorary membership of the Japan Academy, which has a similar role to the Royal Society but is even more exclusive since the number of distinguished members is limited to 150. Moreover, it covers both science and the humanities. The Japan Academy extended an official invitation to Aaron to visit so that he could receive the Academy Medal and make a lecture tour. A visit was planned for May 2000 but had to be abandoned on account of Adam's illness. Since the allocated funds would vanish at the end of the Japanese Financial Year, the visit took place in March 2001. Strictly this was no longer a Presidential visit, but even so it carried a Presidential aura.

Setsuro Ebashi, an internationally famous biochemist, member of the Japan Academy, and a foreign member of the Royal Society, was entrusted with organizing the visit. Ebashi discovered the role of calcium ions in stimulating muscle contraction. On account of chronic ill health, he delegated the organisation to Masashi Susuki, a biochemist who had been a visiting scientist at the LMB. Zinc finger nucleases were now proving to be effective for incorporating DNA sequences at specific sites in the genome. The future of gene therapy looked assured, and Aaron was enthusiastic to talk about these developments. On arrival in Tokyo, Aaron was taken to the National Cancer Research Centre where he held a lecture on zinc fingers. This was followed by lunch at the Japan Academy and the presentation of the Medal. Two days later, Aaron visited the National Institute for Physiological Sciences at Okasaki, as Ebashi was President of the Okasaki National Institutes. Aaron gave a talk on zinc fingers and afterwards met Fumio Oosawa from

---

[19] 'Carthage must be destroyed.'

neighbouring Toyota, who earlier had worked out the mechanism of actin polymerisation. In the evening, Aaron and Liebe were invited to a shabu shabu restaurant by the Ebashis (Plate 7). The tour continued with lectures in Nagoya, Kyoto, Osaka and Tsukuba. On returning to Kyoto, Aaron and Liebe were invited for tea by the Emperor Akihito and Empress Michiko.

In August 2002, George Porter died. This was a shock and a sad loss for Aaron. Over the years, a deep friendship had grown up between them, reinforced by David Klug's very successful scientific apprenticeship with Porter at the Royal Institution. David later moved with Porter to Imperial College, where he is now Chair of the Institute of Chemical Biology. Porter had been elevated to the House of Lords, as Baron Porter of Luddenham, and been awarded the Order of Merit.

Next to Westminster Abbey stands St Margaret's Church, a beautiful Anglican Parish Church often referred to as the Church of the House of Commons. The Royal Society sometimes uses St Margaret's for memorial services, and a Service of Thanksgiving for George Porter was held there on 21st January 2003. Aaron gave the Address, in which he emphasised Porter's discovery of flash photolysis that led to his Nobel Prize and, among many things, to our understanding of the ozone/CFC story. Porter was also very successful in popularising science, an important aspect of his 20 years as Head of the Royal Institution. He was an innovative and engaged President of the Royal Society and fought hard for science funding. Aaron concluded[20]:

> I have tried to convey the breadth of George Porter's public life and the vigour with which he pursued it, but it should be said that, all along, he never gave up the race against time. He won it, by dissecting the minute divisions of time in chemical and biological processes. He retired full of honours and with the satisfaction of having fulfilled his early ambition of 'advancing our understanding of the natural world'. We mark the passing of a great figure in twentieth-century chemistry and in British science. We give thanks for his life.

---

[20] Klug, A. (2003) *Notes Rec. Roy. Soc. Lond.* **57**, 261–264.

20

Ben Gurion University

When Adam Klug left school, he took a 'gap year' (1972–1973) in Israel with a group of 20–30 youngsters who had all belonged to the Habonim youth movement. Deborah (Debbie) Davis, who had been at school in London, was one of the group. During this year, Adam and Debbie started going out together. Rather like Aaron and Liebe a generation before, they were both fired with the idea of Aliyah ('ascent'). Unlike Aaron and Liebe, they actually ended up in Israel.

On the basis of his results in the Oxford entrance exam, Adam was awarded a 'demyship'[1] at Magdalen College. He went up in the autumn of 1973 to read history and ended up with a first-class degree. One doubts if the political atmosphere at Magdalen was in line with Adam's decidedly left-wing views. He apparently chose Magdalen because of the Deer Park. At the same time, Debbie studied medicine at Birmingham. The couple had made up their minds to emigrate to Israel as soon as Debbie had qualified.

After Oxford, Adam studied for a Masters Degree at the London School of Economics. When Debbie qualified, they moved to Israel and lived first in Haifa and then Tel Aviv, where Adam did his PhD and Debbie did a specialty in Family Medicine. They married in 1982. Their

---

[1] A demyship (or demy) is a scholarship at Magdalen College Oxford instituted by the founder, Bishop William of Waynflete. It is derived from *demi-socii* or half-fellows. Oscar Wilde, Lord Denning and T. E. Lawrence were among famous recipients.

first son Yoel was born in April 1984 and their second son Omri in February 1987. In the mean time, Debbie's parents had also emigrated to Israel.

In 1990, Adam, Debbie and the children moved to Princeton. Adam was a visitor in the International Finance Section at Princeton for two years and a visiting professor at Rutgers for two years[2]. Debbie worked in the University Hospital. Adam's field of interest was financial history, and he published several important articles on German Reparations and Sovereign Debt default in the 1930s, as well as an article on the Suez crisis of 1956. He collaborated with Michael Bordo on a project on the Sterling crisis of 1967, with Eugene White on leading indicators of the Great Depression in the United States, and with Doug Irwin on the political economy of tariffs. Adam had a broad range of expertise; his American colleagues told Liebe and Aaron that in a short while Adam's work would make 'a big splash'. Nevertheless, the parents thought that they had better go back before their boys completely forgot their Hebrew, and in August 1994 the family returned to Israel. Adam was offered a post in the Economics Department of Ben Gurion University (BGU) of the Negev in Beer-Sheva, a new university with a lively department. They settled in Omer, a pleasant modern town to the northeast of Beer-Sheva. Omer, originally founded as a Moshav (a cooperative agricultural settlement), had been developed by the BGU for its members. Debbie took a partnership in a medical practice with another UK trained medic. The boys settled into school.

When the University of the Negev was founded in 1969, David Ben Gurion, Israel's first Prime Minister, expressed the hope that this would become Israel's Oxford, perhaps overlooking differences in climate and the fact that the BGU is built entirely out of concrete. One of the dreams of the Zionist founders was to settle the Negev – to make the desert bloom. When Ben Gurion withdrew from politics he lived in the Sde Boker, about 50 km south of Beer-Sheva. Sde Boker is perhaps the best known of all the kibbutzim. Ben Gurion died while the University was just beginning. He would have been delighted that the BGU has been so successful: there are now over 20,000 students, an unprecedented rate of growth for Israel.

---

[2] The following is from a eulogy at Adam's funeral by Michael Bordo, Hugh Rockoff and Eugene White.

From the time that Adam and Debbie settled in Beer-Sheva, Aaron and Liebe became frequent visitors. Liebe pointed out to Aaron that he did a lot for the older universities in Israel – the Hebrew University, the Weizmann Institute, the Haifa Technion – so why not do something for the BGU? While Aaron agreed that this was a good idea, he felt he was already overloaded. He asked Liebe to initiate contacts. Back in London, Liebe telephoned the President of the Governing Body of the University, Hyman Kreitman, to offer Aaron's expertise and experience.

Hyman Kreitman was, for some years, chairman of Tesco, the British supermarket chain that was founded by his father-in-law Sir John (Jack) Cohen. However, Kreitman was too gentle a character to cope with his father-in-law, and in September 1973 he resigned to devote his life to the Arts, both as collector and benefactor. He and his wife Irene were important benefactors of the Tate Gallery. Moreover, they both became ardent supporters of the BGU. The Kreitman School of Advanced Graduate Studies was established in 1996, serving as the framework for all graduate studies. In the same year, Hyman and Irene Kreitman endowed an annual lecture series at the BGU.

Kreitman introduced Aaron to Avishay Braverman, the President of the BGU, and they quickly developed a rapport. For Aaron, this initiated a ten-year involvement with the university. Hyman and Irene Kreitman set up the Kreitman Fellowships, doctoral and postdoctoral, to encourage Israelis to stay on in Israel. For a number of years, Aaron chaired the Fellowship Selection Committee. As mentioned before, on 28th November 1995, two days before he became President of the Royal Society, Aaron addressed the festive session of the Senate of BGU to celebrate 25 years since the founding of the University. Aaron's lecture was on 'Some Reflections on Science and Science Policy'. In the autumn of 1996, Aaron and Liebe undertook two more journeys to Israel. In September Aaron spent a week at the BGU where he gave the first of the Kreitman lectures. This was also a welcome opportunity to visit Adam, Debbie, Yoel and Omri in Omer. The family celebrated the Jewish New Year together.

In November 1996, Aaron gave the 18th Katzir-Katchalsky Memorial Lecture at the Weizmann Institute. Aharon Katzir-Katchalsky, who was a physical chemist of renown, had been assassinated in a terrorist attack on Lod Airport in 1972. He was the designated first President of the BGU but was killed before he could take over the position. The occasion of the memorial lecture was a deeply moving moment for Aaron because he and Aharon had become good friends during Aharon's sabbatical at Birkbeck College in 1959. They had worked together on the structure of polyadenylic acid, which Aharon had thought might be a progenitor molecule to the origin of life. Aaron was invited to give the lecture by Aharon's younger brother Ephraim, a well-known biochemist who became the fourth President of Israel. The next day Aaron was awarded an honorary doctorate by the Weizmann. The irony of the situation was not lost on Aaron as he remembered the brusque refusal even to contemplate offering him a position at the Weizmann in 1950.

When Adam became ill, Aaron and Liebe decided to buy a house nearby. Liebe spent a long weekend in Omer and found a corner house[3] that she liked in the area called 'The Vineyard' just across the green from Adam and Debbie. Adam died on 8th August 2000, just three days before Aaron's 74th birthday. Aaron and Liebe were in Omer at the time.

Adam was very much like Aaron in that he was a born scholar with an extraordinary wealth of knowledge. He had little patience with opinionated people who had no facts on which to base their opinions. When Aaron had to choose a piece of music to be played at the Nobel Prize ceremony, Adam joked that his father knew everything about everything except for two subjects: sports and music. Adam, in contrast, loved all types of music and was an ardent football fan. At Adam's funeral, his colleague Jimmy Weinblatt, now Rector of the BGU, expressed everyone's feelings:

> *Adam's death is untimely, in all respects. He was young, he leaves a young family who needs his presence, he leaves parents who expected to see him flourish, and he leaves us and an unfinished work that would have been superb[4]. Adam was a brilliant economist or rather a brilliant scholar. . . Adam, on top of being a wonderful source of knowledge, was also a great*

---

[3] 15 Ha Karem Street, Omer 84965, Beer-Sheva, Israel.

[4] Adam Klug was an economic historian. His method, known as cliometrics, was to use quantifiable data, such as tax records, to illuminate historical developments. Adam's last paper (published posthumously) was 'Why Chamberlain failed and Bismarck succeeded: the political economy of tariffs in British and German Elections', in the *European Review of Economic History* (2001) **5**, 219–250. He also left behind the draft of a book that was also published posthumously: *Theories of International Trade*, Routledge.

human being who was always willing to share his knowledge and help
people with all his heart. Adam! Dear friend and colleague, we will badly
miss you, your wisdom and kindness.

After Adam's death, Aaron and Liebe visited often to be in touch with
Debbie and their two grandsons. At the time of writing, the elder of
Adam's sons, Yoel, is at the Weizmann Institute doing research for his
doctorate. Omri studied Mechanical Engineering at Tel Aviv University.

At the end of March 2002, Aaron put his name to a statement
published in the Jewish Chronicle drawn up by the 'British Friends of
Peace Now' that called for an end to the West Bank settlements. The
statement accused Israel of a 'betrayal of its moral foundations' which
'affects the whole Jewish people.' While condemning 'the use of terror
by extremist Palestinian groups', it said that 'this should not blind us to
the deeper issues posed by Israel's occupation of, and current behaviour
in, the Palestinian territories... In order to save both the Jewish state
and the spirit of the Jewish people, we all have a duty to speak out
against the occupation.' The developments in Israel were so far from the
socialistic visions of his youth in the Hashomer Hatzair that Aaron felt
impelled to speak out. Moreover, it had been Adam's dying wish that
Aaron should make such a statement. Perhaps unsurprisingly, the
statement led to a total break of the friendship with Norman
Podhoretz. Aaron remarked that one's point of view is influenced by
whether you live in the relative security of Manhattan or in Beer-Sheva,
where sooner or later one has to come to terms with one's neighbours.
A little later, Aaron gave his support to a committee that was trying to
convince British university groups not to boycott Israeli universities on
account of the Israeli Government policy of West Bank settlements. He
felt that the British university protest was aimed in the wrong direction.

The Institute for Applied Biosciences had been set up on the BGU
campus and was inaugurated in 2000. This was the vision of the Swiss
banker and hedge-fund pioneer Edgar D. de Picciotto. Under its director, Shoshana Arad, the Institute specialised in polysaccharide
research. An important member of the Scientific Advisory Committee
was Raymond Dwek from Oxford, an international authority on poly-
saccharides. Dwek and the President of the BGU, Avishay Braverman,
thought that the programme of the Institute was too narrow. Moreover,
the situation was politically fraught. Because Aaron had no personal axe

to grind, they requested his help and advice. Together they set up a new Institute framework with a wider base, incorporating the Institute of Applied Biosciences, to be named the National Institute for Biotechnology in the Negev (NIBN). On 22nd May 2001, Israel's Prime Minister, Ariel Sharon, officially announced the founding of the NIBN. This was to provide an applied research platform for biotechnology-based industries in the Negev and Israel. On becoming Chairman of its International Advisory Committee, Aaron recruited Phillip Needleman, head of the pharmaceutical research and development at Monsanto Company, as a science adviser. The centre now was effectively in the charge of Aaron, Dwek and Needleman. In 2002, the philanthropist Morris Kahn agreed to support the establishment of the Morris Kahn Human Molecular Genetics Laboratory, and had the foresight to move this entity under the umbrella of the NIBN. The NIBN rapidly expanded in both research and faculty.

In November 2005, the Israeli Government announced its commitment to provide $30 million to help defray the costs of establishing the NIBN as part of a $3.6 billion, 10-year master plan targeted to developing the Negev region. Aaron, now 80 years old, was named Acting Executive Director of the NIBN, a post he held for two years. He was able to recruit Irun Cohen, a specialist in auto-immune diseases from the Weizmann Institute, as Executive Director, but Cohen did not stay long. He was succeeded by Varda Shoshan-Barmatz, who had been his deputy.

Aaron made a great difference to BGU, not only in the big things, but also in the way his example gradually changed the culture. In recognition of Aaron's dedication, in April 2013 the NIBN established the Aaron Klug Integrated Centre for Biomolecular Structure and Function (AKIC-BSF). It was opened in the presence of Debbie, Yoel and Omri. Aaron, now nearly 87, felt too frail to travel; he sent a lecture on a CD. The AKIC-BSF was to be a National Centre of Excellence for structural studies. The ceremony was opened by Raymond Dwek who introduced the British Ambassador Matthew Gould. Gould gave a very warm introductory address to the participants in which he explained why he was delighted to be at this dedication. He commented:

> ...*any ambassador would be delighted to pay tribute to Prof. Klug and to be at an event that sets in stone the link between Britain and Israel. It is a*

particular pleasure to be here for three reasons. First, the opening of the centre ties together two of my favourite universities, BGU [Gould received an honorary doctorate from BGU in December 2011] and Cambridge University and specifically Peterhouse, which is Prof. Klug's college and where I studied. Second, it is a celebration of the relationship between Britain and Israel in science. Both are scientific superpowers with complementary strengths. Third, because it pays tribute to Prof. Klug. He is universally known as kind, fair and decent. He is a world class scientist without the ego to match. He has shown us what science can achieve and should be doing to unlock Nature's secrets to provide pathways to healing.

21

Prizes

One winter's day in 1972, Liebe had a phone call from Aaron. In bemused tones, he said, 'I've been awarded something called the Heineken Prize. I don't know what it is, or who nominated me.' Liebe passed on the good news to David, who was somewhat incredulous. How could his dad, who mostly appeared rather vague, possibly get an award? Apparently Aaron's work and achievements had never encroached on family life.

More information soon arrived. Alfred Heineken had created the first Heineken Prize in honour of his father Henri Pierre Heineken, who was a biochemist. The first prize in 1964 was awarded to Erwin Chargaff, renowned for discoveries concerning base ratios in DNA. Alfred Henry 'Freddy' Heineken was for many years the CEO of the Heineken International Brewing Company, founded in 1864 by his grandfather. At the time of his death in 2002, Freddy Heineken was one of the richest people in the Netherlands.

In the spring of 1973, Aaron visited Holland for a week to give lectures at all the major universities. He talked about his work on tobacco mosaic virus assembly. At the end of the week, Prince Claus of the Netherlands presented the prize in a ceremony at the Beurs van Berlage in Amsterdam. Aaron was put up in the Grand Hotel, a splendid Art Deco building. As a tangible representation of the prize Aaron was presented with a reproduction of Leeuwenhoek's microscope, mounted on a large glass crystal. At the Grand Dinner, Freddy Heineken was in a jolly mood. He was intrigued by the question:

What Motivates Mankind?

Prize winners were expected to respond to this question. Aaron's answer was characteristically uncomplicated:

What motivates Mankind is survival. Once survival has been assured, curiosity is the most motivating force.

In September 1981, the Klugs were at a Summer School in Spetsai, an island in the Peloponnese favoured as a yacht harbour by rich Athenians and for package holidays by the British. Aaron and Liebe had attended this School every summer for five years in a row. The subject of this summer's meeting was 'Control and Processing in the Biosynthesis of Macromolecules', co-organised by Brian Clarke and Hermann Bujard. Aaron presented a paper on the structure of chromatin. During

Figure 21.1 Liebe, Aaron and Prince Claus of the Netherlands at the Award of the Heineken Prize to Aaron on 15th April 1973 in the Beurs van Berlage, Amsterdam. (© MRC Laboratory of Molecular Biology.)

the meeting, a phone call came through from Jenny Brightwell, secretary to the Structural Studies Division of LMB, to tell Aaron that he had been awarded the Louisa Gross Horwitz prize.

The Louisa Gross Horwitz Prize was established under the will of the late S. Gross Horwitz through a bequest to Columbia University, and is named to honour the donor's mother. Each year, since its inception in 1967, the prize has been awarded by Columbia University for outstanding basic research in the fields of biology or biochemistry. Ten years previously, the Louisa Gross Horwitz Prize had been awarded to Hugh Huxley. In 1980, before Aaron's award, it had been given to César Milstein, Head of the Protein and Nucleic Acid Chemistry Division of the LMB. In 2008, the prize was awarded posthumously to Rosalind Franklin.

Thus, in November 1981, the Klugs set off for New York. They stayed at the Plaza Hotel on the corner of Fifth Avenue and 59th Street in a suite on the 15th floor overlooking the Park. The presentation was at Columbia University followed by a dinner in the library. Aaron's student friends, Vivian Rakoff, Ralph Hirschowitz, Bennie Kaminer and Freda Kaminer, Norman Podhoretz and Midge Decter, all came along. For the occasion, Aaron had visited his trusted outfitters on Cherry Hinton Road in Cambridge and bought himself a pair of new black shoes, but without trying them on. It turned out that they were far too small. Fortunately, Aaron was able to borrow a pair from Norman Podhoretz. When the presentation and formal dinner were over, the Klugs and friends retired to the Plaza, which with its 'Great Gatsby' ambience seemed appropriately decadent for strawberries and cream. The Klugs subsequently stayed for a few more days with the Podhoretzes, in their elegant flat on the West Side.

The Louisa Gross Horowitz Prize is often cited as the harbinger of a Nobel Prize, and so indeed it was. On Monday 11th October 1982, Aaron phoned Liebe at home. He reported that he had had a phone call from Stockholm. The caller had said, 'Are you sitting down?' before telling Aaron that he had won the Nobel Prize in Chemistry. When Liebe asked Aaron with whom, he replied, 'On my own!'

This was the sixth Nobel Prize awarded to a member of the LMB. The news came during the week of the LMB talks, which had been instigated in the 1960s by Francis Crick to help Perutz (as Chairman and *de jura* Director) keep abreast of what was going on in the laboratory.

They had become an important part of the LMB tradition, and Aaron was loath to disrupt them. Nevertheless, the word quickly spread through the lecture theatre, whereupon the atmosphere brightened. A reporter from the *London Times* phoned up for details and happened to get hold of Sydney Brenner. Having asked Brenner all the standard things about the science and its significance, he tentatively asked Brenner, 'Is he British?' to which Brenner replied in his broadest South African, 'He's as British as I am!'

Then the mail began to arrive, piles and piles of it. Instructions arrived from Stockholm, including a list of what clothes one would have to wear (Vivien Perutz told Liebe that Perutz had had to arrange an overdraft to pay for the family's Nobel outfits) followed by a celebratory lunch at Peterhouse hosted by the Swedish ambassador. The press came – of all kinds and from all manner of places – to a degree that Aaron found overwhelming. He tried to continue working but never got round to writing his Nobel Lecture, which he actually sketched out in Stockholm on the night before the lecture.

The Klugs arrived in Stockholm with seven suitcases. They were met at the airport with flowers, photographers and journalists and conveyed to the Grand Hotel, flames leaping from the pediment displaying the flags of all nations. The Swedish Krona had just been devalued so that the prize was not quite as valuable as it might have been. As a small recompense, the speaker of the Riksdag arranged for a bicycle to be presented to Aaron, it being well known that Aaron generally rode a bike to the lab. Unfortunately, it was a racing bike, not quite Aaron's style. In the end Aaron gave it to David, who used it a lot in London.

A Nobel Laureate is expected to give a summary of his or her work at the Nobel Lecture two days before the prize-giving. Aaron's Nobel Lecture was held in the Beijer Hall of the Swedish Academy of Sciences in the afternoon on 8th December. He was introduced by Bo G. Malmström from Göteborg, chairman of the Chemistry committee at that time. Since Aaron had not managed to write his lecture, he spoke unscripted (this is fairly common for scientific lectures where speakers tend to rely on the sequence of slides as a prompt). He described the work on tobacco mosaic virus, TMV, which had been started with Rosalind Franklin, and he could not help surmising that she would have been sharing the podium had she had not died so young. He recounted how a combination of crystallography, chemistry and electron microscopy had led to a detailed understanding of the structure and function of this simple virus. He then described the work with David DeRosier that had led to 3D reconstructions of structures from electron micrograph images, and explained how mathematical manipulations of the image could remove the effects of underfocus in electron micrographs. He finally showed how the approach that had been developed for TMV could be successful in unravelling the structure of chromatin, the storage form of DNA in the nucleus, which at the beginning had appeared to be an intractable problem. In the written form of his lecture[1] Aaron concluded with these remarks:

> *I particularly wanted to outline the chromatin work because it may serve as a contemporary paradigm for structural studies which try to connect the cellular and the molecular. One studies a complex system by dissecting it out physically, chemically, or in this case enzymatically, and then tries to obtain a detailed picture of its parts by X-ray analysis and chemical studies, and an overall picture of the intact assembly by electron microscopy. There is, however, a sense in which viruses and chromatin, which I have described in this lecture, are still relatively simple systems. Much more complex systems, ribosomes, the mitotic apparatus, lie before us and future generations will recognise that their study is a formidable task, in some respects only just begun. I am glad to have had a hand in the beginnings of the foundation of structural molecular biology.*

The literature prize for 1982 went to the Columbian novelist Gabriel García Márquez. Aaron would very much have liked to hear Márquez's impassioned speech calling for an end of 100 years of South American solitude. However, since it overlapped with his own speech (and was in Spanish), he had to be content with the printed version. Márquez's humanitarian views commended themselves highly to Aaron.

The Nobel Foundation invites members of the family to participate in the jollifications. Thus Aaron's brother Bennie, sister Robin, David, Adam and his wife Debbie, and Aaron's cousins from Cleveland accompanied Aaron and Liebe to Stockholm. The ceremony is held on 10th

---

[1] 'Aaron Klug – Nobel Lecture: From Macromolecules to Biological Assemblies'. *Nobelprize. org*. Nobel Media AB 2013. http://www.nobelprize.org/nobel_prizes/chemistry/laureates/1982/klug-lecture.html

December (the anniversary of Alfred Nobel's death). It starts with a rehearsal in the morning, followed by time to get into the finery, then off to the Stockholm Concert Hall. The King presents each Laureate with the Gold Medal of the Royal Swedish Academy of Sciences (Plate 8) and leather-bound folder containing the prize citation.

The subsequent Nobel Banquet, which is held in Stockholm City Hall, is opened by the King who proposes the toast of Alfred Nobel. During the meal, each Nobel Laureate goes to the top of the staircase to give an expression of thanks for the recognition of his work.

Aaron eloquently praised the prize[2]:

> *Your Majesties, Your Royal Highnesses, Your Excellencies, Ladies and Gentlemen, I wish to express my profound gratitude for the distinction with which you have honoured me today. Ever since I heard the news of the Royal Swedish Academy's decision, and even since arriving in Stockholm, with its flags flying and flames leaping, I have not been able to shake off a feeling of unreality. It is as though I have been cast as an actor invited to take part in a production and not quite knowing his lines. But, the ceremony today, the dignity of the proceedings, the magnificence of the surroundings, this glittering company, have made it all real. Moreover, the formality of this great occasion is combined with a friendliness and a hospitality which makes it delightfully easy to take part.*
>
> *I am deeply conscious that though the Prize has been awarded to me, it is a Prize also to my field of the study of biological machinery. This field is not necessarily glamorous, nor does it often produce immediate results, but it seeks to increase our basic understanding of living processes. The work requires a moderately large investment in technological and theoretical developments and long periods of time to carry them out, without the pressure to achieve quick or short term results. This is, of course, in the gift of our fellow citizens and we very much appreciate the freedom to follow our instincts and to try to solve what we think can be solved.*
>
> *People often ask what is the use of it. In a world where there are pressing problems, why doesn't one devote one's efforts to the practical*

---

[2] 'Aaron Klug – Banquet Speech'. *Nobelprize.org*. Nobel Media AB 2013.
   <http://www.nobelprize.org/nobel_prizes/chemistry/laureates/1982/klug-speech.html>

benefits of mankind. I need only recall the answer of the great Michael Faraday, when at a public lecture he was demonstrating the production of electricity. 'Of what use is your invention, Mr. Faraday?' demanded an important lady. 'Madam', he replied, 'of what use is a new born child?' If – quoting freely from François Jacob – basic science has emerged from its original, and perhaps necessary, obscurity, if the public at large has come to understand its role in the evolution of our culture and society, then this is a large part due to the manner in which you, in this country of Sweden, have interpreted and realized the will of Alfred Nobel. By their independence, by their rigorous work, the Nobel Committees have given the Prize a unique position and prestige.

The Prize has not only marked discoveries of obvious benefit to mankind, but it has also set standards of excellence in fundamental work, which may bear fruit only in the distant future. In these days, when there are constant calls for research devoted to particular ends – and, I do not wish to minimize the importance of these policies – yet there should always be left room for apparently unguided research on problems that seem to have no practical application at the time. One cannot plan for the unexpected. Human curiosity, the urge to know, is a powerful force and is perhaps the best secret weapon of all in the struggle to unravel the workings of the natural world.

It is the celebration of this spirit which, I think, formed part of the intention of Alfred Nobel and of the significance of the Prizes he has created. I am privileged and honoured to have been included. For this day, for this night, I thank you all.

In the old Julian calendar, 13th December was the Winter Solstice, which in dark northern Europe is a serious reason for a celebration with lights. Somehow this pagan festival of lights was transformed into Saint Lucia's Day. The Nobel guests were warned that they would be awakened on that day by singing maidens clad in white with red sashes and crowned with candles, bearing coffee and cinnamon cakes. David was particularly enthusiastic about this happening. What had not been said was that the maidens would be accompanied by a television crew and photographers. That evening, the Laureates joined the students at the Saint Lucia Ball where the literature Laureate, Gabriel García Márquez, crowned Santa Lucia. The Laureates were expected to take part in various charades and games. Aaron did something incorrectly and was awarded the Order of the Frog, of which he was duly proud.

In March 1983, Aaron was awarded the Chancellors Gold Medal of Merit by the University of Cape Town. The two previous recipients of this award were Allan Cormack and Chris Barnard, who performed the first heart transplant. The award was presented by the Vice Chancellor, Stuart Sanders. Sanders had become Vice Chancellor two years before and remained in that post for another 23 years. Aaron and Liebe became firm friends of Sanders and his wife Anita. During this turbulent time, in opposition to official policy, the University of Cape Town, under Sanders' leadership, embarked on a successful programme of recruiting black students. In addition to constant legal battles and police harassment, there was also the problem of financing such students. In 1990, Aaron became a trustee and later chairman of the University of Cape Town Trust in England. On his retirement from the chairmanship in 2010, at a ceremony in London, Aaron was awarded the Vice-Chancellor's Medal (Plate 9). During his 17 years as chairman, the Trust had raised £17 million mostly used for the support of black and mixed-race students at the University of Cape Town.

After the award in 1983, Aaron flew to Durban to visit his old school, Durban High School, that he had left in 1941. There was one main prize at the end of the school year for the most highly regarded student, called the Dux (Plate 10). Aaron had finished school during the Second World War when the custom of awarding prizes was held in abeyance. Now, some 40 years later, Aaron was awarded his Dux. After the ceremony, the boys were asked to write an essay about his visit. One 13-year-old wrote: 'Aaron Klug visited us and told us about his work. He got the *novel* prize for *genital engineering*.'

The next day, Aaron was entertained to dinner in the Durban Jewish club, where he and Bennie had played tennis and their father Lazar had enjoyed playing the card game *Klaberjass*.

In 1985, Aaron was awarded the Copley Medal of the Royal Society and thereby joined a phalanx of great eminence. This is the oldest and most prestigious Royal Society medal, having first been given in 1731 for 'outstanding achievements in research in any branch of science'. At the Anniversary meeting on 30th November, Aaron received the award from the Biological Secretary, David Smith. The citation read:

In recognition of his outstanding contributions to our understanding of complex biological structures and the methods used for determining them.

In June 1988, Aaron received a letter from the Palace asking him if he would accept a knighthood. He had been chosen as a beneficiary of the Queen's Birthday Honours. The citation was: 'For services to molecular biology'. Aaron had been put up by Dai Rees on behalf of the MRC. A tradition had grown up at the Laboratory of Molecular Biology of refusing such offers. John Kendrew had indeed accepted a knighthood but rather in recognition of his work for the Ministry of Defence than for science. By this time, Aaron was Director of the LMB and deeply involved in negotiations on all levels. He accepted the knighthood because it apparently helped when arguing with civil servants.

On 24th October 1995, the Queen appointed Aaron a member of the Order of Merit (Plate 11), one of the highest orders in the Queen's personal gift. The Order of Merit, founded in 1902 on the occasion of the coronation of King Edward VII, is a special mark of honour for persons of exceptional distinction. Membership is limited to 24 individuals. Aaron was elected to replace Dorothy Hodgkin, who had died the previous year. Since both Dorothy and Aaron were crystallographers, and both had won the Nobel Prize for Chemistry, this seemed an appropriate succession. Membership is accompanied by receipt of the Badge of the Order. To Aaron's delight, Nelson Mandela was appointed to honorary membership in the same year (full membership is restricted to UK nationals).

In 2005, Aaron was awarded the Order of Mapungubwe, South Africa's highest honour, 'for achievements in the international area, which have served South Africa's interest'.

About a millennium ago, Mapungubwe Hill was settled by an iron-age people with much skill in metal working: one of the artefacts they left behind is a splendid little rhinoceros made of gold. Although the site was discovered in the 1930s, knowledge of the finding remained very restricted because the fundamental tenet of the Apartheid policy was that the indigenous population was primitive and uncultured. The people who lived on Mapungubwe Hill were clearly cultured and had trade links extending as far as China. In recognition of their culture, the Mapungubwe badge consists of an oval frame above an inverted trapezium. Inside the oval frame sits a golden rhinoceros, with the sun rising above Mapungubwe Hill in the background (Plate 12).

The award has four classes (platinum, gold, silver and bronze). Aaron was awarded gold 'for exceptional achievements in medical science'.

Nobel Laureates appear to qualify for gold: Aaron's colleague Sydney Brenner received the award in gold in 2004, as did Allan Cormack in 2002 and Doris Lessing in 2008. In 2005, the Literature Nobel Laureate J. M. Coetzee was a co-recipient. The award is given by the President of South Africa, who at this time was Thabo Mbeki. Mbeki had been a very competent executive during Mandela's presidency. He did much to get the South African economy rolling. However, when he became President, his continued tolerance of Robert Mugabe and strange (and very damaging) views that HIV does not cause AIDS diminished his reputation. Moreover, from within South Africa, his style of government was thought by some to be remote and academic. Archbishop Desmond Tutu was particularly critical, pointing out that a culture of 'sycophantic, obsequious conformity' was emerging under Mbeki[3]. Aaron was well aware of these problems: his second cousin, Mark Gevisser, was writing Mbeki's biography[4]. Mbeki's irrational HIV denial policy had cost many lives. Aaron wondered about taking an award from Mbeki but he reasoned that, whatever his views, Mbeki was the representative of South Africa and Aaron was proud of the award.

Thus, one fine day in April 2005, Aaron and Liebe were in Pretoria for the award ceremony. It was the antithesis of the Nobel award ceremony. The awardees and guests sat in a large amphitheatre. The President sat on a slightly raised dais. People, mostly black, performed and danced in the centre in a light-hearted atmosphere. The presentation was organised in groups according to the class of the award, with the bronze awards first. Widows and mothers received the award on behalf of partners or sons distinguished for their bravery. Mbeki was very relaxed and offered a helping hand if needed. Each award was accompanied by a laudation: some were in English, some in Xhosa. Interspersed were performances from a string quartet, a choir and an indigenous dance group. Last came the two gold awards to the two Nobel Laureates.

Then there was lunch in an open tent. Aaron and Liebe sat with Thabo Mbeki and his wife Zanele. Mbeki is a charming man with an interest in Shakespeare, and for the Klugs it turned out to be an agreeable meeting. Apartheid was now just unpleasant history.

---

[3] 'Special Report: Thabo Mbeki: A man of two faces' *The Economist* (20th January 2005).
[4] Gevisser, M. *Thabo Mbeki: The Dream Deferred.* Jonathan Ball (2007)

22

Envoi

Liebe's summary of Aaron's life:

Aaron is basically a conventional person who has led an unconventional life, both personally and scientifically. He didn't choose it, but that is how it turned out.

True, Aaron is conventional, but his implementation of conventionality led to a number of revolutionary changes in the way structural cell biology is done. Blessed with an excellent memory and a fine analytical mind, Aaron was driven by an unbridled and intense curiosity. Few human activities escaped his enquiries except, as his son Adam would remark, music and football. He was keen to share his insights, which made him an excellent teacher.

Aaron is a private person. He never said much about his mother's death, but the trauma seems to have reinforced his natural reserve. Not that he was without emotion: his love for his family was intense; he felt deeply about literature. Rosalind Franklin's death hit him hard. He inherited Rosalind's car, but it was more than a year before he could bring himself to drive it.

Aaron appreciated loyalty, supported his co-workers and derived pleasure from their success. He carried out little experimental work himself. John Finch (Plate 13), a gifted experimentalist, was his very effective collaborator for 40 years. All new data were subject to Aaron's Talmudic

307

scrutiny. The method worked well and led to an enormously successful research group. Usually of a generous nature, he could be angered by what he judged to be attempts to steal his results. Nor did arrant thick-headedness escape his public excoriation.

Aaron himself thought that zinc fingers were his most important discovery. Rows of zinc fingers with the appropriate protein sequences would allow the specific recognition of any DNA sequence. Aaron foresaw the ability to target specific DNA sequences as critically important for developmental biology and gene therapy. Unfortunately, specific zinc fingers are difficult to make. In 2012, another gene editing system (CRISPR/Cas9) derived from bacteria was discovered. It was much easier and cheaper to set up than zinc fingers and appears to be the method of choice. Nevertheless, recent progress owes much to Aaron's vision.

The State of Israel, with Jewish traditions and Hebrew as the spoken language, is a manifestation of secular Judaism that fascinated Aaron, but he was deeply disappointed by the politics of the present State of Israel. Once, when interviewed on Israeli television, Aaron was asked whether he believed in God, to which he replied, 'You should live your life as if there were a God.'

While waiting on Cambridge station for the train to London, where Aaron was to receive his Order of Merit, he and Liebe sat next to a classics scholar who was reading a history of Rome. Aaron turned to Liebe and said, 'That's what I really wanted to do,' to which Liebe responded:

'You've left it a bit late now.'

Appendices

Appendix A: Aaron's Birthplace – Zelva, Lithuania
By Mark Gevisser

Mark Gevisser is Aaron Klug's cousin and a South African author. His biography *A Legacy of Liberation: Thabo Mbeki and the Future of the South African Dream* won the South African Alan Paton Prize in 2008. His last book, *Lost and Found in Johannesburg: A Memoir*, includes an account of a visit to Zelva in Lithuania, the birthplace of his grandfather Morris Gevisser, and of Aaron Klug. Klug is Zelva's most famous son, and there is a statue commemorating him in the village square. After visiting Zelva, Gevisser wrote Aaron the following account.

To Aaron Klug   19 August 2010

Per email

Dear Aaron

I have just returned from my visit to Lithuania, which included a trip to Zelva; an experience which I found unexpectedly moving. This was in part because of the physical beauty of the village – it is nestled in a valley along a small riverbed, in an undulating landscape of forests and fields – and in part because there has not been any development there, so the village is, physically at least, very much as it was before the Second World War; a hamlet of brightly-coloured low-roofed wooden houses built and painted in the Baltic-Nordic style, most with well-tended gardens bursting with vegetables, surrounded by agricultural fields.

There are, of course, several definitive changes from the pre-war era. The one that strikes you immediately is the lack of any commerce in this formerly Jewish market town, save for a small general shop in a corner of the ugly two-story concrete block that was built by the Soviet administration (and that now has, inexplicably, a small plaster angel sitting on its roof watching over your statue!). Then, of course, the indicators of a smashed civilisation: the shul is abandoned and derelict, having been pillaged in the Nazi occupation and then turned into the offices of the collective farm during the Soviet era. The Jewish cemetery, on a west-facing slope above the village, is similarly derelict: you have to fight through the undergrowth to find a memorial to the Jews who were marched there and shot in 1941.

The Gevisser/Silin and the Klug houses no longer stand, although I was shown the sites of both. Where the Gevisser house once stood, along the river as you enter Zelva from Ukmerge (Wilkomir) there is now a ramshackle homestead. Yosef Gevisser and his wife Beila-Gittel lived at this site since at least 1877, at which point he was registered as a taxpayer. Yosef's children were Bere-Leib, Morris, Issy and Pesl, and his sister was Ethel Silin, your grandmother. The patriarch was Moyshe Mendel Gevisser, who also lived in Zelva; it seems that he was born in 1816 in Shavel (Siauliai), a major Jewish centre in the west of Lithuania, and that he moved with his wife Berel (née Gene, born in 1819) and family sometime between 1865 and 1877 (the 1865 census records a Movsha Mendel Gividzer and Berel Gividzer in Shavel).

The records in the Kaunas archives state that Moyshe Gevisser and his son Yosef were 'able to pay taxes'; hence they had some income – presumably from their leather and fur business – and Issy Gevisser's memoir suggests a life that, while austere, was secure until the death of his father in 1890. According to the tax records of 1877, the property was leased from the local Polish landlord, and measured 9 by 3.17 sajen (about 100 m^2). The house was low-roofed – an indicator of its inhabitants' modest means – and sand-floored, and backed onto a pasture; the living quarters were behind the leather and fur business and had one of the few matzoh-ovens in the shtetl, which made it a hive of activity before the holidays. According to Issy's memoirs, Yosef died of pneumonia after having been forced to sleep in the open following one of the perennial fires which tore through the village; his wife, in penury, was compelled to marry her competitor, 'Uncle Velvel', and send Bere-Leib, Morris and Pesl away (they were taken in by her Moss relatives, and sent to study in Vilna). Issy named his memoir 'The Village Orphan'. When Beila-Gittel moved into Velvel's house in the centre of the village, the Gevisser house was taken over by your grandparents, Ethel and Reuben Sillin[1]; this is where your mother and her sisters were raised.

---

[1] Spellings of the family names vary.

I was fascinated to read, in Ken Holmes' manuscript, that your grandfather had been a landowner and that Leizer's older brother Yudel inherited the land and was deported to Siberia in 1940 as a member of the Kulak class. Jews were indeed allowed to own land in independent Lithuania, but there was so little of it going (most of it belonged to the local Polish aristocracy and was worked by the Lithuanian peasantry) that the number of Jewish landowners was tiny. In one of the crazy twists of this troubled part of the world, those Jews who were deported by the Soviets in 1940 – as Kulaks, as wealthy townspeople, or as leftists – were ironically spared the death of the Holocaust. While I was in Zelva, I was not told about the Klug land, but I was shown the site of the Klug house in the village, right in the centre of town, built on a rise above the market square opposite the church. I assumed that this is where you had been born. All that remains of it now are some stone steps in densely overgrown bush.

According to the village legend your father Leizer was given this plot on which to build his house after he exhibited an act of great bravery protecting the village from Bolshevik marauders during the 1917 Revolution. The story, in village legend, differs somewhat from the one you told me: in this version, the Bolsheviks had plundered the church and were attempting to cut off the priest's finger so they could steal his ring. Your father and his brother (I assume Yudel?) came to the priest's rescue on their horses; they repelled the Bolsheviks, but in the process Leizer incurred a serious injury and was at grave risk of death. The village tended to him; all the local peasant women came with their traditional herbs; due to their ministrations he eventually recovered.

I have no idea of the provenance of this story; it may well have come from your cousin Sarah Klug, whom I assume was Yudel's daughter and who lived in Kaunas and visited Zelva regularly: she married a Lithuanian, had three children – the two sons became policemen; the daughter is an artist; she died earlier this year but her family all live in Kaunas. I was told the story by Zita Kriauciuniene, the retired teacher who is responsible for the statue commemorating your birth: she is the custodian of the village's history and she is passionate about conserving and retelling the story of its Jewish past. In her telling, your father Leizer is a village hero: the story of his bravery is clearly told to show Lithuanians that Jews were fighters and patriots, and were an integral part of the community. This is an important counter to the general perception that the Jews were all communists who welcomed the Russians, and that they subsequently went like lambs to the slaughter under the Nazis.

It was Zita's passion that ultimately made the trip to Zelva so moving. She has created a precise and beautiful little museum in the local high school, fitted with cabinets decorated in the local style, in which Lithuanian folkloric artefacts coexist harmoniously with records of the town's Jewish history. The museum's centerpiece is a village map made by her students: the Jewish houses are green and the Lithuanian houses blue; the point being to show that the town was overwhelmingly Jewish before the war. With her students, she has also collected oral histories of all the elders of the community, detailing in chilling and intimate detail how the Jews of Zelva were killed – by Lithuanians who joined the fascist 'white-armband' brigades. When I asked her about these people, she knew, precisely, the number of participants in the killings: thirteen. The rest of the village, she said, were passive bystanders. The details of the deaths of about 200 villagers are available on various websites, and also in a chapter written by one of Zita's students in a book of essays by students about Lithuanian Jews, which I have had translated, and which I will send to you if you are interested. About 60 people were killed *in situ*, and another 125 transferred to the Wilkomir jail before being shot in the Pivonija Forest.

You feature prominently in the museum: there are several photocopies of photographs of you, your biography downloaded from the internet, and even the letter you sent to her is neatly tacked onto the wall alongside its envelope. She has kept all the designs for the monument in one file: rendered with care and integrity by a local artist – the son of the school's caretaker – they include other options, besides the magnifying glass that was eventually chosen, to place on the granite plinth: a molecular structure, and a pair of spectacles like the ones you wear in your bio-pic. Also in a display cabinet is a large nail which builders found when they were laying the foundations for a new house on the old Klug property: Zita believes that this is one of your father's work implements. When she found out that I was related to the Klugs, she reacted with intense emotion, and when I showed her the photo of Beila-Gittel with you and Bennie as little boys she was awestruck. We were treated like visiting royalty: she took us on a tour through the town announcing to anyone we passed that we were 'Klug's relatives, from South Africa'. We roused the village's sleepy dogs as we walked around; my companions joked that even they seemed to be barking 'Klug, Klug, Klug!' There was some comedy to this all, but it was also very moving, perhaps because of the way your family history is being used to educate the village about a Jewish history that has otherwise been obliterated.

I asked Zita why she had dedicated herself to this task, and her answer had the moral clarity of a fable. She had been born in prison; her mother was released when she was three weeks old, and she was walking from Wilkomir to Gedrevich with her new-born baby, who was in danger of being snatched by the wolves. In deep distress, she passed through Zelva, where she was taken in by a Jewish family. The family only sheltered her for one night, but it saved the infant girl's life, and Zita now sees her work as a repayment of this debt. I sensed that this story was shorthand for a deeper and more complicated set of emotions, but I did not doubt for a moment her sincerity or the worth of her project: particularly after having visited the killing sites at Ponary outside Vilnius, there was something redemptive about her passion for the work of commemoration, and the care she took in doing it. I understood her attachment to the Klug story and her commitment to its retelling to be twofold: firstly, through your father's story, to demonstrate that Jews had been a much-valued part of the community, and secondly, through your own story, to demonstrate how much could be achieved even by a boy from Zelva if one was ambitious and hard-working; two positive Jewish traits.

Simon Davidovitch, the wonderful Lithuanian-Jewish guide who took us to Zelva, told us that there were people like Zita, unheralded, all over the country. Even so, there was something special about Zita and Zelva. By sheer coincidence, the woman who took us on a tour of Vilnius – a Jewish engineer named Roza who had set up the Jewish museum in the city – had Zelva connections: her mother was from there. She confirmed that it was a special place, unusually open about the past, and she attributed this to Zita and the work she had done. The village is clearly proud but poor, and as we walked through the village at dusk on a hot summer's evening, it was hard not to notice the inebriated, aimless young people gathering in open spaces and tearing through the village in souped-up old cars. Zita commented with anger about the ransacking of both the administrative offices and the synagogue-cum-collective-farm-office following the fall of the Soviet Union; not least because of history and the current recession which has hit the Baltic states particularly severely, Zelva certainly seemed to have its problems. Still, in the end, I felt immense gratitude for the fact that my family comes from a place that is not only more beautiful than I had ever imagined, but that is unusually in touch with its very traumatic past.

Best regards

Mark

Appendix B: Diffraction

B.1 Diffraction and the Phase Problem

If we use light to look at objects with a microscope, then what first happens is that the illuminating light is scattered by the object. This scattered light is the diffraction pattern of the object. In general, the diffraction pattern looks nothing like the original object. The diffraction pattern is mathematically the Fourier transform of the object. One can observe a diffraction pattern by, for example, looking at a distant source of light (such as a distant street lamp) through a stretched handkerchief. You see a small lattice of spots in place of the street lamp. This is the diffraction pattern of the hand-kerchief. Because the Fourier transform of a square lattice (the weave) is also a square lattice, the diffraction pattern of the handker-chief is a square pattern of spots. You can even play games; if you stretch the handkerchief so as to make the weave spacing larger in one direction, you will see the spots in the diffraction pattern move together (the observed lattice becomes smaller) in this direction. Because of this property, the diffraction pattern of a lattice is often referred to as the 'reciprocal lattice', and the diffraction is said to be in 'reciprocal space'.

In a light microscope, the objective lens gathers together all the scattered light and bundles it to form a magnified image. This operation of gathering together all the scattered light is the physical equivalent of a second Fourier transform. There is a very important mathematical result, namely that the Fourier transform of a Fourier transform is the original object (that is, going through the operation twice cancels out). That is really how a microscope works.

If we were to measure all the scattered light (that is, record the diffraction pattern) rather than use a lens, then we would be able to calculate the shape of the object by putting the measured diffraction pattern through a Fourier transform.

Atoms are too small to be seen by normal light (the wavelength and detail in the object have to be of similar size to see anything of the detail), so we have to use light of very short wavelength (5000 times shorter than visible light) to see anything. 'Light' of the appropriate wavelength is actually X-radiation; X-rays of the right wavelength are conventionally produced by accelerating a beam of electrons (in vacuum) to about 50 kilovolts and letting them slam into a cooled metal target, typically copper. The problem with X-rays is that there are no materials that can be used as X-ray lenses. Therefore, to look at structures (molecules) with X-rays, one is forced to measure the scattered X-rays and compute the image by means of a Fourier transform rather than using a lens.

It is very difficult to hold a single molecule in an X-ray beam. A convenient way round this problem is to use crystals. In crystals, all the molecules arrange themselves in the same way (which makes the analysis much easier) and the scattering adds together, which makes it much stronger. In any crystal, there are thousands of millions of molecules. Furthermore, since crystals are built on a lattice, the scattering gets bundled into narrow beams (the lattice effect we noticed with the handkerchief). When these beams hit a film, they make a pattern of spots that are called Bragg reflections after their discoverer William Lawrence Bragg. As the crystal is rotated in the X-ray beam, different Bragg reflections hit the film. In principle, by measuring all the reflections and using a Fourier transform, the 3D structure of the molecule can be reconstructed.

But to do this, one has to solve the *phase problem*. Each scattered wave has two important properties: its strength, which one can measure with a photographic film or appropriate detector; and its phase. Phase is the amount of relative shifting (in time) of the wave crest of the X-ray or light wave compared with some standard wave. Some waves have further to go than other waves, which retards their phase. The amount of retardation in turn depends on which part of the specimen scatters strongly. Thus the phase information is critically necessary to compute the image via a Fourier transform. If one uses a microscope, the phase information is automatically carried through by the lens to form an image. However, in the *no lens* kind of experiment (X-rays), when the scattered waves hit a film this phase information is lost. Therefore, without a lens one is actually in deep trouble. This deep trouble is what makes X-ray crystallography a non-trivial technology and keeps crystallographers in employment.

Max Perutz discovered a general method for solving the phase problem for macro-molecules by using heavy atoms as markers (the method of 'isomorphous replacement'). Max demonstrated that the differences between the diffraction patterns of the same protein crystals but with and without an added heavy atom (such as mercury) can be used to calculate the missing phases. This method is of enormous importance. Max worked it out for crystals. Soon afterwards Don Caspar, (and independently Rosalind Franklin and Ken Holmes) showed that it would work for non-crystalline samples (tobacco mosaic virus) as well.

B.2 Waves and Phases

Diffraction is concerned with how waves add together. This is true for light, X-rays or indeed any kind of wave. Waves are characterised by a wavelength, usually referred to as lambda (λ), and an amplitude.

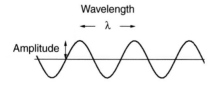

For light, λ is about 500 nm (0.5 micrometres or microns), for X-rays λ is around 0.1 nm (1 Å).

How waves add up depends upon their *phase relationship*, i.e. the displacement of one wave with respect to the other. This displacement can be referred to as fractions of λ – if you displace the wave by λ, you are back where you started; if you displace the wave by $\lambda/2$ the wave is turned upside down – but because of its periodic nature (it repeats every λ), phase is referred to in circular measure (radians or degrees). One full wavelength is one full circle, which is 2π radians or $360°$ (one radian = $57.2958°$). In the following, the phase angle is represented by the symbol ϕ.

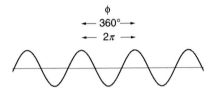

π (pi) is the ratio between half the circumference of a circle and the radius of a circle. It is an irrational number equal to about 3.14159. It is approximately 22/7, although this ratio is not thought to have any cosmic significance. The simplest example of adding waves is when their relative phase is zero. If A and B are *in phase* (i.e. the relative phase angle ϕ is zero), then you just add the amplitudes of the waves point by point along their length.

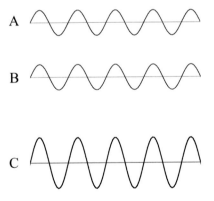

Wave A has amplitude 1, wave B has amplitude 1 and wave C has amplitude 2.

The next simplest example is when the phase angle $\phi = 180°$ (or π radians). This is called *antiphase:*

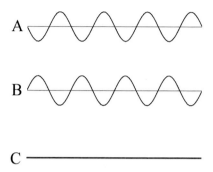

then by adding the waves point by point along their length at each point positive and negative values cancel out and the answer is zero at all points.

Wave A has amplitude 1, wave B has amplitude 1 and wave C has amplitude 0.

This phenomenon is also called interference: *waves can add together to produce no output.* Thus you can get dark stripes (interference fringes) in diffraction patterns.

All other cases call for a special treatment. The result of adding waves with any values of amplitude and phase can be found by means of a *phase vector diagram* (in electrical engineering they are called phasor diagrams). In a phase vector diagram, each wave is represented by a line equal to the amplitude of the wave drawn at an angle to the horizontal equal to the phase angle. The sum is obtained by placing each new line (vector) on the end of the previous line. The end result is the line drawn between the middle (origin) of the diagram and the end point of the lines strung together. The length gives the amplitude and the angle gives the phase.

For example, if for two waves the phase relationship is 90°

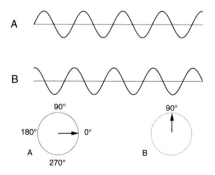

the phase vector diagram is

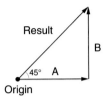

The resulting wave has an amplitude of 1.414 ($\sqrt{2}$) and a phase of 45°. If for two waves the phase relationship is 270°

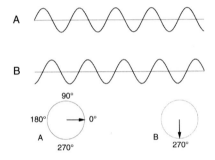

the phase vector diagram is

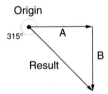

The resulting wave has an amplitude of 1.414 ($\sqrt{2}$) and a phase of 315°. Sometimes, because of symmetry in the diffracting object, a diffracted ray cannot have just any old phase – it has to be 0° or 180°, which is another way of saying that it has to be + or –. Clearly, it is a lot easier to determine a phase if you know it can only be + or –. The phase problem is replaced by the simpler sign problem. This is why the early attempts to solve the phase problem by computation (David Sayre's first paper, for example) were limited to solving the sign problem. Equally, Max Perutz's efforts to determine the phases of haemoglobin experimentally with heavy atoms were initially limited to the signs you get in the diffraction looking down a two-fold crystal axis (the two-fold axis imposes signs rather than the general phases). Sometimes, because of high symmetry, all the diffraction from small molecules can be limited to signs. Unfortunately, this is certainly not true for proteins.

B.3 Fourier Transforms and Resolution

Jean-Baptiste Joseph Fourier[1] (1768–1830) was a French mathematician and engineer best known for the investigation of Fourier series and their applications to problems of heat transfer and vibrations, for example sound waves. Both the Fourier series and the Fourier transform, which is a generalisation of the idea of a Fourier series to objects that do not repeat regularly, are named in his honour.

The Fourier transform of an object is a way of representing all possible sine and cosine waves that can be fitted into the object. In the transform, the direction from the point in the middle of the transform gives the direction of a particular wave. The distance from the middle is inversely proportional to the wavelength (long wavelength – close to the middle; short wavelength – far from the middle). The strength at that point tells you how much of that particular sine or cosine wave you need to make up the object. One can think of a Fourier transformation (FT) as an *app* that can be applied in a computer to a digital rendering of any *object* to yield the Fourier transform (indeed, there are many Fourier transform apps available on the web). As mentioned in the main text, a very important property of Fourier transformations is that *if you calculate the Fourier transform of the Fourier transform of an object you end up with the object you started out with.* If the object is some function f, then more formally, in terms of the Fourier transform app FT

$$\text{FT}\{\text{FT}(f)\} = f$$

---

[1] In the 1820s, Fourier calculated that if the Earth is only warmed by incoming solar radiation it should be considerably colder than it actually is. He considered the possibility that the Earth's atmosphere might act as an insulator to hold in warmth. This was the first proposal of the Greenhouse Effect.

In a microscope the *object,* usually mounted on a glass cover slip, is illuminated with a parallel beam of light. The scattered light is the Fourier transform of the object.

In the following example the object f is a drawing of a duck[2]:

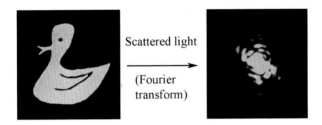

The objective lens gathers up the scattered light and focuses it to form an image. Since the object is usually rather close to the objective lens and the image is some distance away, the image is usually much bigger than the object, which is the purpose of the microscope (in the following example it is the same size). First we show an ideal case where the objective lens has a big diameter so that it can gather up essentially all the scattered light.

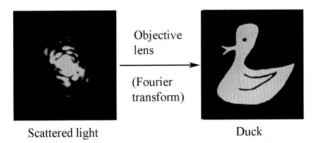

What happens if you have a microscope with only a small-aperture objective lens? Then only the centre region of the scattered light gets included in the Fourier transform (i.e. the long-wavelength components

---

[2] The *intensity* of the Fourier transform should be centro-symmetric (i.e. look the same at equal distances from the middle of the transform). The example shown is actually an optical transform rather than a computed transform, and errors in the optical system have rendered this centro-symmetry imperfect.

are included and the short-wavelength components are left out). This results in image detail being lost. The resolution in the image depends on the size of the aperture at the objective lens. The following shows the effect of limiting the scattered light to just that region contained within the white circle — a low-resolution duck with no detail.

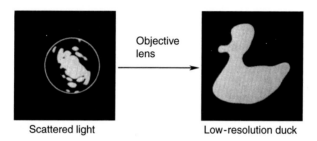

Scattered light   Low-resolution duck

If one makes the limiting circle even smaller, then one obtains just a duck-shaped blob.

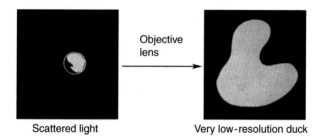

Scattered light   Very low-resolution duck

B.4 Convolution

The convolution of two functions involves putting the first function down at every point in the second function, giving it a weight equal to the value of the second function at that point and then adding them all up. In general, this is clearly a complicated procedure, but if the second function is a lattice then the idea of convolution becomes rather easy to visualise:

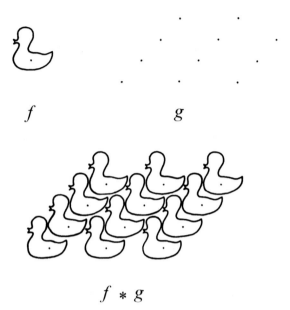

$$f \qquad\qquad g$$

$$f * g$$

Consider crystallised ducks, as in the figure. This can be described as a duck f convoluted with a lattice g. The symbol * is often used to represent convolution. Convolution is a common phenomenon: for example, in the out-of-focus image from a camera, the sharp image has become convoluted with something close to a Gauss function – see below.

Convolution becomes very useful in combination with Fourier transforms because the Fourier transform of two functions convoluted with each other is just the Fourier transform of the first function multiplied with the Fourier transform of the second function. In terms of the Fourier transform app FT, we have

$$FT(f * g) = FT(f) \cdot FT(g)$$

This is a wonderful way of disentangling functions that have become convoluted. For example, for an out-of-focus photo you compute the Fourier transform of the photo and divide this by the Fourier transform of a Gauss function (which is also a Gauss function) to generate the sharp image transform:

$$FT(f) = FT(f * g)/FT(g)$$

If you now calculate the Fourier transform of the sharp image transform, you regenerate the required sharpened (deconvoluted) photo. This procedure is actually unstable because the values of the Gauss function quickly become very small; dividing by very small numbers is not a good idea (it amplifies up all the noise) and you have to stop somewhere. The fact that the procedure has to be stopped because of noise limits the resolution of the photo.

B.5 The Patterson Function

Consider the diffraction pattern of the duck f shown above. This diffraction pattern is not the Fourier transform of the duck but is actually the square of the Fourier transform.

If we call the Fourier transform of the duck F, i.e.

$$F = FT(f)$$

what we actually observe in the diffraction pattern is F^2, the square of the length of F at each point. X-ray crystallography also measures F^2. Using the convolution theorem, if we compute the Fourier transform of F^2, rather than F, we end up with $f * f$ (the duck convoluted with itself). This function is the auto-correlation of the object (in crystallography, it is known as the Patterson function after Arthur Lindo Patterson, an American crystallographer). To find the Patterson function of an atomic object, one needs to make a diagram in which each of the vectors joining atoms is placed with one end at the origin. The outer end is given a value equal to the product of the weights of the atoms. Vectors joining an atom to itself (with zero length) also count. In the following example we consider a molecule containing three atoms, A, B and C:

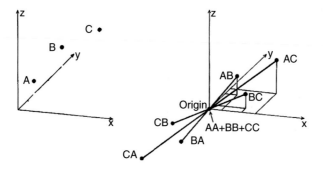

Thus a structure made of three point-like atoms gives rise to nine peaks. Three of these (the self-peaks AA, BB, CC) are at the origin. The total number of peaks is $N^2 - N$ where N is the number of atoms in the structure. One sees that the Patterson function of even this simple object is already quite complicated. The chance of working out a structure from its Patterson decreases abruptly with the size of N (up to $N = 100$ it can be done). Dorothy Hodgkin was awarded the Nobel Prize for working out the structure of vitamin B12 from a Patterson map, but that was approaching the limit. Nevertheless, the Patterson function of a helix is so distinctive that one can dig out the helical parameters (that is, the radius and pitch of the helix) by inspection of the Patterson.

B.6 Fibre Diagrams and Bessel Functions

Molecules that are more or less round (called globular) can often be crystallised. However, long fibrous molecules present a special problem. Their geometry makes crystallisation difficult. If the molecules are really long (good-quality DNA, for instance) then they can be physically pulled out parallel with tweezers. Then the parallel long molecules may spontaneously crystallise. If this happens, you end up with a mass of parallel *micro-crystals* sharing a common axis but randomly orientated about this axis. This is called a *crystalline fibre*. It diffracts as a crystal (so it gives Bragg reflections) but since all orientations of crystal are present at once, the diffraction pattern you get is cylindrically averaged and consists of lots of overlapping Bragg reflections. Because the object is periodic along its length, the Bragg reflections are arranged in lines on the film known as layer lines.

Otherwise, if the molecular symmetry (which will be some sort of helix) does not allow crystallisation, you end up with a sample consisting of parallel *molecules* sharing a common axis but where each molecule is in any orientation. This is a *non-crystalline fibre*. The diffraction pattern is essentially the Fourier transform of the individual molecule (no Bragg reflections are present because there is no lattice). It will be broken into layer lines because the molecule is periodic along its length.

X-ray diffraction pictures obtained from pulled fibres or orientated gels, where the X-ray beam is at right angles to the axis of orientation, are known as X-ray fibre diagrams. The observed X-ray fibre diagram is equivalent to the diffraction (the square of the Fourier transform) from a single microcrystal or molecule spun around the axis of orientation and thereby averaged over all possible orientations. Because of the cylindrical averaging, it is essential to use cylindrical coordinates rather than Cartesian coordinates to evaluate the Fourier transform. In cylindrical coordinates the position of an atom is specified by quoting its radial distance from the axis of the cylindrical system r, the angle ϕ that the radius vector makes with some fixed origin and the distance along the axis z. In cylindrical coordinates the cylindrical averaging is simple: it is just the average as ϕ takes all values between $0°$ and $360°$.

Using cylindrical coordinates to evaluate the Fourier transform, one ends up with Bessel functions in place of the more familiar sine and cosine functions. Bessel functions are a family of functions that are characterised by their 'order' n and are written J_n. Zero-order Bessel functions (J_0) look like a damped cosine wave. A higher order Bessel function J_n starts out at zero and then rises to a first broad maximum. The position of the first maximum depends upon the order – the higher the order the further out you have to go – and then J_n oscillates like a cosine wave.

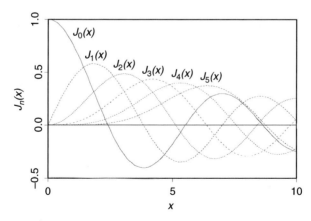

The subscript n is related to the symmetry of the particle. Thus if the particle has four-fold symmetry the Bessel function J_4 in the Fourier transform will be strong. As before, because the particle repeats along z (is periodic along z), the Fourier transform is limited to layer lines. In 1953,

Cochran, Crick and Vand[3] showed how to calculate the diffraction pattern to be expected from a helix using Bessel functions. In particular, they showed how the helical symmetry determines which Bessel functions could show up on which layer lines. They called this the helical selection rule. They showed that for the simplest helix (like the coiled-wire spring in the diagram), the order of Bessel function you get on any layer line is the same as the order of the layer line (i.e. J_0 on layer line 0, J_1 on layer line 1 and -1 etc.), which gives rise to the characteristic 'helix cross'.

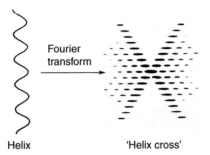

Helix   'Helix cross'

Bessel functions are the natural way to represent a wave with circular symmetry. The order n is related to the symmetry of the wave (that is, how many times it repeats as you go round the axis). Thus the diffraction pattern from objects with circular symmetry (such as a ring of 100 points) can be expressed in Bessel functions.

[3] Cochran, W., Crick, F.H.C. and Vand, V. (1952) *Acta Cryst.* **5**, 581–586.

The optical transform (Fourier transform) from the ring of 100 points is shown below:

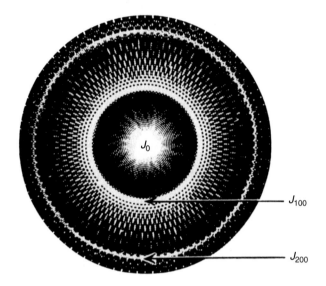

The diffraction pattern divides into zones according to the resolution. The innermost zone of the diffraction pattern (labelled J_0) is circularly symmetric – it does not vary as you go round the middle of the diagram[4]. If we were to make an image from this diffraction pattern (as in a microscope) and if we put a stop in the microscope to let only this inner part of the diffraction pattern into the final image, then the image would look like a continuous ring. In other words, you do not have enough resolution to see that the ring is made of points, but you can see the ring. This inner part of the diffraction pattern has a relatively simple form. In the radial direction it is a zero-order Bessel function[5] J_0. In the azimuthal (ϕ) direction it is continuous. The $J_0(x)$ function, like a cosine, starts off with the value unity and then oscillates.

In the optical transform shown above, there is an outer zone where a new Bessel function occurs. This is the start of J_{100}. At this radius in the optical transform there is enough resolution to know that the ring is

---

[4] If this were a perfect optical transform free from errors it would not vary as you go round.

[5] Since this is a diffraction pattern one actually observes $J_0{}^2$.

made of 100 blobs. The J_{100} is modulated by sines and cosines in the azimuthal (ϕ) coordinate to make the 100-fold pattern. This combination of Bessel functions with sines and cosines with a symmetry equal to the order of the Bessel function is known as a cylinder function.

If we set the microscope aperture to include this part of the diffraction pattern, we would produce an image showing a ring of 100 not very well resolved blobs. Even further out, J_{200} shows up. The full aperture would include J_{200}. Including this higher-order term would produce an image showing 100 quite well resolved points.

B.7 The Gaussian Function

If you make repeated measurements of some quantity, say air pressure, then because of small perturbations such as temperature the values will not all be the same, but the probabilities of a certain value turning up will be distributed around some average value. The average value acquired from a large number of readings will be the most probable answer, but the distribution will have long tails. The shape of this distribution is given by the bell-shaped Gaussian function[6]. The average value is often adjusted to zero (as in the example below).

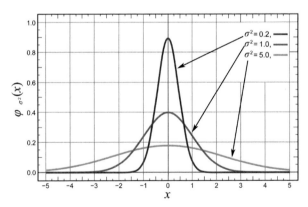

(Diagram from http://commons.wikimedia.org/wiki/File:Normal_Distribution_PDF.svg)

---

[6] Carl Friederich Gauss (1777–1855) lived and worked for most of his life in Göttingen. He is ranked as one of history's most influential mathematicians.

Gaussian functions can be wide or narrow depending on how accurate the measurements are. The width is given by the parameter σ, the standard deviation. A large σ means a wide distribution (not accurate). If the standard deviation is unity, the distribution is referred to as the normal distribution.

Gaussian functions have some interesting properties:

1. The square of a Gaussian function is a Gaussian function (David Sayre made use of this in his equation for determining signs of diffraction pattern amplitudes).
2. The Fourier transform of a Gaussian function is a Gaussian function (used in deconvolution calculations).
3. The convolution of a Gaussian function with a Gaussian function is also a Gaussian function.

Appendix C: Aaron Klug's Scientific Publications

1. Klug, A. 1947. 'Crystal structure of para-bromochlorobenzene' *Nature* **160**, 570.
2. Klug, A. 1950. 'The crystal and molecular structure of triphenylene, C18H12' *Acta Cryst.* **3**, 165–175.
3. Klug, A. 1950. 'The application of the Fourier-transform method to the analysis of the structure of triphenylene C18H12' *Acta Cryst.* **3**, 176–181.
4. Klug, A. 1952. *The Kinetics of Phase Changes in Solids.* Dissertation for the PhD degree, University of Cambridge,
5. Klug, A., Roughton, F.J.W. and 50 others. 1955. 'General discussion' *Faraday Soc. Disc.* No. 20, 278.
6. Franklin, R.E. and Klug, A. 1955. 'The splitting of layer lines in X-ray fibre diagrams of helical structures; application to tobacco mosaic virus' *Acta Cryst.* **8**, 777.
7. Klug, A., Kreuzer, F. and Roughton, F.J.W. 1956. 'Simultaneous diffusion and chemical reaction in thin layers of haemoglobin solution' *Proc. Roy. Soc. B* **145**, 452–472.
8. Klug, A., Kreuzer, F. and Roughton, F.J.W. 1956. 'The diffusion of oxygen in concentrated haemoglobin solutions' *Helv. Physiol. Pharmacol. Acta* **14**, 121–127.
9. Franklin, R.E. and Klug, A. 1956. 'The nature of the helical groove on the tobacco mosaic virus particle' *Biochim. Biophys. Acta* **19**, 403–419.
10. Franklin, R.E., Klug, A. and Holmes, K.C. 1956. 'X-ray diffraction studies of the structure and morphology of tobacco mosaic virus' *CIBA Found. Symp.*, 39–52.
11. Klug, A., Finch, J.T. and Franklin, R.E. 1957. 'Structure of turnip yellow mosaic virus' *Nature* **179**, 683–684.
12. Gibson, Q.H. and Roughton, F.J.W. with an Appendix by A. Klug. 1957. 'The determination of the velocity constants of the four successive reactions of carbon monoxide with sheep haemoglobin' *Proc. Roy. Soc. B* **146**, 205–224.

13. Klug, A. and Franklin, R.E. 1957. 'The reaggregation of the A-protein of tobacco mosaic virus' *Biochim. Biophys. Acta* **23**, 199–201.

14. Klug, A., Finch, J.T. and Franklin, R.E. 1957. 'The structure of turnip yellow mosaic virus; X-ray diffraction studies' *Biochim, Biophys. Acta* **25**, 242–252.

15. Klug, A., Crick, F.H.C. and Wyckoff, H.W. 1958. 'Diffraction by helical structures' *Acta Cryst.* **11**, Pt. 3, 199–213.

16. Klug, A. 1958. 'Joint probability distributions of structure factors and the phase problem' *Acta Cryst.* **11**, 515–543.

17. Klug, A. and Franklin, R.E. 1958. 'Order-disorder transitions in structures containing helical molecules' *Farad. Soc. Disc.* No. 25.

18. Franklin, R.E., Klug, A., Finch, J.T. and Holmes, K.C. 1958. 'On the structure of some ribonucleoprotein particles' *Faraday Soc. Disc.* No. 25.

19. Finch, J.T. and Klug, A. 1959. 'Structure of poliomyelitis virus' *Nature* **183**, 1709–1714.

20. Klug, A. 1959. 'A reply to some comments by Karle and Hauptman' *Acta Cryst.* **12**, 943.

21. Klug, A., Franklin, R.E. and Humphreys-Owen, S.P.F. 1959. 'The crystal structure of Tipula Iridescent virus as determined by Bragg reflection of visible light' *Biochim. Biophys. Acta* **32**, 203–219.

22. Franklin, R.E., Caspar, D.L.D. and Klug, A. 1959. *Plant Pathology: Problems and Progress, 1908–1958* (Golden Jubilee Volume of the American Phytopathological Society) (University of Wisconsin Press).

23. Finch, J.T., and Klug, A. 1960. 'The form of crystals of mahoney poliovirus grown in phosphate-saline' *Biochim. Biophys. Acta* **41**, 430–433.

24. Klug, A. and Finch, J.T. 1960. 'The symmetries of the protein and nucleic acid in turnip yellow mosaic virus; X-ray diffraction studies' *J. Mol. Biol.* **2**, 201–215.

25. Klug, A. and Caspar, D.L.D. 1960. 'The structure of small viruses' *Adv. Virus Res.* **7**, 225–325.

26. Finch, J.T. and Klug, A. 1960. 'X-ray "powder" diagrams of crystals of an artificial top component from turnip yellow mosaic virus' *J. Mol. Biol.* **2**, 434–435.

27. Klug, A., Holmes, K.C. and Finch, J.T. 1961. 'X-ray diffraction studies on ribosomes from various sources' *J. Mol. Biol.* **3**, 87–100.

28. Caspar, D.L.D. and Klug, A. 1962. 'Physical principles in the construction of regular viruses' *Cold Spring Harb. Symp. Quant. Biol.* **27**, 1–24.

29. Caspar, D.L.D. and Klug, A. 1963. 'Structure and assembly of regular virus particles' *MD Anderson Symp.: Viruses, Nucleic Acid and Cancer* 27–39.

30. Klug, A. and Berger, J.E. 1964. 'An optical method for the analysis of periodicities in electron micrographs, and some observations on the mechanism of negative staining' *J. Mol. Biol.* **10**, 565–569.

31. Finch, J.T., Klug, A. and Stretton, A.O.W. 1964. 'The structure of the "polyheads" of T4 bacteriophage' *J. Mol. Biol.* **10**, 570–575.

32. Finch, J.T. and Klug, A. 1965. 'The structure of viruses of the papilloma-polyoma type. III. Structure of rabbit papilloma virus' *J. Mol. Biol.* **13**, 1–12.

33. Klug, A. and Finch, J.T. 1965. 'Structure of viruses of the papilloma-polyoma type. I. Human wart virus' *J. Mol. Biol.* **11**, 403–423.

34. Klug, A. 1965. 'Structure of viruses of the papilloma-polyoma type. II. Comments on other work' *J. Mol. Biol.* **13**, 424–431.

35. Klug, A. and Finch, J.T. 1965. 'Structure of viruses of the papilloma-polyoma type' *J. Mol. Biol.* **11**, 961–962.

36. Klug, A., Finch, J.T., Leberman, R. and Longley, W. 1966. 'Design and structure of regular virus particles'. In: *CIBA Found. Symp. 1966: Principles of Biomolecular Organisation* (ed. G.E.W. Wolstenholme and M. O'Connor), 158–189.

37. Grimstone, A.V. and Klug A. 1966. 'Observations on the substructure of flagellar fibres' *J. Cell Sci.* **1**, 351–362.

38. Klug, A., Longley, W. and Leberman, R. 1966. 'Arrangement of protein subunits and the distribution of nucleic acid in turnip yellow mosaic virus. I. X-ray diffraction studies' *J. Mol. Biol.* **15**, 315–343.

39. Finch, J.T. and Klug, A. 1966. 'Arrangement of protein subunits and the distribution of nucleic acid in turnip yellow mosaic virus. II. Electron microscope studies' *J. Mol. Biol.* **15**, 344–364.

40. Klug, A. and DeRosier, D.J. 1966. 'Optical filtering of electron micrographs: reconstruction of one-sided images' *Nature* **212**, 29–32.

41. Finch, J.T., Leberman, R., Chang, Y.-S. and Klug, A. 1966. 'Rotational symmetry of the two turn disk aggregate of tobacco mosaic virus protein' *Nature* **212**, 349–351.

42. Harrison, B.D. and Klug, A. 1966. 'Relation between length and sedimentation coefficient for particles of tobacco rattle viruses' *Virology* **30**, 738–740.

43. Finch, J.T. and Klug, A. 1967. 'Structure of broad bean mottle virus. I. Analysis of electron micrographs and comparison with turnip yellow mosaic virus and its top component' *J. Mol. Biol.* **24**, 289–302.

44. Finch, J.T., Klug, A. and van Regenmortel, M.H.V. 1967. 'The structure of cucumber mosaic virus' *J. Mol. Biol.* **24**, 303–305.

45. Klug, A. 1967. 'The design of self-assembling systems of equal units.' In *Formation and Fate of Cell Organelles. Symp. Int. Soc. Cell Biol.* Vol. 6, 1–18 (Academic Press).

46. Finch, J.T., Klug, A. and Nermut, M.V. 1967. 'The structure of the macromolecular units on the cell walls of bacillus polymyxa' *J. Cell Sci.* **2**, 587–590.

47. DeRosier, D.J. and Klug, A. 1968. 'Reconstruction of three-dimensional structures from electron micrographs' *Nature* **217**, 130–134.

48. Klug, A. and Finch, J.T. 1968. 'Structure of viruses of the papilloma-polyoma type. IV. Analysis of tilting experiments in the electron microscope' *J. Mol. Biol.* **31**, 1–12.

49. Kiselev, N.A., DeRosier, D.J. and Klug, A. 1968. 'Structure of the tubes of catalase: analysis of electron micrographs by optical filtering' *J. Mol. Biol.* **35**, 561–566.

50. Klug, A. 1968. 'Rosalind Franklin and the discovery of the structure of DNA' *Nature* **219**, 808–810.

51. Clark, B.F.C., Doctor, B.P., Holmes, K.C., Klug, A., Marcker, K.A. and Morris S.J. 1968. 'Crystallisation of transfer RNA' *Nature* **219**, 1222–1224.

52. DeRosier, D.J. and Klug, A. 1969. 'Positions of ribosomal subunits' *Science* **163**, 1470.

53. Kiselev, N.A. and Klug, A. 1969. 'The structure of viruses of the papilloma-polyoma type. V. Tubular variants built of pentamers' *J. Mol. Biol.* **40**, 155–171.

54. Klug, A. 1968. 'Symmetry and function of biological systems at the macromolecular level. Point groups and the design of aggregates' *Nobel Symp.* **11**.

55. Finch, J.T. and Klug, A. 1969. 'Two double helical forms of polyriboadenylic acid and the pH-dependent transition between them' *J. Mol. Biol.* **46**, 597–598.

56. Fowle, L.G., Juritz, J.F.W., Klug, A. and Stephen, A.M. 1970. 'Crystallinity of the gummy polysaccharide from *Watsonia pyramidata* corm-sacs' *SA Med. J.* **44**, 152.

57. Crowther, R.A., DeRosier, D.J. and Klug, A. 1970. 'The reconstruction of a three-dimensional structure from projections and its application to electron microscopy' *Proc. Roy. Soc. Lond. A* **317**, 319–340.

58. Crowther, R.A., Amos, L.A., Finch, J.T., DeRosier, D.J. and Klug, A. 1970. 'Three dimensional reconstruction of spherical viruses by Fourier synthesis from electron micrographs' *Nature* **226**, 421–425.

59. Finch, J.T., Klug, A. and Leberman, R. 1970. 'The structures of turnip crinkle and tomato bushy stunt viruses. II. The surface structure; dimer clustering patterns' *J. Mol. Biol.* **50**, 215–222.

60. Erickson, H.P. and Klug, A. 1970. 'The Fourier transform of an electron micrograph; effects of defocussing and aberrations, and implications for the use of underfocus contrast enhancement' *Ber. Bunsen-Gesselsch. Phys. Chem. (früher Z. Electrochem.)* **74**, 1129–1137.

61. Erickson, H.P. and Klug, A. 1971. 'Measurement and compensation of defocusing and aberrations by Fourier processing of electron micrographs' *Phil. Trans. Roy. Soc. B* **261**, 105–118.

62. Klug, A. 1971. 'Applications of image analysis technique in electron microscopy' *Phil. Trans. Roy. Soc. B* **261**, 173–179.

63. Finch, J.T. and Klug, A. 1971. 'Three-dimensional reconstruction of the stacked-disk aggregate of tobacco mosaic virus protein from electron micrographs' *Phil. Trans. Roy. Soc. B* **261**, 211–219.

64. Durham, A.C.H., Finch, J.T. and Klug, A. 1971. 'States of aggregation of tobacco mosaic virus protein' *Nature New Biol.* **229**, 37–42.

65. Durham, A.C.H. and Klug, A. 1971. 'Polymerisation of protein subunits and its control' *Nature New Biol.* **229**, 42–46.

66. Butler, P.J.G. and Klug, A. 1971. 'Assembly of the particles from RNA and disks of protein' *Nature New Biol.* **229**, 47–50.

67. Crowther, R.A. and Klug, A. 1971. 'ART and science or conditions for three-dimensional reconstruction from electron microscope images'. *J. Theor. Biol.* **32**, 199–203.

68. Klug, A. and Durham, A.C.H. 1971. 'The disk of TMV protein and its relation to the helical and other modes of aggregation' *Cold Spring Harb. Symp. Quant. Biol.* **36**, 449–460.

69. Klug, A. 1971. 'Interpretation of the rotation function map of satellite tobacco necrosis virus; octahedral packing of icosahedral particles' *Cold Spring Harb. Symp. Quant. Biol.* **36**, 483–486.

70. Barrett, A.N., Barrington-Leigh, J., Holmes, K.C., Leberman, R., Mandelkow, E. von Sengbusch, P. and Klug, A. 1971. 'An electron density map of tobacco mosaic virus at 10 Å resolution' *Cold Spring Harb. Symp. Quant. Biol.* **36**, 433–448.

71. DeRosier, D.J. and Klug, A. 1972. 'Structure of the tubular variants of the head of bacteriophage T4 (polyheads). I. Arrangement of subunits in some classes of polyheads' *J. Mol. Biol.* **65**, 469–488.

72. Yanagida, M., DeRosier, D.J. and Klug, A. 1972. 'The structure of the tubular variants of the head of bacteriophage T5 (polyheads). II. Structural transition from a hexamer to a 6 + 1 morphological unit' *J. Mol. Biol.* **65**, 489–499.

73. Klug, A. 1972. 'Assembly of tobacco mosaic virus' *Fed. Proc.* **31**, 1.

74. Durham, A.C.H. and Klug, A. 1972. 'Structures and roles of the polymorphic forms of tobacco mosaic virus protein; III. A model for the association of A-protein into discs' *J. Mol. Biol.* **67**, 315–332.

75. Mirzabekov, A.D., Rhodes, D., Finch, J.T., Klug, A. and Clark, B.F.C. 1972. 'Crystallization of tRNAs as cetyltrimethylammonium salts' *Nature New Biol.* **237**, 27–28.

76. Crowther, R.A., Amos, L.A. and Klug, A. 1972. 'Three-dimensional image reconstruction using functional expansions' *Proc. Fifth Eur. Congress on Electron Microscopy*, Manchester, p. 539.

77. Amos, L.A. and Klug, A. 1972. 'Image filtering by computer' *Proc. Fifth Eur. Congress on Electron Microscopy*, Manchester, p. 580.

78. Klug, A. 1972. 'The polymorphism of tobacco mosaic virus protein and its significance for the assembly of the virus' *CIBA Found. Symp. 7*.

79. Butler, P.J.G. and Klug, A. 1972. 'Assembly of tobacco mosaic virus in vitro: effect of state of polymerisation of the protein component' *Proc. Natl Acad. Sci. USA*. **69**, 2950–2953.

80. Klug, A. and Crowther, R.A. 1972. 'Three-dimensional image reconstruction from the viewpoint of information theory' *Nature* **238**, 435–440.

81. Mellema, J.E. and Klug, A. 1972. 'Quaternary structure of gastropod haemocyanin' *Nature* **239**, 146–150.

82. Ladner, J.E., Finch, J.T., Klug, A. and Clark, B.F.C. 1972. 'High resolution X-ray diffraction studies on a pure species of transfer RNA' *J. Mol. Biol.* **72**, 99–101.

83. Brown R.S., Clark, B.F.C., Coulson, R.R., Finch, J.T., Klug, A. and Rhodes, D. 1972. 'Crystallisation of pure species of bacterial tRNA for X-ray diffraction studies' *Eur. J. Biochem.* **31**, 130–134.

84. Butler, P.J.G., Durham, A.C.H. and Klug, A. 1972. 'Structures and roles of the polymorphic forms of tobacco mosaic virus protein. IV. Control of mode of aggregation of tobacco mosaic virus protein by proton binding' *J. Mol. Biol.* **72**, 1–18.

85. Finch, J.T. and Klug, A. 1972. 'The helical surface lattice of bacterial flagella'. In: *Proc. First John Innes Symp., Norwich. The Generation of Subcellular Structures* (ed. R. Markham and J. Bancroft).

86. Butler, P.J.G. and Klug, A. 1973. 'Effect of state of polymerisation of the protein component on the assembly of tobacco mosaic virus' *Molec. Gen. Genet.* **120**, 91–93.

87. Klug, A. and Butler, P.J.G. 1973. 'Dislocations in tobacco mosaic virus' *Nature* **244**, 115–116.

88. Amos, L.A. and Klug, A. 1974. 'Arrangement of subunits in flagellar microtubules' *J. Cell Sci.* **14**, 523–549.

89. Klug, A. 1974. 'Rosalind Franklin and the double helix' *Nature* **248**, 787–788.

90. Leberman, R., Finch, J.T., Gilbert, P.F.C., Witz, J. and Klug, A. 1974. 'X-ray analysis of the disk of tobacco mosaic virus protein' *J. Mol. Biol.* **86**, 179–182.

91. Finch, J.T., Gilbert, P.F.C., Klug, A. and Leberman, R. 1974. 'X-ray analysis of the disk of tobacco mosaic virus protein. II. The packing arrangement in the crystal' *J. Mol. Biol.* **86**, 183–192.

92. Gilbert, P.F.C. and Klug, A. 1974. 'X-ray analysis of the disk of tobacco mosaic virus protein. III. A low resolution electron density map' *J. Mol. Biol.* **86**, 193–207.

93. Unwin, P.N.T. and Klug, A. 1974. 'Electron microscopy of the stacked disk aggregate of TMV protein. I. 3-dimensional image reconstruction' *J. Mol. Biol.* **87**, 641–656.

94. Finch, J.T. and Klug, A. 1974. 'The structural relationship between the stacked disk and helical polymers of TMV protein' *J. Mol. Biol.* **87**, 633–640.

95. Robertus, J.D., Ladner, J., Finch, J.T., Rhodes, D., Brown, R., Clark, B.F.C. and Klug, A. 1974. 'Structure of yeast phenylalanine tRNA at 3 Å resolution' *Nature* **250**, 546–551.

96. Robertus, J.D., Ladner, J., Finch, J.T. Rhodes, D., Brown, R., Clark, B.F.C. and Klug, A. 1974. 'Correlation between three-dimensional structure of tRNA and chemical reactivity' *Nucleic Acids Res.* **1**, 7.

97. Klug, A., Robertus, J.D., Ladner, J., Brown, R, and Finch, J.T. 1974. 'Conservation of the molecular structure of yeast phenylalanine transfer RNA in two crystal forms' *Proc. Natl Acad. Sci. USA* **71**, 3711–3715.

98. Klug, A., Ladner, J. and Robertus, J.D. 1974. 'The structural geometry of coordinated base changes in transfer RNA' *J. Mol. Biol.* **89**, 511–516.

99. Crowther, R.A. and Klug, A. 1974. 'Three dimensional image reconstruction on an extended field – a fast, stable algorithm' *Nature* **251**, 490–492.

100. Branton, D. and Klug, A. 1975. 'Capsid geometry of bacteriophage T2: a freeze-etching study' *J. Mol. Biol.* **92**, 559–565.

101. Sperling, R., Amos, L.A. and Klug, A. 1975. 'A study of the pairing interaction between protein subunits in the tobacco mosaic

virus family by image reconstruction from electron micrographs' *J. Mol. Biol.* **92**, 541–558.

102. Wakabayashi, T., Huxley, H.E., Amos, L.A. and Klug, A. 1975. 'Three-dimensional image reconstruction of actin-tropomyosin complex and actin–tropomyosin–troponin T–troponin I complex' *J. Mol. Biol.* **93**, 477–497.

103. Sperling, R. and Klug, A. 1975. 'States of aggregation of the Dahlemense strain of tobacco mosaic virus protein and their relation to crystal formation' *J. Mol. Biol.* **96**, 425–430.

104. Crowther, R.A. and Klug, A. 1975. 'Structural analysis of macromolecular assemblies by image reconstruction from electron micrographs' *Ann. Rev. Biochem.* **44**, 161–182.

105. Ladner, J.E., Jack, A., Robertus, J.D., Brown, R.S., Rhodes, D., Clark, B.F.C. and Klug, A. 1975. 'Atomic co-ordinates for yeast phenylalanine tRNA' *Nucleic Acids Res.* **2**, 1629–1637.

106. Crick, F.H.C. and Klug, A. 1975. 'Kinky helix' *Nature* 255, 530–533.

107. Amos, L.A. and Klug, A. 1975. 'Three-dimensional image reconstructions of the contractile tail of T4 bacteriophage' *J. Mol. Biol.* **99**, 51–73.

108. Ladner, J.E., Jack, A., Robertus, J.D., Brown, R.S., Rhodes, D., Clark, B.F.C. and Klug, A. 1975. 'Structure of yeast phenylalanine transfer RNA at 2.5 Å resolution' *Proc. Natl Acad. Sci. USA* **72**, 4414–4418.

109. Clark, B.F.C. and Klug, A. 1975. 'Structure and function of tRNA with special reference to the three dimensional structure of yeast phenylalanine tRNA' *Proc. 10th FEBS Meeting*, 183–206.

110. Champness, J.N., Bloomer, A.C., Bricogne, G., Butler, P.L.G. and Klug, A. 1976. 'The structure of the protein disk of tobacco mosaic virus to 5 Å resolution' *Nature* **259**, 20–24.

111. Laemmli, U.K., Amos, L.A. and Klug, A. 1976. 'Correlation between structural transformation and cleavage of the major head protein of T4 bacteriophage' *Cell* **7**, 191–203.

112. Jack, A., Klug, A. and Ladner, J.E. 1976. '"Non-rigid" nucleotides in tRNA: a new correlation in the conformation of a ribose' *Nature* **261**, 250–251.

113. Finch, J.T. and Klug, A. 1976. 'A solenoidal model for superstructure in chromatin' *Proc. Natl Acad. Sci. USA* **73**, 1897–1901.

114. Butler, P.J.G., Bloomer, A.C., Bricogne, G., Champness, J.N., Graham, J., Guilley, H., Klug, A. and Zimmern, D. 1976. 'Tobacco mosaic virus assembly – specificity and the transition in protein structure during RNA packaging' in *Proc. 3rd John Innes Symp. Structure–Function Relationships of Proteins* (ed. R. Markham and R. W. Horne) 101–110 (Elsevier).

115. Amos, L.A., Linck, R.W. and Klug, A. 1976. 'Molecular structure of flagellar microtubules' *Cold Spring Harb. Symp. Cell Motility,* 847–867.

116. Jack, A., Ladner, J.E. and Klug, A. 1976. 'Crystallographic refinement of yeast phenylalanine transfer RNA at 2.5 Å resolution' *J. Mol. Biol.* **108,** 619–649.

117. Crick, F.H.C., Brenner, S., Klug, A. and Pieczenik, G. 1976. 'A speculation on the origin of protein synthesis' *Origins Life* 7, 389–397.

118. Jack, A., Ladner, J.E., Rhodes, D. Brown, R.S. and Klug, A. 1977. 'A crystallographic study of metal-binding to yeast phenylalanine transfer RNA' *J. Mol. Biol.* **111,** 315–328.

119. Sperling, L. and Klug, A. 1977. 'X-ray studies on "native" chromatin' *J. Mol. Biol.* **112,** 253–263.

120. Finch, J.T., Lutter, L.C., Rhodes, D., Brown, R.S. Rushton, B., Levitt, M. and Klug, A. 1977. 'Structure of nucleosome core particles of chromatin' *Nature* **269,** 29–36.

121. Klug, A., Lutter, L.C., Rhodes, D., Brown, R.S., Rushton, B. and Finch, J.T. 1977. 'X-ray crystallographic and enzymatic analysis of nucleosome cores' *FEBS 11th Meeting,* Copenhagen, 1977. *Gene Expression* **43,** Symp. A2, 233–235.

122. Klug, A. 1978. 'I. Structure of chromatin. Introductory remarks' *Phil. Trans. Roy. Soc. Lond. B* **283,** 233–239.

123. Finch, J.T. and Klug, A. 1978. 'X-ray and electron microscope analysis of crystals of nucleosome cores' *Cold Spring Harb. Symp. Quant. Biol.* **42,** 1–9.

124. Butler, P.J.G. and Klug, A. 1978. 'The assembly of a virus' *Scient. Am.* **239,** 52–59.

125. Bloomer, A.C., Champness, J.N., Bricogne, G., Staden, R. and Klug, A. 1978. 'Protein disk of tobacco mosaic virus at 2.8 Å resolution showing the interactions within and between subunits' *Nature* **276,** 362–368.

126. Finch, J.T., Lutter, L.C., Rhodes, D., Brown, R.S., Rushton, B. and Klug, A. 1978. 'X-ray and electron microscope studies on nucleosome structure' *FEBS 12th Meeting*, Dresden, *Gene Functions* **51**, 193–197.

127. Klug, A., Jack, A., Viswamitra, M.A., Kennard, O., Shakked, Z. and Steitz, T.A. 1979. 'A hypothesis on a specific sequence-dependent conformation of DNA and its relation to the binding of the lac-repressor protein' *J. Mol. Biol.* **131**, 669–680.

128. Prunell, A., Kornberg, R.D., Lutter, L.C., Klug, A., Levitt, M. and Crick, F.H.C. 1979. 'Periodicity of deoxyribonuclease I digestion of chromatin' *Science* **204**, 855–858.

129. Thoma, F., Koller, Th. and Klug, A. 1979. 'Involvement of histone Hl in the organization of the nucleosome and of the salt dependent superstructures of chromatin' *J. Cell Biol.* **83**, 403–427.

130. Klug, A. 1979. 'From Macromolecules to Biological Assemblies'. Address on award of Heineken Prize: *Proc. Roy. Neth. Acad. Sci.*, April 1979.

131. Klug, A. 1978/79. 'Image analysis and reconstruction in the electron microscope of biological macromolecules' *Chem. Scripta* **14** (*Proc. 47th Nobel Symp.*) 245–256.

132. Klug, A. 1978/79. 'Direct imaging of atoms in crystals and molecules. Status and prospects for biological sciences' *Chem. Scripta* **14** (Proc. 47th Nobel Symp.) 291–293.

133. Klug, A. 1979. 'The assembly of tobacco mosaic virus: Structure and specificity' *Harvey Lecture*, 1978. *The Harvey Lectures*, Series 74, 141–172.

134. Rhodes, D. and Klug, A. 1980. 'Helical periodicity of DNA determined by enzyme digestion' *Nature* **286**, 573–578.

135. Klug. A., Rhodes, D., Smith, J., Finch, J.T. and Thomas, J.O. 1980. 'A low resolution structure for the histone core of the nucleosome' *Nature* **287**, 509–515.

136. Finch, J.T., Brown, R.S., Rhodes, D., Richmond, T.J., Rushton, B., Lutter, L.C. and Klug, A. 1980. 'X-ray diffraction study of a new crystal form of the nucleosome core showing higher resolution' *J. Mol. Biol.* **145**, 757–769.

137. Bloomer, A.C., Graham, J., Hovmoller, S., Butler, P.J.G. and Klug, A. 1981. 'Tobacco mosaic virus: Interaction of the protein disk

with oligonucleotides and its implications for virus structure and assembly'. In: *Structural Aspects of Recognition and Assembly in Biological Macromolecules. Proc. 7th Aharon Katzir-Katchalsky Conf.*, 851–864.

138. Kornberg, R.D. and Klug, A. 1981. 'The nucleosome' *Scient. Am.* **244**, 52–64.

139. Klug,A. and Lutter, L.C. 1981. 'The helical periodicity of DNA on the nucleosome' *Nucleic Acids Res.* **9**, 4267–4283.

140. Richmond, T.J., Klug, A., Finch, J.T. and Lutter, L.C. 1981. 'The organization of DNA in the nucleosome core particle'. In: *Biomolecular Stereodynamics*, Vol. II (ed. R.H. Sarma), 109–124 (Adenine Press).

141. Lomonossoff, G.P., Butler, P.J.G. and Klug, A. 1981. 'Sequence-dependent variation in the conformation of DNA' *J. Mol. Biol.* **149**, 745–760.

142. Rhodes, D. and Klug, A. 1981. 'Sequence dependent helical periodicity of DNA' *Nature* **292**, 378–380.

143. Hingerty, B.E., Brown, R.S. and Klug, A. 1982. 'Stabilization of the tertiary structure of yeast phenylalanine tRNA by $[Co(NH_3)_6]$. X-ray evidence for hydrogen bonding to pairs of guanine bases in the major groove' *Biochim. Biophys. Acta* **697**, 78–82.

144. Klug, A. Lutter, L.C. and Rhodes, D. 1982. 'Helical periodicity of DNA on and off the nucleosome as probed by nucleases' *Cold Spring Harb. Symp. Quant. Biol.* **XLVII**, 285–292.

145. Richmond, T.J., Finch, J.T. and Klug, A. 1982. 'Studies of nucleosome structure' *Cold Spring Harb. Symp. Quant. Biol.* **XLVII**, 493–501.

146. Klug, A. 'Structures of DNA: A summary' *Cold Spring Harb. Symp. Quant. Biol.* **XLVII**, 1215–1223.

147. Brown, R.S., Hingerty, B.E., Dewan, J.C. and Klug, A. 1983. 'Pb(II)-catalysed cleavage of the sugar-phosphate backbone of yeast tRNAPhe implications for lead toxicity and self-splicing RNA' *Nature* **303**, 543–546.

148. Klug, A. 1983. 'Architectural design of spherical viruses' *Nature* **303**, 378–379.

149. Klug, A. 1983. 'From macromolecules to biological assemblies'. In: The Nobel Foundation, 93–125. Les Prix Nobel en 1982. Nobel lecture also published in: (a) *Biosci. Rep.* **3**, 395–430 (1983);

(b) *Angew. Chem.* **22**, 565–582 (1983) (Int edn), **95**, 579–596 (German edn).

150. Cockell, M., Rhodes, D. and Klug, A. 1983. 'Location of the primary sites of micrococcal nuclease cleavage on the nucleosome core' *J. Mol. Biol.* **170**, 423–446.

151. Klug, A. 1983. 'Nucleosome structure and chromatin superstructure'. In: *Nucleic Acids Res. Proc. AMBO Symp.*, 91–112 (Academic Press).

152. Klug, A. and Butler, P.J.G. 1983. 'The structure of nucleosomes and chromatin'. In: *Genes: Structure and Expression* (ed. A.M. Kroon), 1–41 (Wiley).

153. de Bruijn, M.H.L. and Klug, A. 1983. 'A model for the tertiary structure of mammalian mitochondrial transfer RNAs lacking the entire 'dihydrouridine' loop and stem' *EMBO J.* **2**, 1309–1321.

154. de Bruijn, M.H.L. and Klug, A. 1984, 'A model for the structure of a small mitochondrial tRNA'. In: *Gene Expression, Alfred Benzon Symp. 19* (ed. B.F.C. Clark and H.U. Petersen), 259–278 (Munksgaard).

155. Westhof, E., Altschuh, D., Moras, D., Bloomer, A.C., Mondragon, A. Klug, A. and Van Regenmortel, M.H.V. 1984. 'Correlation between segmental mobility and the location of antigenic determinants in proteins' *Nature* **311**, 123–126.

156. Richmond, T.J., Finch, J.T., Rushton, B., Rhodes, D. and Klug, A. 1984. 'Structure of the nucleosome core particle at 7 Å resolution' *Nature* **311**, 532–537.

157. Brown, R., Dewan, J.C. and Klug, A. 1985. 'Crystallographic and biochemical investigation of the lead(II)-catalyzed hydrolysis of yeast phenylalanine tRNA' *Biochemistry* **24**, 4785.

158. Miller, J., McLachlan, A.D. and Klug, A. 1985. 'Repetitive zinc-binding domains in the protein transcription factor III A from Xenopus oocytes' *EMBO J.* **4**, 1609–1614.

159. Widom, J. and Klug, A. 1985. 'Structure of the 300 Å chromatin filament: X-ray diffraction from oriented samples' *Cell* **43**, 207–213.

160. Klug, A. 1985. 'The higher order structure of chromatin'. In: *The Robert A. Welch Foundation Conf. Chem. Res. XXXIX. Genetic Chemistry: The Molecular Basis of Heredity*, 133–160.

161. Klug, A., Finch, J.T. and Richmond, T.J. 1985. 'Crystallographic structure of the octamer histone core of the nucleosome' *Science* **229**, 1109–1110.

162. Van Regenmortel, M.H.V., Altschuh, D. and Klug, A. 1986. 'Influence of local structure on the location of antigenic determinants in tobacco mosaic virus protein' *CIBA Found. Symp.* **119**, 76–84.

163. Rhodes, D. and Klug, A. 1986. 'An underlying repeat in some transcriptional control sequences corresponding to half a double helical turn of DNA' *Cell* **46**, 123–132.

164. Fairall, L., Rhodes, D. and Klug, A. 1986. 'Mapping of the sites of protection on a 5S RNA gene by the Xenopus transcription factor IIIA: A model for the interaction' *J. Mol. Biol.* **192**, 577–591.

165. Diakun, G.P., Fairall, L. and Klug, A. 1986. 'EXAFS study of the zinc-binding sites in the protein transcription factor IIIA' *Nature* **324**, 698–699.

166. Altschuh, D., Lesk, A.M., Bloomer, A.C. and Klug, A. 1987. 'Correlation of co-ordinated amino acid substitutions with function in viruses related to tobacco mosaic virus' *J. Mol. Biol.* **193**, 693–707.

167. Nelson, H.C.M., Finch, J.T., Luisi, B.F. and Klug, A. 1987. 'The structure of an oligo(dA) · oligo(dT) tract and its biological implications' *Nature* **330**, 221–226.

168. Travers, A.A. and Klug, A. 1987. 'The bending of DNA in nucleosomes and its wider implications' *Phil. Trans. Roy. Soc. Lond.* B **317**, 537–561.

169. Travers, A.A. and Klug, A. 1987. 'Nucleoprotein complexes: DNA wrapping and writhing' *Nature* **327**, 280–281.

170. Klug, A. and Rhodes, D. 1987. 'Zinc fingers, a novel protein motif for nucleic acid recognition' *Trends Biochem. Sci.* **12**, 464–469.

171. Rhodes, D. and Klug, A. 1988. '"Zinc fingers": a novel motif for nucleic acid binding'. In: *Nucleic Acids and Molecular Biology* Vol. 2 (ed. F. Eckstein and D.M.J. Lilley), 149–166 (Springer).

172. Klug, A. and Rhodes, D. 1987. 'Zinc fingers: A novel protein fold for nucleic acid recognition' *Cold Spring Harbor Symp. Quant. Biol.* **LII**, 473–482.

173. O'Halloran, T.V., Lippard S.J., Richmond T.J. and Klug, A. 1987. 'Multiple heavy-atom reagents for macromolecular X-ray structure determination. Application to the nucleosome core particle' *J. Mol. Biol.* **194**, 705–712.

174. Goedert, M., Wischik, C.M., Crowther, R.A., Walker, J.E. and Klug, A. 1988. 'Cloning and sequencing of the cDNA encoding a core protein of the paired helical filament of Alzheimer disease: identification as the microtubule associated protein tau' *Proc. Natl Acad. Sci. USA* **85**, 4051–4055.

175. Wischik, C.M., Novak, M., Thogersen, H-C., Edwards, P.C., Runswick, M.J., Jakes, R., Walker, J.E., Milstein, C., Roth, M. and Klug, A. 1988. 'Isolation of a fragment of tau derived from the core of the paired helical filament of Alzheimer disease' *Proc. Natl Acad. Sci. USA* **85**, 4506–4510.

176. Wischik, C.M., Novak, M., Edwards, P.C., Klug, A., Tichelaar, W. and Crowther, R.A. 1988. 'Structural characterization of the core of the paired helical filament of Alzheimer disease' *Proc. Natl Acad. Sci. USA* **85**, 4884–4888.

177. Klug, A. & Travers, A.A. 1989. 'The helical repeat of nucleosome-wrapped DNA' *Cell* **56**, 9–11.

178. Klug, A. 1989. 'Zinc fingers: ubiquitous protein modules for nucleic acid recognition' *S. Afr. J. Sci.* **85**, 576–581.

179. Sundquist, W.I. & Klug, A. 1989. 'Telomeric DNA dimerizes by formation of guanine tetrads between hairpin loops' *Nature* **342**, 825–829.

180. Rhodes, D., Brown, R.S. and Klug, A. 1989. 'Crystallisation of nucleosome core particles' *Methods Enzymol.* **170**, 420–428.

181. Spillantini, M.G., Goedert, M., Jakes, R. and Klug, A. 1990. 'Different configurational states of b-amyloid and their distributions relative to plaques and tangles in Alzheimer disease' *Proc. Natl Acad. Sci. USA* **87**, 3947–3951.

182. Spillantini, M.G., Goedert, M., Jakes, R. and Klug, A. 1990. 'Topographical relationship between b-amyloid and tau protein epitopes in tangle-bearing cells in Alzheimer disease' *Proc. Natl Acad. Sci. USA* **87**, 3952–3956.

183. Travers, A.A. and Klug, A. 1990. 'Bending of DNA in nucleoprotein complexes'. In: *DNA Topology and its Biological*

Effects (ed. N.R. Cozzarelli and J.C. Wang), 57–106 (Cold Spring Harbor Laboratory Press).

184. Churchill, M.E.A., Tullius, T.D. and Klug, A. 1990. 'Mode of interaction of the zinc finger protein TFIIIA with 5S RNA gene of Xenopus' *Proc. Natl Acad. Sci. USA* **87**, 5528–5532.

185. Neuhaus, D., Nakaseko, Y., Nagai, K. and Klug, A. 1990. 'Sequence-specific ^{1}H NMR resonance assignments and secondary structure identification for 1- and 2-zinc finger constructs from SWI5. A hydrophobic core involving four invariant residues.' *FEBS Lett.* **262**, 179–184.

186. Bondareff, W., Wischik, C.M., Novak, M., Amos, W.B., Klug, A. and Roth, M. 1990. 'Molecular analysis of neurofibrillary degeneration in Alzheimer's disease' *Am. J. Path.* **137**, 711–723.

187. Klug, A. 1990. 'Reminiscences of Sir Lawrence Bragg'. In: *Selections and Reflections: The Legacy of Sir Lawrence Bragg* (ed. J.M. Thomas and Sir D. Phillips), 129–133 (Science Reviews Limited).

188. Klug, A. 1991. 'Order in molecular biology'. In: *Evolutionary Trends in the Physical Sciences. Springer Proceedings in Physics*, Vol. 57 (ed. M. Suzuki and R. Kubo), 225–237 (Springer).

189. Struck, M-M., Klug, A. and Richmond, T.J. 1992. 'Comparison of X-ray structures of the nucleosome core particle in two different hydration states' *J. Mol. Biol.* **224**, 253–264.

190. Nakaseko, Y., Neuhaus, D., Klug, A. and Rhodes, D. 1992. 'Adjacent zinc finger motifs in multiple zinc finger peptides from SWI5 form structurally independent, flexibly linked domains' *J. Mol. Biol.* **228**, 619–636.

191. Neuhaus, D., Nakaseko, Y., Schwabe, J.W.R. and Klug, A. 1992. 'Solution structures of two zinc finger domains from SWI5, obtained using two-dimensional ^{1}H NMR spectroscopy: a zinc finger structure with a third strand of β-sheet' *J. Mol. Biol.* **228**, 637–651.

192. Klug, A., Neuhaus, D., Schwabe, J., Nakaseko, Y., Fairal, L., Nagai, K. and Rhodes, D. 1992. 'The structures of two classes of zinc-finger domains and their modes of interaction with DNA' *Sci. Technol. Japan* **33**, 60–80.

193. Rhodes, D. and Klug, A. 1993. 'Zinc fingers' *Scient. Am.* **268**, 56–65.

194. Klug, A. 1993. 'Transcription: Opening the gateway' *Nature* **365**, 486–487.

195. Choo, Y. and Klug, A. 1993. 'A role in DNA binding for the linker sequences of the first three zinc fingers of TFIIIA' *Nucleic Acids Res.* **21**, 3341–3346.

196. Klug, A. 1993. 'Protein designs for the specific recognition of DNA' *Gene* **135**, 83–92.

197. Schwabe, J.W.R. and Klug, A. 1994. 'Zinc mining for protein domains' *Nature Struct. Biol.* **1**, 345–349.

198. Klug, A. 1994. 'Macromolecular order in biology' (a) *Phil. Trans. Roy. Soc. Lond. A* **348**, 167–178; (b) 4th William and Mary Lecture, Leiden University, 30th October 1996.

199. Jakes, R., Harrington, C.R. Spillantini, M.G. Goedert, M. and Klug, A. 1995. 'Characterisation of an antibody relevant to the neuropathology of Alzheimer disease' *Alzheimer Dis. Assoc. Disord.* **9**, 47–51.

200. Choo, Y. and Klug, A. 1994. 'Towards a code for the interactions of zinc fingers with DNA: selection of randomised fingers displayed on phage' *Proc. Natl Acad. Sci. USA* **91**, 11163–11167.

201. Choo, Y. and Klug, A. 1994. 'Selection of DNA binding sites for zinc fingers using rationally randomised DNA reveals coded interactions' *Proc. Natl Acad. Sci. USA* **91**, 11168–11172.

202. Choo, Y., Sánchez-García, I. and Klug, A. 1994. '*In vivo* repression by a site-specific DNA-binding protein designed against an oncogenic sequence' *Nature* **372**, 642–645.

203. Klug, A. and Schwabe, J.W.R. 1995. 'Zinc fingers' *FASEB J.* **9**, 597–604.

204. Scott, W.G., Finch, J.T., Grenfell, R., Fogg, J., Smith, T., Gait, M.J. and Klug, A. 1995. 'Rapid crystallization of chemically synthesized hammerhead RNAs using a double screening procedure' *J. Mol. Biol.* **250**, 327–332.

205. Scott, W.G., Finch, J.T. and Klug, A. 1995. 'The crystal structure of an all-RNA hammerhead ribozyme: A proposed mechanism for RNA catalytic cleavage' *Cell* **81**, 991–1002.

206. Choo, Y. and Klug, A. 1995. 'Designing DNA-binding proteins on the surface of filamentious phage' *Curr. Opin. Biotechnol.* **6**, 431–436.

207. Klug, A. 1995. 'Gene regulatory proteins and their interaction with DNA'. In: *DNA: The Double Helix. Perspective and*

Prospective at Forty Years (ed. D.A. Chambers) *Ann. NY Acad. Sci.* **758**, 143–160.

208. Klug, A. 'Protein designs for the specific recognition of DNA' (a) 1995. In: *Molecular Biology and Biotechnology. A Comprehensive Desk Reference* (ed. R.A. Meyers), 746–753 (VCH); (b) 1996. In: *Encyclopaedia of Molecular Biology and Molecular Medicine* (ed. R.A. Meyers), 127–135(VCH)

209. Scott, W.G., Finch J.T. and Klug, A. 1995 'The crystal structure of an all-RNA hammerhead ribozyme'. In: *Proc. 22nd Symp. Nucleic Acids Chemistry. Nucleic Acids Symp. Series* **34**, 214–216 (Oxford University Press).

210. Goedert, M., Spillantini, M.G., Hasegawa, M., Jakes, R., Crowther, R.A. and Klug, A. 1996, 'Molecular dissection of the neurofibrillary lesions of Alzheimer's disease' *Cold Spring Harb. Symp. Quant. Biol.* **61**, 565–573.

211. Scott, W.G. and Klug, A. 1996. 'Ribozymes: structure and mechanism in RNA catalysis' *Trends Biol. Sci.* **21**, 220–224.

212. Scott, W.G., Murray, J.B., Arnold, J.R.P., Stoddard, B.L. and Klug, A. 1996. 'Capturing the structure of a catalytic RNA intermediate: The hammerhead ribozyme' *Science* **274**, 2065–2069.

213. Choo, Y. and Klug, A. 1997. 'Physical basis of a protein-DNA recognition code' *Curr. Opin. Struct. Biol.* **7**, 117–125.

214. Isalan, M., Choo, Y. and Klug, A. 1997. 'Synergy between adjacent zinc fingers in sequence-specific DNA recognition' *Proc. Natl Acad. Sci. USA* **94**, 5617–5621.

215. Choo, Y., Castellanos, A., García-Hernández, B., Sánchez-García, I. and Klug, A. 1997. 'Promoter-specific activation of gene expression directed by bacteriophage-selected zinc fingers' *J. Mol. Biol.* **273**, 525–532.

216. Nowak, M., Krakauer, D., Klug, A. and May, R. 1998. 'Prion infection dynamics' *Integr. Biol.* **1**, 3–15.

217. Spillantini, M.G., Murrell, J.R., Goedert, M., Farlow, M.R., Klug, A. and Ghetti, B. 1998. 'Mutation in the tau gene in familial multiple system tauopathy with presenile dementia' *Proc. Natl Acad. Sci. USA* **95**, 7737–7741.

218. Isalan, M., Klug, A. and Choo, Y. 1998. 'Comprehensive DNA recognition through concerted interactions from adjacent zinc fingers' *Biochemistry* **37**, 12026–12033.

219. Klug, A. 1999. 'Commentary on the two papers by Caspar and Franklin'. In *Tobacco Mosaic Virus: One Hundred Years of Contribution to Virology* (ed. K-B.G. Scholthof, J.G. Shaw and M. Zaitlin) 119–121 (American Phytopathology Society Press).

220. Klug, A. 1999. 'The tobacco mosaic virus particle: structure and assembly' *Phil. Trans. Roy. Soc. Lond.* **354**, 531–535.

221. Varani, L., Hasegawa, M., Spillantini, M.G., Smith, M.J., Murrell, J.R., Ghetti, B., Klug, A., Goedert, M. and Varani, G. 1999. 'Structure of tau exon 10 splicing regulatory element RNA and destabilization by mutations of frontotemporal dementia and parkinsonism linked to chromosome 17' *Proc. Natl Acad. Sci. USA* **96**, 8229–8234.

222. Klug, A. 1999. 'Zinc finger peptides for the regulation of gene expression' *J. Mol. Biol.* **293**, 215–218.

223. Goedert, M. and Klug, A. 1999. 'Tau protein and the paired helical filament of Alzheimer's disease. Special Millennium Issue. A hundred years of neuroscience' *Brain Res. Bull.* **50**, 469–470.

224. Searles, M.A., Lu, D. and Klug, A. 2000. 'The role of the central zinc fingers of transcription factor IIIA in binding to 5SRNA' *J. Mol. Biol.* **301**, 47–60.

225. Moore, M., Choo, Y. and Klug, A. 2001. 'Design of polyzinc finger peptides with structured linkers' *Proc. Natl Acad. Sci. USA* **98**, 1432–1436.

226. Moore, M., Klug, A. and Choo, Y. 2001. 'Improved DNA binding specificity from polyzinc finger peptides by using strings of two-finger units' *Proc. Natl Acad. Sci. USA* **98**, 1437–1441.

227. Klug, A. 2001. 'The Human Genome Project' *IUBMB Life* 51, 1–4.

228. Klug, A. 2001. 'RNA polymerase II: A marvellous machine for making messages' *Science* **292**, 1785–1952.

229. Isalan, M., Klug, A. and Choo, Y. 2001. 'A rapid, generally applicable method to engineer zinc fingers illustrated by targeting the HIV-1 promoter' *Nature Biotechnol.* **19**, 656–660.

230. Klug, A. 2002. 'Retrospective: Max Perutz (1914–2002)' *Science* **295**, 2382–2383.

231. Klug, A. 2002. 'Chris Calladine and biological structures: a personal account'. In *New Approaches to Structural Mechanics, Shells and Biological Structures* (ed. H.R. Drew and S. Pellegrini), 413–419 (Springer).

232. Klug, A. 2002. 'The development of image analysis and
     3D-reconstruction in the electron microscopy of biomolecular
     assemblies'. Paper presented at ICEM-15, Durban.
233. Reynolds, L., Ullman, C., Moore, M., Isalan, M., West, M.J.,
     Clapham, P., Klug, A. and Choo, Y. 2003. 'Repression of the
     HIV-1 5 LTR promoter and inhibition of HIV-1 replication by
     using engineered zinc-finger transcription factors' *Proc. Natl
     Acad. Sci. USA* **100**, 1615–1620.
234. Papworth, M., Moore, M., Isalan, M., Minczuk, M., Choo, Y. and
     Klug, A. 2003. 'Inhibition of herpes simplex virus 1 gene
     expression by designer zinc-finger transcription factors'
     Proc. Natl Acad. Sci. USA **100**, 1621–1626.
235. Klug, A. 2003. 'The life of George Porter OM FRS' *Notes Rec. Roy.
     Soc. Lond.* **57**, 261–264.
236. Lu, D., Searles, M.A. and Klug, A. 2003. 'Crystal structure of a zinc
     finger RNA complex reveals two modes of molecular
     recognition' *Nature* **426**, 96–100.
237. Klug, A. 2004. 'The discovery of the DNA double helix' *J. Mol. Biol.*
     335, 3–26.
238. Klug, A. 2004. 'The discovery of zinc fingers and their practical
     applications in gene regulation: A personal account'. In:
     *Zinc Finger Proteins: From Atomic Contact to Cellular
     Function* (ed. S. Iuchi and N. Kuldell), 1–6 (Kluwer Academic/
     Plenum).
239. Klug, A. 2005. 'Obituary: Francis Crick (8 June 1916–28 July 2004):
     A Memoir.' *FEBS Lett.* **579**, 852–854. An expanded version of
     the Obituary which appeared in the September (2004) issue of
     Nature Cell Biology.
240. Klug, A. 2005. 'Introduction to eukaryotic transcription and
     chromatin' *FEBS Lett.* **579**, 890–891.
241. Klug, A. 2005. 'Towards therapeutic applications of engineered
     zinc finger proteins' *FEBS Lett.* **579**, 892–894 (*Proc. Nobel Symp.*
     130).
242. Klug, A. 2005. 'Review: The discovery of zinc fingers and their
     development for practical applications in gene regulation' *Proc.
     Japan Acad.* **81**, 87–102.
243. Minczuk, M., Papworth, M.A., Kolasinska, P., Murphy, M.P. and
     Klug, A. 2006. 'Sequence-specific modification of

mitochondrial DNA using a chimeric zinc finger methylase'
Proc. Natl Acad. Sci. USA **103**, 19689–19694.

244. Goedert, M., Klug, A.K., Crowther, R.A. 2006. 'Tau protein, the paired helical filament and Alzheimer's disease' *J. Alzheimer's Dis.* **9**, 195–207.

245. Goedert, M., Grazia Spillantini, M., Ghetti, B., Crowther, R.A. and Klug, A. 2006. 'Discovery of the tangle'. In *Alzheimer: 100 Years and Beyond* (ed. M. Jucker *et al.*), 297–304 (Springer),

246. Grazia Spillantini, M., Murrell, J.R., Goedert, M., Farlow, M., Klug, A. and Ghetti, B. 2006. 'Mutations in the tau gene (MAPT) in FTDP-17: the family with multiple system tauopathy with presenile dementia (MSTD)' *J. Alzheimer's Dis.* **9**, 373–380.

247. Lu, D. and Klug, A. 2007. 'Invariance of the zinc finger module: A comparison of the free structure with those in nucleic-acid complexes' *Proteins* **67**, 508–512.

248. Santiago, Y., Chan, E., Liu, P.Q., Orlando, S., Zhang, L., Urnov, F.D., Holmes, M.C., Guschin, D., Waite, A., Miller, J.C., Rebar, E. J., Gregory, P.D., Klug, A. and Collingwood, T.N. 2008. 'Targeted gene knockout in mammalian cells using engineered zinc-finger nucleases' *Proc. Natl Acad. Sci. USA* **105**, 5809–5014.

249. Minczuk, M., Papworth, M. A., Miller, J. C., Murphy, M. P. and Klug, A. 2008. 'Development of a single-chain, quasi-dimeric zinc-finger nuclease for the selective degradation of mutated human mitochondrial DNA' *Nucleic Acids Res.* **36**, 3926–3938.

250. Klug, A. 2010. 'The discovery of zinc fingers and their applications in gene regulation and genome manipulation' *Annu. Rev. Biochem.* **79**, 213–231.

251. Klug, A. 2010. 'From virus structure to chromatin: X-ray diffraction to three-dimensional electron microscopy' *Annu. Rev. Biochem.* **79**, 1–35.

252. Klug, A. 2010. 'The discovery of zinc fingers and their development for practical applications in gene regulation and genome manipulation' *Quart. Rev. Biophys.* **43**, 1–21.

Index